the evolving earth

the evolving earth

A Text in Physical Geology

FREDERICK J. SAWKINS
University of Minnesota, Minneapolis

CLEMENT G. CHASE
University of Minnesota, Minneapolis

DAVID G. DARBY
University of Minnesota, Duluth

GEORGE RAPP, JR.
University of Minnesota, Minneapolis

Macmillan Publishing Co., Inc.
NEW YORK

Collier Macmillan Publishers
LONDON

Macmillan Publishing Co., Inc.
866 Third Avenue, New York, New York 10022
Collier-Macmillan Canada, Ltd.

Library of Congress Cataloging in Publication Data
Main entry under title:

The Evolving earth.

   1. Physical geology.   2. Plate tectonics.
I. Sawkins, Frederick J.
QE28.2.E86   551   73-7678
ISBN 0-02-406500-5

Printing: 1 2 3 4 5 6 7 8   Year: 4 5 6 7 8 9 0

For Professor Strathmore B. Cooke,
in recognition of his dedication to teaching
and his unstinting help to the authors
during preparation of this manuscript.

preface

Within the last decade there has been a revolution in the earth sciences. Most of the first-order features of the outermost portions of the earth can now be related to a set of concepts, grand and far-reaching in their effects and elegant in their simplicity—plate tectonics.

We feel it is only right that someone studying introductory geology should share fully in the excitement of this new approach to understanding the earth. This book uses the concept of plate tectonics to develop a framework within which the student can integrate the facts he learns about geological processes, their causes, and their results.

Though virtually all recent introductory geology textbooks deal with plate tectonics, they make no real attempt to demonstrate the extent to which these mechanisms influence or control geological environments. This text departs from the traditional format of most physical geology books. The recent revolution in earth science is so fundamental that such a departure is as necessary in introductory texts as in advanced research programs. Thus, once the scene is set, this book shows in progressive steps how plates interact, why the outlines of some continents appear to fit together, why sea floors are geologically young, and by what mechanisms the continents with their great mountain ranges and extensive plains are formed.

Most colleges and universities require a natural science course, with lab, as one of the general requirements for graduation. The course (or sequence) is often the only natural science a student will take in his college career. This book was written with the non-science student in mind. Attention is given to the broad aspects of scientific inquiry and the nature of science, viewed from the framework of investigating the earth.

In the interest of the student, we have attempted to write a tightly constructed text in which every element is of significance. The result is a relatively short book, which nevertheless covers those aspects of physical geology we believe to be of central importance to the evolution of the surface of our planet.

In the process of writing this book we have received considerable guidance from our colleagues of the University of Minnesota, in particular S. R. Cooke, P. Hudleston, R. Murthy, R. Sloan, and J. Stout. Their help is gratefully acknowledged.

ACKNOWLEDGMENTS

During the earlier stages of manuscript planning and preparation we received invaluable input and constructive criticism from the following distinguished geologic educators.

John J. Rusch	*University of Oklahoma*
Claude Albritton	*Southern Methodist University*
John E. Allen	*Portland State University*
Don Groff	*Western Connecticut State College*
James Gunderson	*Wichita State University*
O. T. Hayward	*Baylor University*
Thomas E. Hendrix	*Indiana University*
Phil Hewitt	*State University of New York, Brockport*
Larry Lattman	*University of Cincinnati*
Charles Matsch	*University of Minnesota, Duluth*
William H. Matthews, III	*Lamar State College of Technology*
Richard Paull	*University of Wisconsin, Milwaukee*
Charles P. Thornton	*Pennsylvania State University, University Park*
Don Zenger	*Pomona College*

contents

one

two

three

four

five

six

seven

eight

nine

ten

Rocks in the Surface Environment 287

eleven

Transport and Topography 321

twelve

Geology and Man   **397**

one

Introduction

WHY STUDY GEOLOGY?

The surface of the earth is your physical environment, and geology is related to the most basic factors of this environment. It is clear that, if we are to live in harmony with our physical environment, we must attempt to understand it as fully as possible. The inevitable conflict that develops between modern technological man and the natural environment is perhaps one of the most pressing issues of our day. Until we, and in particular the planners and decision makers amongst us, show a distinct sensitivity in our dealings with this natural environment, the conflict can only be expected to become more intense.

This factor alone provides enough justification for the study of geology, but we can add another—the vital resources man obtains from the earth. Winning these resources requires an understanding of the various geologic processes involved in their formation.

Finally, in addition to these important practical factors, there is the matter of man's inherent curiosity. This encompasses both his quest for his origins and his innate desire to understand the processes that shaped, and continue to shape, his physical world. To the layman, rocks are merely inanimate objects; to the geologist, they provide insights into the dynamic processes involved in their formation and clues to the geologic history of the area in which they occur. This book attempts to help the reader gain an understanding of the earth and how it has evolved to its present state.

HISTORICAL PERSPECTIVES IN GEOLOGY

The science of geology has developed into its present form by innumerable small, and some large, advances in our knowledge of the earth. Some concepts and models of the physical world developed along the way have proven inadequate to explain later observations, and new models and concepts have been generated. Discarding old and developing new concepts inevitably generates major scientific controversies that often persist for decades. The fact that this occurs and is almost always finally resolved in favor of the model that best explains the phenomena points to the significant characteristic of science, namely, that it is a self-correcting system for thinking about the world. Our review of the early history of geologic thought is brief, but it should provide you with some helpful examples of earlier attempts to understand a few of the more basic principles of earth science.

Eratosthenes' Attempt to Measure the Earth

Several centuries before Christ, Greek astronomers had made a number of observations that suggested to them the earth was not flat. They had observed, for example, that the stars visible in the night sky were not all the same at widely separated places. They had noticed that the shape of the earth's shadow on the moon during an eclipse was rounded and that ships putting out to sea gradually disappeared over the horizon.

About B.C. 250 a Greek astronomer named Eratosthenes made a set of observations that demonstrated that the earth's surface was indeed not flat. He knew that at noon on June 21st of any year the sun was directly over the city of Syene, in Egypt. He had deduced this fact by observing that the buildings at that time cast no shadows and that the sun's reflection could be observed in the water at the bottom of a deep well. Eratosthenes himself lived in Alexandria, a city 800 kilometers (km) north of Syene. He demonstrated that at

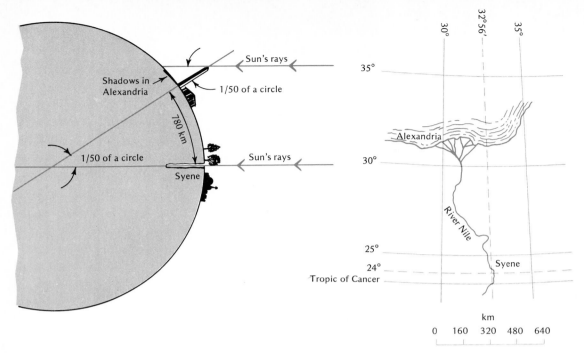

1-1. *Eratosthene's measurement of the earth's circumference.*
What assumptions were made in this method?
1. Are rays from the sun parallel? If not would this method be valid?
2. Is Alexandria 800 km north of Syene?
3. Do buildings and wells lie along lines that point to the center of the earth?

noon, the 21st of June in any year, the sun was not overhead in Alexandria but slightly to the south (Figure 1-1). By measuring the length of noontime shadows cast by buildings, Eratosthenes was able to determine that the sun was approximately 7.2° to the south at its meridian passage.

Eratosthenes had proved that the earth was not flat. *Assuming* it was spherical, he calculated the circumference of the earth by multiplying the distance between the two cities by 50 (7.2° = 1/50 of a full circle, 360°). His result, 40,000 km, was surprisingly close to the true distance. This experiment stands as a major scientific accomplishment.

Steno's Laws

Real progress in understanding the earth and the geologic processes that shape it was limited until the seventeenth century. Prior to that time most of the valid observations made on geologic phenomena, such as fossils in rocks, earthquakes, and volcanoes, were explained in terms of supernatural causes, and it was widely believed that most of the features of the earth such as mountains were formed by sudden, catastrophic events.

Nicholas Steno was born and educated in medicine in Copenhagen but later lived in Italy, where he began making observations of fossils in rocks. He interpreted them correctly as the remains of organisms preserved in accumulating sediment. His most important contribution to the development of geologic thought, however, lies in the laws he formulated regarding the deposition and relative ages of layers of sedimentary rocks. The most important of these are

1. *The law of original horizontality.* Water-laid sediments are deposited in layers (strata) that are not far from horizontal and are parallel or nearly parallel to the surface on which they are accumulating.
2. *The law of superposition.* In any succession of sedimentary strata that has not been disrupted since its deposition, the youngest stratum is at the top and successively older strata lie below it.

These concepts may seem inherently obvious to us today, but they form the key to understanding sedimentary rocks. Furthermore, the two laws introduced the idea of time and sequence into geologic processes and indicated that sedimentary rocks could not all be identical in age (that is, have been created at the same time). From these beginnings a systematic science of the earth started to take form.

For the next hundred years or so most of the geological investigation that took place centered on the problem of fossils, and Steno's view that fossils must represent the remains of organisms preserved by burial in accumulating sediments was gradually accepted. This conclusion brought the early pioneers of geology into an inevitable conflict with the Christian Church of the day because the theories and concepts they evolved were strongly at odds with the story of Genesis as expounded in the Bible.

Werner and Neptunism

In 1787, a teacher at the Freiberg Mining Academy in Saxony, named Abraham Werner, published a general theory of the origin of crustal rocks. His observations in the area where he lived led him to postulate that the crust had a layered structure and everywhere consisted of hard crystalline rocks overlain successively by rocks of a less crystalline nature. Werner rejected ideas of catastrophism and believed, rather, that all rock was formed in a universal ocean. To the lowermost crystalline rocks, which lacked fossils, Werner ascribed an origin involving chemical precipitation. To the uppermost fossiliferous rocks, Werner ascribed a mechanical origin involving the deposition of sands and muds derived by erosion from uplifted portions of the underlying crystalline crust (Figure 1-2).

This broad theory, with its emphasis on deposition in some primeval ocean, soon became known as "neptunism." Although the neptunist scheme is at odds with many simple observations that can be made on rocks formed from molten material, such as volcanics, it dominated geologic thought for many decades. Werner himself was an extremely persuasive and influential teacher, and, furthermore, the concept of a universal sea could be fitted to the Biblical flood and thus appealed to the theologians of the time.

Werner himself never travelled outside of Saxony but eventually some of his students, when confronted with evidence of volcanic

1-2. *Simplified cross section of the typical rock layers near Freiburg, Germany. The first column describes the rock types, the second column gives the names ascribed to the rock units, and the third column describes the Wernerian interpretation of their origin. As with all geologic columns, the series of events should be read from the bottom up to preserve chronological order.*

	Rock Types	Names	Origins
	Loose unconsolidated gravel, boulders, peat bogs; nonmarine	Tertiary or Alluvial	Formed during recession of ocean to its present position
	Sandstone, limestone, basalt, coal; fossils present, mostly marine	Secondary	Most biologically and fragmentally formed during oscillations of sea level
	Hard shales and slates; first appearance of fossils, all marine	Transition	Mostly chemically deposited with some materials transported as fragments
	Hard crystalline rocks, granites, schists, gneisses; no fossils	Primary or Primitive	Chemically precipitated from ocean to form original surface of earth

activity in other parts of Europe, found themselves compelled to seriously question his teachings.

Hutton and Plutonism

Near the end of the eighteenth century scientific enquiry was coming of age and a general belief in rational order had evolved. In this philosophical climate, James Hutton began making observations and deductions that eventually led people to question neptunist concepts and to question the general idea that the rocks observable at the earth's surface had been created by a series of catastrophic events. Hutton was trained in law and medicine, but he rejected both of these professions and became a geologist and gentleman farmer.

Hutton was an extremely perceptive observer and felt that, as a result of his field studies, he could recognize relationships due to currently operating processes, such as erosion, deposition, and volcanism, in the rocks he studied. His great insight perhaps was into the nature of geologic time and the internal forces at work within the earth. Hutton himself saw no need for the concept of catastrophism in geology. Instead he reasoned that, due to internal forces certain areas could be uplifted and then subjected to erosion (Figure 1-3), whereas other areas could subside to become basins of deposition for sediments, given sufficiently long periods of time.

Hutton's other remarkable accomplishment was his correct interpretation of the nature of granite and basalt. Again through careful observation and reasoning, he deduced that basalts and granitic rocks were once molten and had been emplaced in this condition, recognizably baking the rocks with which they came into contact.

1-3. *An old sketch illustrating some of the concepts formulated by Hutton regarding the upheaval of mountains, folding of strata, and intrusion of granite. This old sketch illustrates concepts that are in many ways remarkably similar to some modern ideas regarding the deformation of rocks. [After G. P. Scrope:* Considerations on Volcanoes, *1825.]*

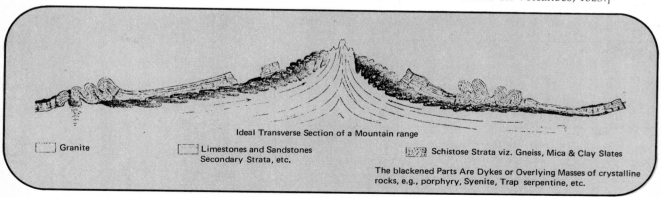

Ideal Transverse Section of a Mountain range

☐ Granite
☐ Limestones and Sandstones Secondary Strata, etc.
▨ Schistose Strata viz. Gneiss, Mica & Clay Slates

The blackened Parts Are Dykes or Overlying Masses of crystalline rocks, e.g., porphyry, Syenite, Trap serpentine, etc.

Hutton's evidence was actually extremely fragmentary, but he was on the right track. However, his insistence on the importance of great subterranean heat to produce rocks such as granite earned his ideas the nickname "plutonism" (for Pluto, god of the lower, infernal world). Thus, concepts of a dynamic earth in which both internal and external processes continue to act over very long periods of time came into being. The foundation stones of modern geology had been firmly laid.

Lyell and Uniformitarianism

Inherent in Hutton's ideas of time and gradual change through existing physical causes was the concept of **uniformitarianism.** It remained for an English geologist, Charles Lyell, to give specific expression to this concept and press for acceptance of the idea. In 1830 Lyell published a book entitled *Principles of Geology* that had tremendous impact on geologic thought. In it he marshalled carefully and lucidly all the observations he could to support the idea that *the present was the key to the past.* In so doing, Lyell established uniformitarianism, at the expense of catastrophism, as the accepted conceptual approach to studies of the earth. The doctrine of uniformity had been incorporated in the writings of earlier workers, but it was Lyell who effectively interpreted and publicized it for society in general.

The concept of uniformity in nature, and hence in geology, is of great importance, but it needs to be approached with caution. To the best of our modern scientific knowledge it is the laws of physics and chemistry that are uniform. Actually the earth changes through an irreversible series of cumulative events, but the physical and chemical laws that govern the processes by which these changes take place are themselves unchanging. Thus, by careful study of the geologic processes taking place today and the laws that govern them, we can, by analogy and inductive reasoning, gain insights into the history of our planet.

Consider for a moment the readily made observation that a rapidly flowing stream can transport coarse sand and even pebbles, whereas a slow flowing stream is able to transport only fine grained material such as mud. This is simply a manifestation of the physical laws that relate the maximum size of particles in transport to the speed and turbulence of the transporting medium, in this case water. A geologist can use this information, together with the concept of uniformity, to interpret the conditions under which water-laid ancient rocks of different grain size were deposited.

MODELS AND HYPOTHESES IN GEOLOGY

The concept of uniformity illustrates two very important aspects of modern geology. These aspects are the central role of ideas in the progress of science and the relationship of geology to other sciences. Science is not composed just of a series of observations and facts; to have any meaning to men, facts must be fitted into some framework of ideas. After all, Werner and Hutton looked at rocks that were quite similar, but they came to very different conclusions about how rocks could be formed. In each case, their ideas about earth history helped to determine what kind of observation they made, just as the observations helped to shape their ideas.

Models and hypotheses are the tools that scientists use in attempting to create some understanding of the endless complexities of nature. Faced with a set of observations about a facet of the natural world, a scientist will form a **hypothesis,** or idea, relating those observations to each other so that they have meaning. Hopefully, the hypothesis will be precise enough to predict the outcome of further observations. Those further observations are then made and compared to the results predicted from the hypothesis. If, in the opinion of the scientist, they match, the hypothesis is reinforced. If it can be shown that the observations do not match the predictions, the hypothesis is in need of refinement or replacement.

In this process, there is no final proof because a limitless number of observations is possible and we have no way of saying whether further facts will fit the hypotheses already formed. A scientific **theory** is simply a hypothesis that has proved to be especially successful in explaining facts. Progress in science is as much a matter of new and improved hypotheses and theories as of more facts and instrumentation.

The word **model** is used in two senses in geology. In the more restricted sense, a model is a physical object that is designed to scale down a natural geological process to a size that can be studied in the laboratory. In a more general sense, a geological model is a set of hypotheses designed to scale down a natural geological process intellectually so that we can understand how the process works and make predictions about it. Often, this kind of model must be deliberately oversimplified to be comprehensible and to emphasize the most important features of the process.

Both Hutton's and Werner's ideas were models of rock formation processes. The concept of uniformitarianism, in stating that the physical and chemical processes we can see operating today have also operated in the past, is definitely a hypothesis. It is not a model because it is not concerned with just how those physical and chemical principles are to be applied.

GEOLOGY IN RELATION TO OTHER SCIENCES

As a science, geology does not stand alone. The concept of uniformity emphasizes that geological processes are controlled by physical and chemical laws that we can observe in action today. Geologists always operate as though the prevailing chemical and physical theories are valid, and all geologic explanation must be consistent with those theories. Thus, to understand geology, we need to apply the results of the entire spectrum of other natural sciences. Physics and chemistry provide insight into the nature and behavior of earth materials, and knowledge of biology is essential to understand how living organisms can modify and control some geological processes. In turn, problems arising in geology have contributed to the progress of physics, chemistry, and biology.

The concept of energy, derived from physical and chemical studies, is particularly important for understanding geology. **Energy** is defined as the capacity to do work, which in turn can be defined in terms of a force acting through a distance. Work is done in geological processes of all scales, from the weathering of a rock to the uplifting of a mountain range. Energy, which is a measure of the capacity of a system for change, can take many forms.

Geological processes can often be usefully thought of in terms of conversion of energy from one form to another. For example, the uplift of a mountain range takes heat energy from within the earth and stores it as gravitational energy of the uplifted rocks. Weathering of those rocks is a matter of interchange of chemical and heat energy. When the products of weathering are washed down to the sea, they give up gravitational energy, which is released into the environment as thermal energy. In many surface environments, organisms can act to influence the flow of energy needed to drive geological processes. For example, lichen on the rocks in the high mountains (Figure 1-4) can greatly speed the weathering process. Biological effects can

1-4. *Lichens, the dark spots on the rocks in the foreground, have helped create the topography in the background. [Photo Courtesy S.R.B. Cooke]*

enter into many other geological processes in both obvious and subtle ways.

Although geology draws heavily upon the sciences of physics, chemistry, and biology, it does differ from these sciences in more than its emphasis on the earth. The other distinctive ingredient is the importance of time. Geology is a historical science, concerned not only with how natural processes work but also with the ordering of past events. Of the natural sciences, only astronomy shares this concern with great stretches of time. A geologist must not only understand something of how processes work today but must also be able to piece together, often from very incomplete evidence, a reasonable picture of what the particular results of these processes have been in the near and distant past.

MODERN PERSPECTIVES IN GEOLOGY

During the twentieth century, geology has undergone important changes. The emphasis of physical and chemical laws and the availability of sophisticated instruments have allowed geologists to become more quantitative in their approach to earth problems. The once separate fields of geophysics and geochemistry have become more central to any geologist's understanding of the earth.

Modern geologic laboratories can simulate the physical and chemical environments encountered within the upper 100 km of the earth, and the studies so made have helped to quantify many of the processes that have been inferred by geologists working on field problems. However, for all their success experimental laboratory studies cannot really come to grips with one of the most fundamental parameters of geology—time. Many of the changes and chemical reactions that occur in the earth proceed extremely slowly by human standards and simply cannot be reproduced in laboratory experiments. Another important advance in techniques for studying the earth has been the development of instrumentation by which the systematic decay of naturally occurring radioactive elements could be studied. Radioactivity studies have provided great insights into the nature of geologic time and the history of rocks formed by internal processes in the earth.

Rocks provide the geologists with only fragmentary evidence of the events that led to their formation. In many areas large volumes of rock have been stripped away by erosion, leaving no physical record of past events. Geologic investigation thus involves piecing together fragmentary scraps of information in an attempt to understand both the nature of geologic processes and the events that have occurred within one particular geographic area.

Inevitably the geologist is attempting to probe backward through the mists of time. In a real sense he is much like a detective, who must reconstruct a crime on the basis of whatever clues he can assemble and his knowledge of human psychology. It is hardly surprising that geology holds such a fascination for those involved in it. It is also hardly surprising that geology as a science is characterized by a wealth of lively controversy. Modern techniques of data gathering and interpretation have finally laid some of the older controversies to rest, but some persist and new ones are constantly being generated.

THE CURRENT STATE OF GEOLOGIC SCIENCE

Within the last decade a series of unifying concepts called **plate tectonics** has truly revolutionized the science of geology. The impact of plate tectonic concepts on earth science has been aptly likened to the impact that atomic theory had on physical science around the turn of the century. This does not mean that work done previous to this revolution in thinking has been invalidated. It means, rather, that the theory of plate tectonics has allowed an integration and correlation of many geologic phenomena that formerly appeared to be hopelessly disconnected and unrelated.

Armed with this new set of concepts, geologists can now understand many of the first-order features of the earth, such as continents, ocean basins, mountain belts, earthquakes, and volcanoes, and those factors that exercise fundamental control of their geographic locations. New insights into the location of mineral resources and even the controls of biologic evolution have been gained. This set of concepts has proved to be so powerful that the vast majority of geologists now accept them readily, and there is a great rush to reinterpret the geology of various geographic regions in terms of plate tectonic theory. This book was conceived with the intention of bringing the full breadth and importance of plate tectonic theory to the beginning student in geology.

The current enthusiasm for plate tectonic concepts among geologists has brought about a need for caution. There is always a danger that the overenthusiastic acceptance of any set of concepts in any field of science will tend to blind its practioners to other important concepts. Although plate tectonics appear to represent the grand scheme of earth evolution, the basic tenets of the theory must constantly be tested and critically appraised in terms of empirical data, and much work still remains to be done.

In our attempt to make this book truly contemporary in terms of earth science, we have of necessity included some interpretations and models that may not stand the test of time. Hopefully we have done so in such a manner that the questioning reader will be able to recognize the degree of uncertainty involved. Geology never was, and never will be, a cut and dried subject and, in keeping with this fact, we have incorporated in this book as much of the dynamic current thinking as possible.

THE WORK AHEAD

As knowledge in any field expands and new insights are gained certain problems are solved, but it is the nature of science that fresh horizons always stretch outward and new challenges arise. Geology is no exception. The rapid progress of the last decade has created more, rather than less, avenues for research. In particular, the dynamic controls of plate tectonics are presently poorly understood, and much remains to be learned about the early history of the earth. Furthermore, geologists will have to play a central role in the discovery and assessment of earth resources that will be urgently needed in the years to come. To be equal to this major and important task, geologists will have to draw on virtually every branch of geologic science.

QUESTIONS

1. What are the two important factors that distinguish geology from the other physical sciences?
2. How did Werner's concepts (Neptunism) differ from those of Hutton (Plutonism)? How was the controversy resolved?
3. What is the basic idea of the concept of uniformitarianism? Explain how this helps geologists understand ancient rocks.
4. The earth is constantly changing and evolving. How can this fact be reconciled with uniformitarianism?

Atoms, Minerals, and Rocks

THE ANATOMY OF ATOMS

Atoms are the basic building blocks of all matter, both organic and inorganic. Until this fact became known to man, he had no real chance of understanding the true nature of matter. Some knowledge of the anatomy of atoms is mandatory, not only to understand the nature of geologic materials but also to gain some insight into the radioactive and isotopic processes that are important factors in modern geologic research.

Atoms are extremely small and measure from less than 1 Ångstrom (Å) to about 2 Å in diameter (1 Å $= 1 \times 10^{-8}$ centimeters (cm) or 1/100,000,000 cm). Despite their extremely small size, the atoms of each chemical element are composed of even smaller, subatomic particles called **electrons, protons,** and **neutrons*** in unique combinations. The unique combination of configuration and specific numbers of electrons, protons, and neutrons of each atomic species (chemical element) defines its special chemical properties. If we could perceive atoms in some way, they would appear to us as spheres of different sizes (Figure 2-1A and B).

---

*Protons and neutrons can be split even further, but a discussion of this subject is beyond the scope of this book.

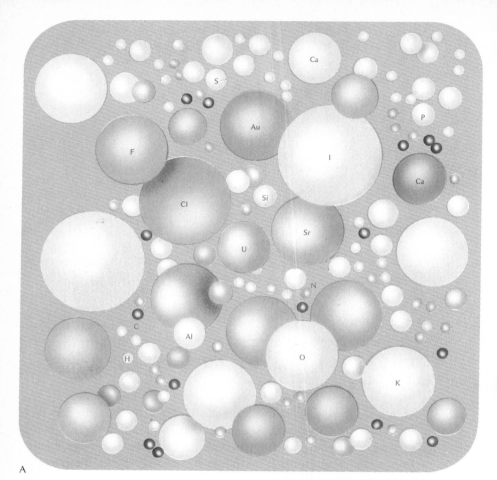

A

B

2-1. **A.** *The chemical elements. Each atom can be considered a sphere. When bound in crystals, these spheres become slightly distorted.* **B.** *The crystal structure of the mineral fluorite, CaF$_2$. The smaller atoms are calcium; the larger are fluorine.*

2-2. *Model of a chlorine atom, magnified about 2.5 thousand million times. Surrounding the nucleus are three shells of electrons; an innermost shell containing 2 electrons, a second shell with 8 electrons, and an outermost shell with 7 electrons.* [*Courtesy of Pennsylvania Salt Manufacturing Co.*]

Although atoms can be conveniently visualized as spheres, each atom consists of a tiny central nucleus surrounded, at some distance, by orbiting electrons organized in specific spherical shells (Figure 2-2). The tiny nucleus, which is made up of protons and neutrons, contains virtually all the mass (matter) of the atom. Protons and neutrons have almost the same mass but differ in their electrical nature. A proton has a positive electrical charge designated as +1, whereas a neutron has no electrical charge. Around the nucleus, but some distance from it, are the extremely small electrons, each of which has a negative electrical charge (designated as −1). The electrons constantly spin around the nucleus in orbits controlled by electromagnetic forces. Atoms in their elemental, uncombined form have a specific number of positively charged protons balanced by a similar number of orbiting electrons. One of the laws of atomic theory states that positive and negative electrical charges will always attract one another in an attempt to achieve electrical neutrality.

An electron has negligible mass in comparison to a proton or neutron; therefore, the **atomic weight** (**mass number**) of an atom is essentially equal to the number of protons and neutrons in its nucleus. The number of protons (**atomic number**) in each nucleus determines whether the atom is gold, iron, hydrogen, and so on. The number of protons also determines the number of electrons in any atom, and it is basically the number of electrons that controls the chemical behavior of the element. A change in the number of protons in an atomic nucleus will cause a change in the identity of the element. This was the goal of the alchemist of the Middle Ages, who, without adequate equipment and knowledge, sought and failed to change base metals into gold.

A change in the number of neutrons in an atomic nucleus is a much less profound matter because it involves no change in either the size or electrical characteristics of the atom. Among the 92 naturally occurring elements, 60 contain different numbers of neutrons in their atoms. For example, one variety of oxygen (^{16}O) contains eight protons and eight neutrons, a second (^{17}O) contains eight protons and nine neutrons, and a third (^{18}O) contains eight protons and ten neutrons. Each variety has the same atomic number (8) but a different atomic weight (16, 17, and 18). These distinctive nuclear forms of the same element are known as **isotopes.** Some elements have many isotopes; tin, for example, has ten.

Not all isotopes are atomically stable. All the isotopes of oxygen are stable, but, in the case of carbon, two isotopes (^{12}C and ^{13}C) are stable and one (^{14}C) is unstable. The consequence of this instability is natural radioactivity, a nuclear process. Radioactivity is the spontaneous emission from the nucleus of certain types of particles or rays. This subject is developed further in the following chapter.

Studies of the distribution of natural isotopes in rocks and minerals have greatly aided our understanding of many geologic processes and events. Some radioactive isotopes are useful as natural geologic clocks (see Chapter 3), and studies of both radioactive and stable isotope distributions in minerals have provided insights into the origin of certain rocks. Natural decay of the relatively abundant radioactive isotopes of potassium, uranium, and thorium is an important mechanism for generating heat within the earth. Many of the dynamic internal processes of the earth are set in motion by this heat energy.

MINERALS AS COMBINATIONS OF ATOMS

The elements in the earth's crust and interior for the most part do not occur as individual atoms but are combined into crystalline substances called **minerals.** The combination of atoms of one element with those of others to form a mineral involves a chemical bonding (**reaction**) in which the outermost electrons of the atoms of the different elements are rearranged. Atoms involved in chemical bonds are generally not in their neutral state but have been converted, by either loss or gain of one or more electrons, into **ions.** Some atoms can lose electrons (negative charges) and thus form an ion carrying a positive charge. Other atoms can gain electrons (negative charges), and thus form ions that have a residual negative charge. Positively charged atoms are called **cations** and negatively charged atoms are called **anions.**

The factors that cause certain atoms to have a tendency to gain one or more electrons and other atoms to tend to lose one or more electrons when they form ions are complex. For our purposes it is sufficient to realize that such changes in the electron population of atoms do routinely occur and, as a result, neutral atoms are changed into electrically charged ions.

A relatively simple example of this phenomenon is the mineral halite (sodium chloride, NaCl), or common table salt. In sodium (Na) atoms the outer electron shell contains only one loosely held electron. Chlorine atoms, on the other hand, contain seven electrons in their

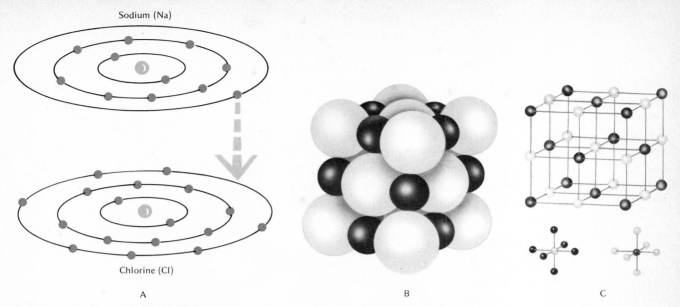

Sodium (Na)

Chlorine (Cl)

A

B

C

2-3. A. *In sodium chloride (halite), sodium becomes Na$^+$ by giving one of its eleven electrons to chlorine, making it Cl$^-$. Then each element has eight electrons in its outer shell, a noble gas configuration. B. The mineral halite, NaCl. Sketch showing relative ion sizes. Chlorine is the larger. C. Expanded view showing the position of the centers of the atoms. Each sodium is surrounded by six chlorine atoms and likewise each chlorine is surrounded by six sodium atoms.*

outermost electron shell. It so happens that eight electrons in a shell form a particularly stable configuration. There is, therefore, a strong tendency for sodium atoms to lose their outermost electron and become Na$^+$ ions, and for chlorine atoms to gain one electron to become Cl$^-$ ions (Figures 2-3A, B and C). The combination of Na$^+$ and Cl$^-$ ions forms a stable, electrically neutral **chemical compound,** NaCl. This compound is routinely used by man in his food, but both sodium and chlorine in their elemental forms are extremely reactive and poisonous substances.

Water, an Important Geologic Substance

Water, although it is not a solid (except when frozen), can be considered as a mineral and provides an excellent example of the influence of atomic properties on the behavior of a material. Water (H_2O) has a **molecular structure;** that is, it consists of atoms bonded together to form molecules. The molecular structure of water (Figure 2-4) is such that the two hydrogen atoms lie on practically the same side of the oxygen atom to which they are linked. This structural arrangement makes the water molecule a **dipole,** which means that one end is positively charged and the other is negatively charged. It is precisely this property of water molecules that gives water the ability to dissolve many chemical compounds, such as minerals. A water molecule can "wedge" its way between surface ions in a mineral and "float" neighboring ions away from one another (Figure 2-5).

Water, although it is a liquid, is not entirely without structure. A water molecule because of its atomic configuration has in effect four electrically charged "arms" extending from the nucleus. Two extend from the positively charged hydrogen atoms, and the other two carry the excess negative charges. When water molecules are packed together, as in a liquid, each negative arm attracts a positive (hydrogen) arm in a neighboring water molecule; the hydrogen atoms then act to join the molecules in what is called a **hydrogen bond** (Figure 2-6). This is an extremely tough bond compared to most bonds that form between *molecules* (atomic bonds are stronger), and it is the strength of the hydrogen bond that gives water its remarkable properties. For example, no other substance can absorb and release more heat than water. The evaporation (or boiling) of water involves breaking these tough intermolecular hydrogen bonds and therefore requires a great deal of heat energy. Approximately five times as much heat energy is needed to change a given volume of water from liquid to vapor as is needed to raise its temperature from the freezing to the boiling point.

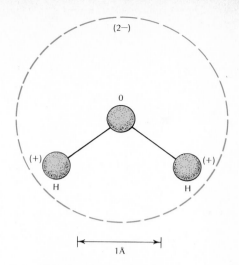

2-4. *Dipolar structure of the water molecule. The hemisphere with the two hydrogens carries a net positive charge; the opposite hemisphere carries a net negative charge.*

2-5. *Schematic illustrations of the mechanism of ionic solution. Dipolar water molecules attach themselves to the ions, overcoming the ionic bonds of the ions in the crystalline solid and "float away" the detached ions in solution.*

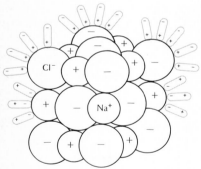

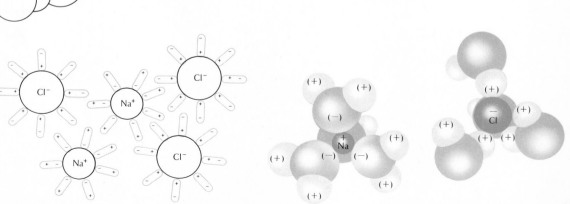

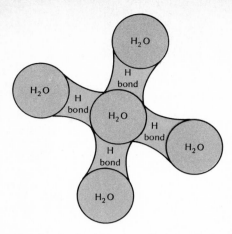

2-6. *In the liquid state each water molecule establishes hydrogen bonds with four of its nearest neighbors, and it is largely due to these bonds that water has many unique qualities.*

It is precisely this effect that your body uses as a cooling system —evaporation of water from your skin. That is why hot humid weather is so much less comfortable than hot dry weather; the increased humidity in the air reduces the rate at which water molecules can evaporate from one's skin. As a result, the efficiency of the human cooling system is impaired. The world's climate is tempered by the ability of water to soak up and store the sun's heat and then release it slowly. Without large bodies of surface water, the earth's surface would be an extremely inhospitable place because the daily variation in temperature would be much more extreme than at present.

Most substances, whether solid, liquid, or gaseous, contract when cooled. Water is no exception over a wide range of temperature. However, close to its freezing point water behaves quite differently. As the temperature of water falls below 4°C, it gradually expands. A rapid expansion occurs at the freezing point during the change from liquid water to solid ice. In fact the volume of ice is about 9% greater than that of liquid water. In this respect water is decidedly abnormal, and the consequences of this abnormality govern many geologic phenomena at the earth's surface. If ice was denser than water, frost action would not be a major factor in the mechanical breakdown of surface rocks in temperate and cold climates. Lakes in high latitudes would freeze from the bottom up during the winter and most of them would remain partially frozen even during the summer. The net effect would be either that life within them would not have evolved at all or that the fish and plants in such lakes would have radically different survival mechanisms. In addition, climates in temperate regions would be much harsher.

Mineral Formation

Each solid mineral (with the exception of a few amorphous minerals) is defined by two fundamental properties:

1. *Crystal structure*—the geometric arrangement of the atoms (or ions) composing the mineral.
2. *Chemical composition*—the proportions of different chemical elements contained in the mineral.

In geologic environments where mineral formation is taking place, such as the cooling of molten rock material, various chemical elements are, in a sense, competing with one another to form stable

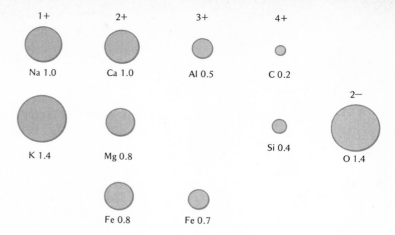

2-7. *Ionic size (in angstroms) and charge of some of the abundant cations in the earth's crust. Also included is the important anion O^{2-}.*

crystalline compounds (minerals). The kinds of minerals that will form in any particular case depend on various factors, such as temperature, pressure, and the mobility and relative abundance of chemical elements at the site of mineral formation. In detail, mineral formation can be exceedingly complex and many aspects of the process are still not completely understood. Nevertheless, we do know that two fundamental properties of each ion, its **size** and electrical **charge**, influence its ability to combine with other ions or atoms to form minerals. Figure 2-7 illustrates the relative size and charge of the most important mineral forming ions. Note that iron (Fe) forms two distinct ions, one with a plus two (+2) charge and one with a plus three (+3) charge. Stripping off two electrons from the outer shell of the original iron atom reduces its size. Removal of a third electron to produce Fe^{3+} makes the ion even smaller. The reduction in size results not only from the loss of another electron but also because the remaining electrons are now more strongly attracted to the protons in the nucleus and pull in closer. Conversely, oxygen ions (O^{2-}) are larger than oxygen atoms because electrons are added in the process of ionization (O (atomic) + 2 electrons = O^{2-}).

THE CRYSTAL CHEMISTRY OF SILICATE MINERALS

About 2000 distinctive mineral types or species have been discovered. Fortunately, the large majority of these are extremely rare. The most important minerals, from the viewpoint of the geologist, can be divided into five chemically distinctive families: silicates, carbonates, oxides, sulfates, and sulfides (Table 2-1). Silicate minerals are by far the most important rock-forming minerals. Of these the

Table 2-1 Common Rock-Forming Minerals

Mineral	Idealized Formula	Color	Hardness, Mohs' scale*	Cleavage
SILICATES				
Quartz	SiO_2	Variable	7	None
Orthoclase	$KAlSi_3O_8$	White to pink	6	Two at right angles
Plagioclase	$CaAl_2Si_2O_8$ $NaAlSi_3O_8$	White to green	6	Two at right angles
Muscovite mica	$KAl_3Si_3O_{10}(OH)_2$	Clear to light green	2–2.5	One perfect
Biotite mica	$K(Mg,Fe)_3Si_4O_{10}(OH)_2$	Black to dark green	2.5–3	One perfect
Clay minerals	$Al_2Si_2O_5(OH)_4$	Light	Very low	
Amphibole	$Ca_2(Mg,Fe)_5Si_8O_{22}(OH)_2$	Black to dark green	5–6	Two at 60°
Pyroxene	$(Mg,Fe)SiO_3$	Black to dark green	5–6	Two at right angles
Olivine	$(Mg,Fe)_2SiO_4$	Green	6.5–7	None
Chlorite	$(Mg,Fe,Al)_6(Al,Si)_4O_{10}(OH)_8$	Green	2–2.5	One perfect
Garnet	$(Ca,Mg,Fe,Al)(SiO_4)$	Green, brown, red	6.5–7.5	None
CARBONATES				
Calcite	$CaCO_3$	Generally white or colorless	2.5–3	Three, not at right angles
Dolomite	$(Ca,Mg)CO_3$	Pinkish or white	3.5–4	Three, not at right angles
OXIDES				
Magnetite	Fe_3O_4	Black	6	None
Hematite	Fe_2O_3	Reddish brown to black	5.5–6.5	None
SULFATES				
Anhydrite	$CaSO_4$	White		
Gypsum	$CaSO_4 \cdot 2H_2O$	White or clear	2	One perfect, other less good
SULFIDES				
Pyrite	FeS_2	Pale brass yellow	6–6.5	None
Pyrrhotite	FeS	Brownish	3.5–4.5	One sometimes distinct

*See glossary for definition of Mohs' scale

Crust		Earth	
Oxygen	= 46.1	Iron	= 34.8
Silicon	= 27.7	Oxygen	= 29.3
Aluminum	= 8.3	Silicon	= 14.7
Iron	= 5.4	Magnesium	= 11.3
Calcium	= 4.9	Sulfur	= 3.3
Magnesium	= 2.6	Nickel	= 2.4
Sodium	= 2.2	Calcium	= 1.4
Potassium	= 1.9	Aluminum	= 1.2

*It is important to realize that, because we can only sample the outermost surface of the earth, these data are merely best estimates of average crustal and earth compositions. Recent studies, for example, have shown that the average potassium content of the earth's crust is probably about 1%. Note that the crust is enriched in silicon, oxygen, aluminum, sodium, and potassium (all relatively light elements) with respect to the earth as a whole.

simplest chemically is quartz (SiO_2). All other important rock-forming silicate minerals are combinations of silicon and oxygen with one or more of the other common elements, such as sodium, potassium, aluminum, calcium, magnesium, or iron. In fact these eight elements make up over 98% by weight of the upper crust of the continents (Table 2-2), that portion of the earth accessible for direct study by geologists. Note that this is not true with respect to the estimated composition of the whole earth or of the universe.

The fundamental unit in silicate crystal structure is the **tetrahedral** (pyramid shaped) arrangement of four O^{2-} ions around a central Si^{4+} ion (Figure 2-8). This tetrahedral packing arrangement (or **coordination**) is a consequence of the tiny size and high charge (see Figure 2-7) of the Si^{4+} ion, and the relatively large size of O^{2-} ions. Cations (positively charged ions) such as iron, calcium, and magnesium share these tetrahedrally bound oxygens in most silicate mineral structures. The basic silicon-oxygen tetrahedral building blocks may thus be linked together by these cations, by sharing oxygen atoms between neighboring tetrahedra, or by some combination of both. The type of linkages that form define the various silicate structural types (see Figure 2-9, for example, rings, chains, sheets, or complex three-dimensional frameworks).

Mineral structures can thus be thought of as geometrically organized **sites** or positions where specific ions are stably bound during mineral formation. Because ions and groups of ions are assembled into defined three-dimensional structures during mineral growth, each site within the **crystal structure** is of a more or less specific

2-8. *The silicon-oxygen tetrahedron, the basic structural unit of the silicates. Black central sphere is silicon, larger surrounding spheres are oxygens (top and side views).*

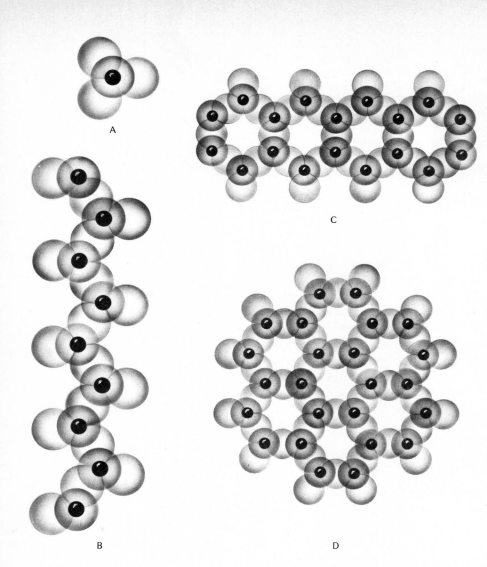

A

C

B

D

2-9. *Types of linkages of the SiO_4 tetrahedra common in silicate minerals. A. Independent tetrahedra (olivine). B. Single chains (pyroxenes). C. Double chains (ampiboles). D. Two dimensional sheets (clays and micas). Three-dimensional networks (feldspars and quartz) not shown. Note that these are top views, looking through "transparent" oxygen atoms.*

size. This explains why different ions of approximately similar size and charge can substitute for one another in the same site in a mineral structure, and why certain specific minerals can display a range of composition. For example, **olivine** can range in composition from pure magnesium silicate (Mg_2SiO_4) to pure iron silicate (Fe_2SiO_4). All naturally occurring olivines lie somewhere between the above end-member compositions and are consequently designated $(Mg,Fe)_2SiO_4$. Clearly then Mg^{2+} and Fe^{2+} can substitute for one another in the olivine structure. This substitution is not a random matter and is controlled mainly by temperature and the relative availability of Mg^{2+} and Fe^{2+} ions during olivine formation. The same is true for

2-10. *Handspecimen of obsidian, a black glass of granitic composition. Note typical fracture pattern. [Photo by M. Kauper]*

└ 5 cm ┘

other rock-forming silicates that contain magnesium and iron, such as some **pyroxenes, amphiboles** and **biotite mica** (see Table 2-1).

Plagioclase feldspars also exhibit a range in composition from a calcium-rich end-member ($CaAl_2Si_2O_8$) to a sodium-rich end-member ($NaAlSi_3O_8$). Due to the difference in charge between Na^+ ions (+1) and Ca^{2+} ions (+2) the substitution of sodium for calcium must be accompanied by a second, coupled substitution that compensates for the difference in charge between Ca^{2+} and Na^+ ions. This is achieved by the substitution of a third Si^{4+} ion for one of the Al^{3+} ions. Thus, the coupled substitution that occurs in the plagioclase series can be written

$$Ca^{2+} + Al^{3+} \longrightarrow Na^+ + Si^{4+}$$

As in the case of olivine, the degree of substitution of sodium for calcium is not random but is governed by the conditions (physico-chemical environment) under which any particular plagioclase crystal forms. In general terms, minerals that can encompass these types of compositional variations are extremely valuable for the reconstruction of environments of mineral formation.

Minerals are crystalline substances; that is, they have an orderly internal structure. There is a class of geologic materials, however, that do not have orderly internal structure—glasses. Glasses (Figure 2-10) are liquids that have been cooled so rapidly that their constituent ions have not had time to organize themselves into completely defined crystalline structures (minerals) before solidification occurred. Glass is a familiar material to us, but naturally occurring glasses are not so common. Their nature, however, immediately tells a geologist that he is dealing with material that was not only once molten but also underwent rapid chilling.

MINERALS AS ENVIRONMENTAL INDICATORS

In the previous section we have seen that many minerals contain keys in their makeup to the conditions under which they originally formed. An understanding of these keys has come largely from studying minerals in the laboratory. Experiments involving the synthetic growth of minerals under controlled laboratory conditions have provided geologists with rather precise data on the pressure, temperature, and chemical conditions under which specific minerals can form.

Certain minerals, for example, quartz (SiO_2), can form under a wide range of environmental conditions. Others fortunately can form only within a narrow range of conditions, and thus provide far more specific information regarding the environments in which they formed. A number of minerals are good indicators of temperature, especially those that undergo the types of compositional variations described in the previous section. Fewer provide us with any information about pressure; however, **diamond** is one of these. A dense crystalline form of carbon (C), diamond can only form at high temperatures and extremely high pressures. **Graphite** is also a crystalline form of carbon, but graphite forms at much lower temperatures and pressures. These forms are interchangeable, but fortunately for the jewelry business the change of diamond (the less stable form) to graphite (the more stable form) at the temperatures and pressure of the earth's surface is so extremely slow it cannot be measured.

A few minerals can also provide us with rather specific chemical information. For example, the presence of a bed of halite (NaCl) in sediments that have clearly formed in a marine environment can indicate that the overlying sea water had become ten times more salty than normal sea water, thus allowing direct precipitation of halite and other salts on the sea floor.

The types of minerals that form from the weathering of rocks at the earth's surface depend to some extent on the local climatic conditions. In some cases these minerals are preserved more or less in place by burial beneath younger sediments. When uplift and erosion again make the minerals available for study, geologists can gain information about past climatic conditions in that area.

Minerals are thus considered by geologists not merely as the building blocks of rocks but as important keys to the history of the rocks in which they occur.

PHYSICAL PROPERTIES OF MINERALS

Under normal working conditions the geologist does not have the facilities to investigate the composition and crystal structure of the minerals he encounters. He must therefore make his mineral identifications in the field on the basis of the external aspects (physical properties) of the samples he encounters. These physical properties are governed by the chemical composition of a mineral and the manner in which these chemical compounds are assembled (crystal structure of the mineral).

In general terms, *color, shape, density, fracture,* and *hardness* are all properties the geologist can use to aid him in field identification of minerals. The external shape of well-formed crystals is a considerable aid in mineral identification, but in most geologic environments mineral growth occurs under conditions of mutual interference so that well-formed crystals (Figure 2-11) are comparatively rare in nature. The beautiful crystals one sees in museums are vir-

2-11. *Well formed quartz crystals showing typical shape. Dark areas are crystals of zinc sulfide (ZnS). [Photo by M. Kauper]*

| 5 cm |

2-12. *As a group the micas are characterized by their perfect two-dimensional cleavage, giving thin, flexible, and elastic cleavage plates. The underlying sheet structure of the micas is illustrated in Figure 2-9.*

tually never encountered in routine geologic work. Color, perhaps the most obvious physical characteristic of minerals, is much less diagnostic than might be imagined. Some minerals have bright and distinctive colors, but many minerals are nondescript in color, and in certain cases different specimens of the same mineral can exhibit a wide range of colors. Probably the most helpful single physical property of minerals, in terms of field identification, is the pattern displayed in broken surfaces. Not only is this how one sees minerals in most rocks but the fracture patterns are controlled by the fundamental internal structure of each mineral.

Cleavage is the tendency of certain minerals to fracture cleanly along certain preferred directions, which are dictated by the internal arrangement of their atoms (crystal structure). These cleavage directions represent planar surfaces within the mineral along which the molecular bonds are relatively weak. The excellent cleavage of micas (Figure 2-12), for example, is a direct reflection of their underlying sheetlike crystal structure: the bonding between the sheets being much weaker than the lateral bonding within each sheet. Some minerals exhibit two, or even three, directions of cleavage (Figure 2-13). As a consequence, these minerals fracture into distinct shapes, or cleavage fragments, dictated by the geometric arrangement of the surfaces along which bonding is weak. In all cases, the way in which a mineral fractures, either along well defined cleavage directions or in a hackly, uneven manner, is a direct manifestation of its internal atomic structure.

Another example of the close relationship between the physical properties of minerals and their crystal structure is provided by diamond and graphite. Diamond is prized not only for its attractive

2-13. *Cleavage fragments of the mineral halite, NaCl. Compare the external shape with the internal atomic structure illustrated in Figure 2-3. Cleavage is always parallel to a possible crystal face and halite crystals also exhibit a cubic form.*

brilliance when cut and faceted but for its extreme hardness. Due to its formation at extremely high pressures, diamond has a compact crystal structure, a density three and a half times that of water, and is harder than any other naturally occurring substance. Graphite, on the other hand, has a less compact, sheetlike structure, a density only slightly more than twice that of water, and is a very soft substance. The bonding between individual graphite sheets is so weak, in fact, that powdered graphite is used as a lubricant. Graphite is also the principal constituent of pencil "lead." As you draw the point of a pencil across paper, the sheets of graphite are easily stripped off onto the paper.

One of the most important physical properties of certain iron-bearing oxide minerals, from the viewpoint of current geologic research, is their magnetism. These minerals, for example **magnetite** (Fe_3O_4), have the ability to act as weak magnets if they become magnetized in some manner. Large specimens of magnetite (called lodestones) were used by ancient navigators as crude compasses long before the nature of magnetism and the earth's magnetic field were understood.

It is the nature of all magnetic materials to lose their magnetism above a certain temperature (Curie temperature or **Curie point**). The Curie point of pure iron is 770°C; of magnetite, 580°C. Many rocks that contain magnetite are formed by the crystallization of once molten material, and this crystallization is completed before the rock has cooled to a temperature of 580°C. The magnetite crystals in these newly formed rocks are therefore formed at temperatures above which they can become magnetized. As the rock cools through the Curie point, the magnetite crystals become magnetized by the earth's magnetic field. The net result is that the cooling rock acquires a magnetism aligned in direction with that of the earth's magnetic field in that area.

The importance of this fact is that the rock freezes in this magnetism. Thereby it retains a record (or memory) of the attitude and direction of the earth's magnetic field at the time and place of the formation of that particular rock. Techniques have now been developed by which the frozen magnetic record of the earth's magnetic field can be measured in oriented specimens from certain rock types that are millions or even billions of years old. Careful compilation of this type of **paleomagnetic data** has provided critical supporting evidence for the revolutionary geologic concepts of plate tectonics, referred to in Chapter 1. We will deal with the subject of rock magnetism in more detail in Chapter 5.

2-14. *Typical hand-specimen of coarse grained granitic rock. Individual crystal grains exhibit interlocking grain boundaries due to mutual interference during growth.* [*Photo by M. Kauper*]

⌞ 5 cm ⌟

THE THREE MAJOR ROCK FORMING ENVIRONMENTS

Rocks, the materials that the geologist encounters and must interpret in the field, are aggregates of minerals. In some instances a rock might consist of grains of only one mineral; in others it might consist of a complex mixture of mineral types. As we stressed in the foregoing sections, minerals can provide information regarding the environment in which they formed. There are also certain aspects of rocks that provide the geologist with information regarding their formation.

The **texture** of a rock is influenced by the manner in which its constituent minerals were assembled. An assemblage of tightly interlocked, irregular mineral grains (Figure 2-14), for instance, indicates a formational environment distinctly different from an assemblage of rounded mineral grains (Figure 2-15) cemented together by some other mineral substance. A rock containing minerals of large **grain size** has obviously experienced different conditions during its formation or subsequent history than one containing minerals of extremely small grain size. Some rocks have their constituent mineral grains arranged in a uniform **fabric** that shows no sign of any directional

2-15. *Handspecimen of a conglomerate showing large pebbles of quartz cemented together by other minerals. [Photo by M. Kauper]*

| 5 cm |

texture. Others exhibit a highly directional fabric, in which all or most of the mineral grains are oriented along a certain preferred direction. By using their knowledge of mineralogy and textural relationships, geologists can classify virtually any rock they encounter in one of the three major families of rock types: igneous, sedimentary, and metamorphic.

Igneous Rocks

Igneous (from the Latin word *ignis,* meaning fire) rocks form from the cooling and crystallization of hot, molten rock material called **magma.** The formation of igneous rocks can take place both within the earth and at the earth's surface. Any traveller fortunate enough to be in Hawaii during one of the volcanic eruptions can watch rocks form from the hot molten material that flows up from within the earth (Figure 2-16).

Rocks of different compositions melt at different temperatures; therefore, natural magmas can exist over a rather wide range of temperatures, depending on their composition. Some information about the temperature of magma has come from direct measurements of lava pools, but most of our knowledge of this subject has been obtained from laboratory studies of the melting of rocks. The whole subject of the crystallization of igneous rocks and their constituent

2-16. *Hawaiian lava cooling as it flows slowly over a step-like feature about 1 m. high.*

minerals is one of considerable complexity, and it has taken a very large number of field studies and laboratory experiments on igneous minerals to come to our present understanding of igneous processes.

The texture and grain size of igneous rocks, two features readily observable in the field, provide considerable insight into the crystallization history of igneous rocks. Those rocks with a glassy (noncrystalline) or fine grained texture can be inferred to have crystallized rapidly as a result of sudden cooling in surface or near-surface environments. Those with coarse grained texture can be inferred to have crystallized more gradually in deeper seated environments where cooling would be slower.

The minerals in certain igneous rocks exhibit a mixture of strongly contrasted grain size. Large crystals or **phenocrysts** of one mineral appear to be floating in a groundmass of fine grained minerals (Figure 2-17). This type of relationship is called **porphyritic texture,** and the igneous rocks that exhibit it are called **porphyries.** A porphyritic texture indicates that a certain species of mineral began to crystallize before, and grew more rapidly than, any other mineral in the cooling melt. These early forming crystals had sufficient time to attain their relatively large size before the rest of the magma started to crystallize. Porphyritic texture is common in igneous rocks formed at or near the surface of the earth. In both cases the inference can be made that crystallization had begun in the magma well before its final solidification.

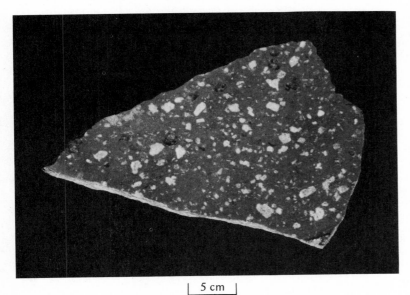

2-17. *Handspecimen of a porphyritic volcanic rock in which phenocrysts of plagioclase are present in a fine grained groundmass of other minerals.* [*Photo by M. Kauper*]

5 cm

2-18. *Two zoned plagioclase crystals in a porphyritic igneous rock. Cores of early formed plagioclase crystals are rich in calcium (Ca). As crystallization proceeded plagioclase crystals became increasingly richer in sodium (Na). Finally the remainder of the magma crystallized as a fine grained assemblage of minerals (matrix).*

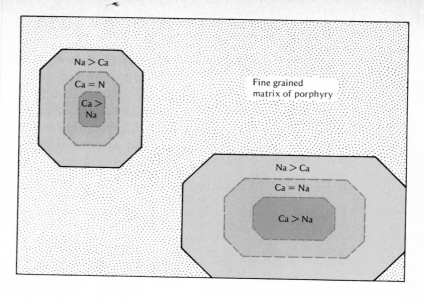

Scale ⊢——————————⊣
1 cm

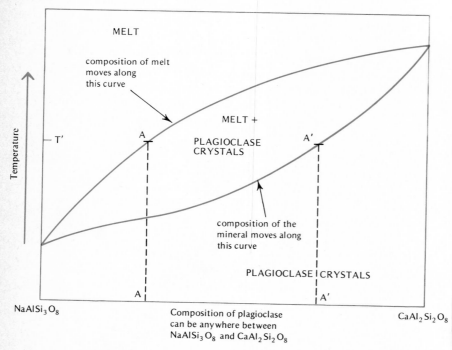

2-19. *Temperature-composition diagram for the plagioclase feldspars. The relationships provide information on the compositions and physical conditions of the various materials at elevated temperatures in the plagioclase system. For example at temperature T' the system will contain melt of composition A plus crystals of composition A'. If the temperature is lowered, the compositions of melt and crystals will slide down their respective curves to new positions.*

Table 2-3 **Composition of Volcanic Gases from Kilauea Volcano, Hawaii**

Constituent	Volume %
H_2O	67.7
CO_2	12.7
N_2	7.65
SO_2	7.03
SO_3	1.86
S_2	1.04
H_2	.75
CO	.67
Cl_2	.41
Ar	.20

2-20. *Dike intruding very young gravel deposit near Lake Mead, Utah. [Photograph courtesy of C. K. Chase]*

The phenocrysts in many igneous rocks that exhibit porphyritic texture are magnesium-rich olivine or calcium-rich plagioclase, as these minerals tend to start crystallizing first from common magma types. Because the phenocrysts differ in composition from the magma in which they are forming, the net result of their crystallization is to change the composition of the remaining melt. If early formed igneous minerals are separated by some process (for example, gravitational settling) from the magma, the separated minerals and the residual magma will form different igneous rocks. Such rocks will not only differ from one another in composition but also differ from the composition of the original magma that produced them.

Phenocrysts of plagioclase are commonly observed to consist of a number of concentric zones, each of slightly differing composition. In many cases the zoning patterns are complex, but in general terms the cores of the phenocrysts contain calcium-rich plagioclase and the margins contain sodium-rich plagioclase (Figure 2-18). These zonal patterns are best understood in terms of the melting relationships of the plagioclase series ($Ca_2Al_2Si_2O_8 - NaAlSi_3O_8$) that have been worked out experimentally (Figure 2-19).

All magmas contain gases in solution. The most important of these is water, but carbon dioxide (CO_2), nitrogen (N_2), sulfur dioxide (SO_2), sulfur trioxide (SO_3), hydrogen (H_2), carbon monoxide (CO), chlorine (Cl_2), and argon (Ar) are generally present in lesser quantities (Table 2-3). Although most of the gas escapes when a magma reaches the surface, some expands, is trapped in the solidifying magma, and forms the cavities or **vesicles** commonly observed in the upper parts of former lava flows. Igneous rocks that crystallize deep under the earth's surface do not develop vesicles, but the amount and composition of the gases in their parent magmas tend to influence the types of igneous minerals that will form and the temperature range of crystallization.

Igneous rocks that crystallize in surface environments (**extrusives**) form either lava flows or layers of fragmental volcanic material. Igneous rocks that crystallize beneath the earth's surface (**intrusives**) can assume a variety of forms. The most common shapes of intrusives formed near the surface are sheets or **pipes**. Sheetlike intrusions that cut across their host rocks are called **dikes** (Figure 2-20), whereas sheets of igneous rock that are emplaced parallel to the layering of stratified rocks are called **sills** (Figure 2-21). Igneous rocks emplaced in deeper crustal environments commonly have the form of somewhat irregular domes, and if these bodies have dimensions of less than 30 km across they are called **stocks**. Intrusive bodies of larger dimensions are called **batholiths** (see Figure 2-21).

2-21. *Sketch to illustrate various igneous rock forming environments and the names given to various shapes and forms of igneous bodies. Note that dikes are sheetlike and crosscut the enclosing rock layers sharply. Sills are also sheetlike but are emplaced parallel to the enclosing rock layers. Stocks are generally cylindrical in shape, whereas the much larger batholiths are generally somewhat irregular in shape. X's along the margins of intrusive igneous bodies indicate baking of adjacent rocks by heat. Volcanic cones consist of varying amounts of lava flows and volcanic ash that have been erupted from a volcanic neck.*

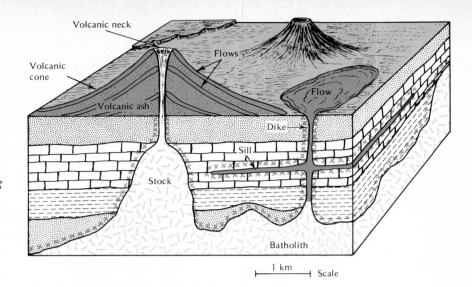

Sedimentary Rocks

Sedimentary rocks are formed in surficial environments from materials derived from the weathering of preexisting rocks. Although they cover about 75% of the surface of the earth, sedimentary rocks comprise only about 5% of the volume of the crust. The various surficial processes that are involved in the breakdown of rocks and in the transport of this debris to areas of sediment deposition will be dealt with in later chapters of this book. Sediments, the starting materials in the formation of sedimentary rocks, can be divided into two broad types: clastic sediments and chemical (nonclastic) sediments. **Clastic** sediments, such as mud and sand, are composed mainly of broken and worn fragments of preexisting minerals, rock particles, and even fossils that have been transported to the site of their deposition by water, wind, or ice. **Chemical** sediments are formed at the site of their accumulation by chemical precipitation. An important subtype of chemical sediments are those of biologic origin; that is, those composed of the skeletons of marine organisms.

Geologists have developed most of their insights regarding ancient surface environments by studying modern sediments and sedimentary environments. By careful comparison of the data obtained in such studies with those obtained from ancient sedimentary rocks, inferences can generally be made about conditions in the area at the time the ancient sediments were formed. Here we have an example of the use of the concept of uniformitarianism. Most sediments are deposited in the sea, but some are deposited in land areas. In fact nonmarine sediments will tend to accumulate in any site where the input of materials is more rapid than their removal by erosion.

The transformation of a sediment, which is unconsolidated material, into a sedimentary rock is an important geological process called **lithification.** Lithification occurs mainly by three processes: cementation, compaction, and recrystallization. The most important cementing agents are calcite ($CaCO_3$), silica (SiO_2), and iron oxides (for example, hematite Fe_2O_3). These cementing agents can be introduced into the sediment by the movement of mineralized water, or they can be derived from the solution and reprecipitation of original mineral constituents of the sediment. The compaction of sediments to form sedimentary rocks generally results from the weight of overlying sediments during burial and from a reduction in volume (especially of muds) due to squeezing out trapped water. Recrystallization is an important mechanism in the lithification of fine grained sediments, especially calcite-rich mud. The new crystals that grow within a sediment and the enlargement of some of those already present result in a mosaic of interlocking grains; as a consequence, the sediment hardens into a rock.

Clastic Sediments. In many cases clastic sedimentary rocks are of great help in deciphering ancient environments. Their grain size (Table 2-4) can tell us something about the energy characteristics of the medium that transported them. Slowly flowing water, for example, can transport clay and possibly silt sized materials but is incapable of transporting coarse sand and gravel. The composition of the mineral grains, and rock fragments if they are present, can provide information regarding the rock types and climatic conditions in the source area. The degree of rounding, especially of the sand sized grains, provides a measure of the degree of abrasion they have suffered during their journey from source rock to site of deposition. Coarse grained clastic rocks such as sandstones and conglomerates

Table 2–4 **Grain Size Classification of Clastic Sedimentary Rocks**

Grain Size, mm	Sediment	Rock
Greater than 2	Gravel	Conglomerate*
1/16–2	Sand	Sandstone*
1/256–1/16	Silt	Siltstone
Less than 1/256	Clay	Shale

*Note: Conglomerates and sandstones commonly contain a certain fraction of grains of smaller grain size. Thus, many conglomerates have a sandy matrix and many sandstones have a silty matrix.

2-22. *Handspecimen of medium grained quartz sandstone. An idea of the granular nature of the rock can be gained by noticing its surface texture. [Photo by M. Kauper]*

| 5 cm |

(Figure 2-22) tend to be more useful for the reconstruction of ancient environments because, in most cases, they are deposited close to, or even on, land areas. Fine grained sedimentary rocks such as shales are generally deposited further away from land areas.

Some minerals break down relatively rapidly in surface environments; consequently they are seldom found in sedimentary rocks, even though they may be important constituents of the source rocks of a particular sediment. Quartz, on the other hand, is the most resistant of all the common rock-forming minerals so that quartz grains form the bulk of most sandstones.

Nonclastic Sediments. The deposition of nonclastic sediments differs markedly from that of clastic sediments. Nonclastic sediment components are formed by chemical and biological agencies from material in solution. The starting materials for virtually all such sediments are the ionic species dissolved in sea water (Cl^-, Na^+, SO_4^{2-}, Mg^{2+}, Ca^{2+}, CO_3^{2-}).

The predominant mineral in most nonclastic sediments is **calcite** ($CaCO_3$, calcium carbonate). A large proportion of the calcite extracted from sea water, which eventually forms nonclastic sedimentary rocks, is of biologic origin. It is extracted from sea water by marine organisms that secrete calcium carbonate to form the hard parts of their skeletons, and these skeletons accumulate on the sea floor after the organisms die. Eventually this material forms sedimentary rocks called **limestones.** Not all limestones are of biologic origin however; under certain circumstances calcite can precipitate from sea water by purely chemical (inorganic) means.

A less important, but still significant nonclastic sedimentary rock type is **chert**, composed of very fine grained silicon dioxide (SiO_2). Much like limestone, chert can result from the precipitation of silicon dioxide by either biological or chemical means.

Evaporites are chemically precipitated rocks resulting from the evaporation of sea water. They comprise about 3% of the total sedimentary rock mass and consist chiefly of the minerals gypsum ($CaSO_4 \cdot 2H_2O$), anhydrite ($CaSO_4$), and halite ($NaCl$), with associated carbonates such as calcite and dolomite [$(Ca,Mg)CO_3$]. Obviously, a rather special combination of conditions involving a supply of sea water, absence of clastic sediments, and arid climates is a prerequisite for the formation of evaporites. Nevertheless, they are found in many sedimentary basins that formed at various stages during the last 500 million years. Environments of large scale evaporite deposition are rare today, but evaporite minerals are forming in parts of the Caspian Sea and along the tidal flats of the Persian Gulf.

Metamorphic Rocks

Metamorphism (from the Greek for "changed form"), the process by which metamorphic rocks are generated, involves the mineralogical and textural reordering of preexisting rocks within the earth's crust. In fact all bodies of rock tend to change their mineralogical and textural character when subjected to new sets of physical and chemical conditions, whether within the earth or at its surface. Many of these processes are discussed elsewhere in this book. Our concern here is with processes that occur at depths greater than approximately 1 km within the earth's crust.

The most important agents of metamorphism are heat and pressure, but stresses due to earth movements and the migration of chemically active fluids can play a significant role in the formation of metamorphic rocks. At relatively low temperatures and pressures, mineralogical reactions in metamorphic rocks are limited mainly to the recrystallization of micas and carbonates, and physical changes such as the development of planar structures (**schistosity,** Figure 2-23) tend to represent the most noticeable effects. At higher temperatures and pressures readjustment of the minerals to the new conditions and recrystallization proceed more readily, although in most cases changes in the *bulk chemical composition* of the total rock are minor. Thus, the original rock and its metamorphosed counter-

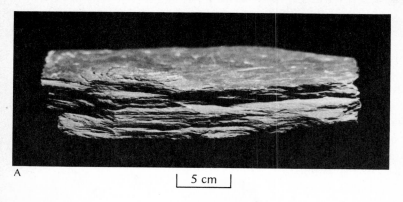

A

| 5 cm |

B

2-23. A. *Handspecimen of a slate to illustrate the planes of schisto-sity. (M. Kauper) B. Rock outcrop showing nearly vertical schistosity planes cutting across inclined beds. [Courtesy U. S. Geological Survey]*

part may resemble each other no more than a lump of dough and the loaf of bread that emerges from the oven after it is baked (Figure 2-24).

An instructive example of metamorphic change is the progressive mineralogical and textural reordering of a shale (itself originally formed from mud) that is subjected to pressure from specific directions and rising temperatures. The first step in response to the increasing temperature and pressure is the formation of **slate,** a hard, fine grained rock that splits readily along smooth, closely spaced parallel surfaces. The development of these surfaces or **schistosity planes** is a consequence of the recrystallization of the original clay minerals in the shale into platy silicates (for example, mica) that now have their flat planar surfaces all lying parallel. This oriented growth occurs in response to the directed pressure imposed upon the minerals and need have no relationship to the original bedding of the shale (see Figure 2-23B). The net result is a hard cohesive rock that, when carefully split apart, can be used for a roofing material.

As temperatures rise still higher (to approximately 300°C) the grain size of the rock will increase and the slate becomes a **phyllite.** Phyllites are smooth, lustrous rocks of high mica content that cleave along wavy surfaces. With further increase in the intensity of metamorphism the phyllite changes into a **schist.** At this stage quartz begins

2-24. *A. Handspecimen of a siltstone showing fine grained nature of rock and layering of thin beds. B. Handspecimen of a gneiss, a strongly metamorphosed rock that could have been reproduced from the rock shown in A. [Photos by M. Kauper]*

A

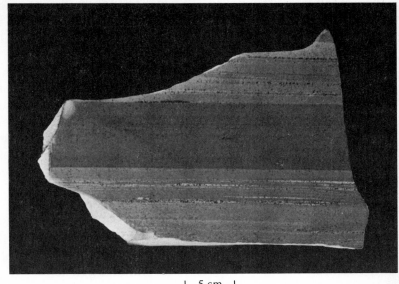

| 5 cm |

B

| 5 cm |

to separate into distinct bands, and the mica minerals undergo a further increase in their grain size. The rock also loses a certain degree of its planar structure and takes on a **foliated** (leafy) texture.

At temperatures in the vicinity of 400°C the mica minerals start to break down and are replaced by feldspar and amphibole. Grain size increases still further, and the minerals tend to segregate themselves into alternating bands of light and dark color (see Figure 2-24B). The light bands are rich in feldspar and quartz, and the dark bands are rich in amphibole and mica minerals that are still stable (for example biotite). The rock is now composed of interlocking mineral grains, and despite its banded nature does not exhibit any pronounced tendency to break into planar fragments. Such rocks are called **gneiss.** Gneisses in most cases represent the culmination of metamorphism and can develop not only from shale and other sedimentary rocks but from volcanic and intrusive rocks as well.

If rocks are subjected to even more extreme conditions deep within the crust, they will either begin to melt or be converted to **granulites,** that is, pyroxene-bearing gneisses. The determination of which of the two alternatives is followed is dictated mainly by the availability of water. Dry assemblages will form granulites, whereas in wet assemblages melting of certain of the minerals will take place.

Monomineralic sedimentary rocks, such as limestones and quartz sandstones, undergo no mineralogical changes during metamorphism and simply recrystallize to coarser and coarser grained assemblages. The metamorphic changes described thus far relate to bodies of rock of regional dimensions and are consequently referred to as **regional metamorphism.** Along the margins of intrusive igneous bodies, the intruded rocks are subjected to strong heating and to the action of chemically active fluids emanating from the crystallizing magma. The metamorphic changes that occur in such localized environments are referred to as **contact metamorphism,** and one of the commonest rock types produced is called **hornfels,** a tough, generally fine grained rock.

Careful work on metamorphic rocks in the field in combination with laboratory studies on minerals common in metamorphic rocks have allowed geologists to define in a general way the temperature and pressure conditions at which different types of metamorphism occur (Figure 2-25). As a result metamorphic rocks can be classified on the basis of their constituent minerals into different, environmentally significant types or **facies.**

The above brief and generalized description of the processes and products of metamorphism may have given the reader the impression that metamorphic rocks are relatively uncomplicated. Nothing could be further from the truth. The study of metamorphic rocks and the reconstruction of metamorphic environments in any detail is an exceedingly complex matter, and many problems remain to be

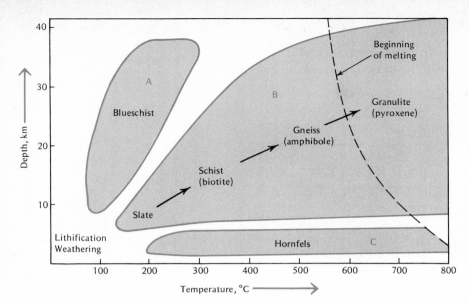

2-25. *Diagrammatic illustration of the approximate pressure-temperature fields of various types of metamorphism. Indicated are major rock types and critical (not necessarily most abundant) minerals. A. Field of high pressure–low temperature metamorphism. B. Field of regional metamorphism. C. Field of contact metamorphism.*

worked out. For example, water is an important catalyst to many metamorphic reactions, and its relative abundance during metamorphism can exert a strong influence on the minimum values of temperature and pressure at which critical metamorphic reactions take place. Furthermore, many metamorphic rocks that have experienced extreme conditions of metamorphism undergo **retrograde** (reverse) metamorphic changes when temperature and pressure subsequently decrease. Fortunately, retrograde effects are generally rather limited, otherwise the mineral assemblages of high-grade metamorphic rocks would never survive the drop in temperature and pressure associated with their eventual unroofing, and all evidence of their former history would be lost.

CONCLUDING STATEMENT

This brief survey of the raw materials of the earth, from the level of atoms to minerals to rocks, contains much background information that will be used in many of the later chapters of this book. In some of these chapters we will be dealing with the various large scale environments in which different types of rocks and their constituent minerals form, and we will have to depend heavily on the material covered in this chapter. It should therefore be read and studied carefully and referred to as necessary.

Before we can deal meaningfully with these large scale environments, however, we must consider one of the most important dimensions of earth science, that of time.

QUESTIONS

1. Which subatomic particles determine the chemical characteristics of an element?
2. What is the relationship between an atom and an ion of a particular element?
3. Many of the unique properties of water are determined by its bipolar molecular structure. Explain this statement.
4. What is the definition of a mineral?
5. What two properties of an ion are most important in determining its role in mineral formation?
6. Why are the silicon and oxygen ions of silicate minerals arranged in tetrahedral building blocks? Explain the ways in which these tetrahedra are combined to build up various kinds of silicate mineral structures.
7. What kinds of information can be gained about geological environments by studying minerals?
8. List the physical properties of minerals that are useful to a geologist? Which one of these is most useful in identifying most minerals, and why?
9. What can a geologist learn from the grain size and texture of an igneous rock?
10. Distinguish between sills, dikes, pipes, stocks, and batholiths.
11. What can a geologist learn from the size and shape of the grains in a clastic sedimentary rock?
12. Almost all non-clastic sedimentary rocks are composed of material precipitated from sea water. What are three processes by which this can occur?
13. Metamorphism involves the mineralogical and textural reordering of rocks within the earth's crust. Why does this happen?
14. Trace the progressive metamorphism of a shale subjected to increasing temperature and directed pressure.

three Time

Time is an abstract concept, and apparently there is no way to measure a concept. The nature of time is most commonly a topic of philosophy; mathematicians and physicists have added ideas but never simplified the always elusive subject. The crux of the matter is the human understanding of time, and we only know of time through observing or experiencing *events*. Such events may be the day by day appearance of the sun, the aging of our bodies, or the movements of hands on a clock, but without such events, perhaps time is impossible to perceive.

Isaac Newton wrote: "[Time] . . . flows equably without relation to anything external." If time flows along, what is flowing? If time is independent of external things, why must it be considered equable (uniform)? Can it just as well be variable? Still, the Newtonian idea of time as an unalterable framework in which sequences of events occur is somewhat understandable and comfortable to live with. Einstein, however, developed a time concept divorced from our common senses and consciousness. Time is not an absolute, but a relative thing and changes according to the velocity of a mass and the position of an observer. Several experiments have confirmed Einstein's ideas. Actually, time is not invariable.

In this chapter we intend to accept the idea that time, whatever it is, has passed and that its passage is measurable (Figure 3-1). The record of prior events held in our minds or in writings serves to tell us about our own past or that of humans before us. To determine the

3-1. *Time. There are three potential time recorders here. The clock, the growth rings of the logs, and the changing ratio between carbon-12 and the radioactive carbon-14 in the wood. [Used by permission of Conzett and Huber, Zurich, Switzerland.]*

events in the earth's past the records must be found in rocks, and the rocks tell two kinds of time, *real* and *relative*.

Let us take real time to mean time that can be described in terms of a particular unit in a more or less regular sequence of events. The 16th of March, 1869, are the particular units of daily, monthly, and yearly sequences that define the time when Einstein was born. In real geologic time, a starting point for time zero such as the birth of Christ is not practical. Most dates of geologic events in the earth's past are confined to yearly units and time zero is considered to be the present. Major events in the earth's history are seldom instantaneous, and our methods of telling time are not accurate enough to assign even a rapid event to any particular year in the remote past. That the dinosaurs became extinct *approximately* 65 million years ago is really a remarkably accurate statement in spite of the qualifying adverb.

Relative time is a concept commonly used in geology. To place an object or event into a framework of relative time does not require that any "time zero" be defined, nor does it require a date in any time unit such as the year. The idea of a sequence is still important, but the sequence need not be part of any regular series of events. Relative time simply defines that an object or event is older or younger than something else. If you found dinosaur skeletons in rock and established that none were alive during the memory of man, it would be accurate to state that dinosaurs were older than you. Before and after are the units of relative time. You would have no idea whether dinosaurs became extinct 10 thousand or 65 million years ago, only that the event of their extinction was before you.

Several principles have been formulated to guide our understanding of earth history in terms of relative time. Some of these now seem so self-evident they hardly need to be stated. Nonetheless, simple though they are, they were not well understood or even followed in the not too distant past (relatively speaking).

PRINCIPLES CONCERNING TIME

Uniformitarianism

Any physical or biological feature of earth was produced by processes acting in accord with the laws of nature that now exist.

"Laws" of nature are not formulated by nature but by people. These laws are simply statements describing our ideas about reality. They are attempts to provide a coherent pattern into which we can fit the data we obtain from our examination of the world; that is, we test the world against our laws so that (if the laws we invented seem true) we can gain a deeper understanding of the universe in which we live.

Scientists may not correctly or fully understand a process, and our "law" or our idea of a "truth" may not be true. This is irrelevant since the process itself is independent of our understanding of it; it operates—and uniformitarianism is the theory that any process has always done so in accord with the laws that now govern any physical, chemical, or biological process. For example, if there was gravity in the past, it acted as a force in the same way it now does; that is, water always flowed downhill, not up. If one were to find a mammoth (a type of elephant) skeleton in a layer of sandstone, uniformitarianism is the basis of the apparently logical conclusion that it was an animal; it had parents; it ate and grew. It was not put into the rock by a devil to confuse us. If we understand how sand gets transported and deposited, we can explain, in a rational way, how the mammoth was buried.

Uniformitarianism does *not* imply that things were always the same on earth. Such seeming invariables as the amount of solar energy, the number of days in a year, or the force of gravity may not have been absolutely uniform throughout time. We know that no physical thing on earth is unchanging; even rock, often a symbol of stability to humans, contains its own record of instability. Uniformitarianism does *not* imply that the rates of processes now operating were invariable through time; there even may have been processes operating in the past that we have no direct knowledge of today. The principle of uniformitarianism encompasses all of these ideas and lets us work with the rational assumption that whatever took place in the past, it followed the same laws of physics, chemistry, and biology that govern processes operating today.

Superposition

In an undisturbed sequence of layered sedimentary rock, each layer was formed after the one beneath it.

This sounds self-evident, and the concept is simple. It is useful not only because it enables us to make statements about the relative ages of rock layers but also, partly, because there is an implication of "up" in the statement. Most of the rock at the surface in the major mountain belts of the world is sedimentary. Most of these sediments were deposited in the sea, later uplifted, and often broken and distorted. The layers may be vertical or even overturned from their original nearly horizontal attitudes. Although determination of the relative age of layers may be difficult, the principles of uniformitarianism and superposition allow a sensible solution.

Figure 3-2 illustrates two conditions where relative time is deter-

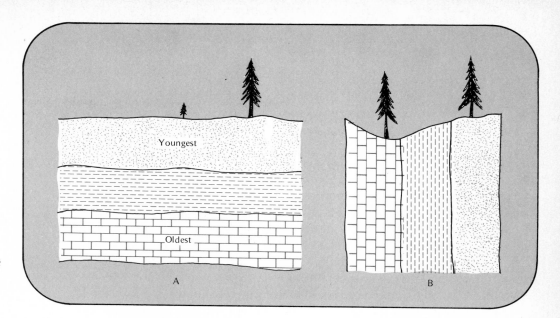

3-2. *Relative time and superposition. In case B the youngest layer is still on "top," but which way originally was "up"?*

minable; the situation A is simple, but B would require some study. In case B perhaps ocean waves produced ripple marks on the sand, and later the grains were cemented together to form sandstone and preserve the ripples (Figure 3-3). These ripple marks can be used to tell whether "up," when the layers were nearly horizontal, is now to the right or left. If the original "up" is now to the right, then the rock layers have the same relative age relations to each other as those in case A.

3-3. *Preserved ripple marks. These ripples formed on sediment by the current action in a shallow ancient sea. After the sediments were cemented in such a way as to preserve the ripples, the rocks were uplifted and tilted. Today the marks serve not only to indicate something about an ancient environment but also that the rocks behind them are older and that the sedimentary rocks on the other side of the road, out of view, are younger. [Courtesy R. Ojakangas.]*

Faunal Succession

Populations of organisms always change through time and once changed will not appear again in their previous form.

Centuries ago people noticed that the plant and animal fossils in one layer of rock could be very different from those in a layer a few meters above or below. Partly because the concept of uniformitarianism was not yet widely articulated or understood, some believed that populations were periodically destroyed worldwide by catastrophes and that the populations in rocks above represented separate creations. As more and more fossils were studied, it became evident that some fossils differed in only minor, but consistent, ways from those in older or younger layers. The groups were separated in different layers, therefore they were separated in time, yet they were related. This observation is a keystone to the idea of biological change through time, better known as evolution.

Evolution usually proceeds gradually, and the appearance of new species of organisms as time progresses is a fact. The definition of what constitutes a separate species and the ramifications of the processes of evolution are things beyond the scope of this chapter. In spite of the apparent gradualness of most biological change, geologists often have no trouble recognizing that one population of fossils, say snails, is very different from another population of snails in a layer of rock a few meters above or below.

Commonly, in a sequence of sedimentary rock layers, most of the *time* between the deposition of the lowest layer and the topmost layer has no physical representative in the form of rock. For example, if a regular sequence of sedimentary layers is 50 m thick, the age difference between the bottommost and topmost layers may be 10 million years of real time. Yet each layer may have been deposited in a short time, ranging from days to perhaps thousands of years. Say that the time required for deposition of the entire rock sequence is totaled and found to be only 100,000 years. If the sequence represents 10 million years, yet 100,000 years were all that seem to be required to *produce* the sequence, what happened during the rest of that time?

It is evident that, in a thick sequence of sedimentary rocks, we are often seeing physical evidence of only a fraction of the time that passed. There may have been long periods of no deposition, or there may have been deposition and later erosion before the sediment was consolidated into rock. These gaps in the sedimentary record are called **unconformities.** In neither case is there always visual evidence of what happened. Much of the time is represented by the planes separating the layers of rock. With this in mind, and with the idea of faunal succession in mind, it is not at all surprising that close or even adjacent rock layers often have very different fossil populations; they may be widely separated in time.

The lack of a sedimentary sequence is one important reason why most of the organisms that inhabited this earth in the past are, and will remain, unknown to us. There are many factors that might result in the nonfossilization of an organism, for example lack of a hardened skeleton, destruction of the hard parts by predators, or break up of the skeletal or shell material by wave action. But many, probably most, types of organisms that have lived have left no record because no rocks remain in which they might have been preserved. By coupling estimates of the quantity of species in our modern world with knowledge of past environments and uniformitarianism, it has been estimated that less than 1% of the types of organisms that ever lived have been recognized!

Since the record of faunal succession is so fragmentary, those geologists whose specialty is the study of ancient ecological and evolutionary patterns are very much inhibited in their work. However, the geologist who wishes to use fossils to indicate relative time relations is not confronted with a series of fossil populations that differ from layer to layer in only small and subtle ways. The fossils in one layer are often easily distinguished from those in near or adjacent layers of rock. This eases the task of comparing rocks by their fossil content. Clearly, a shale deposited 100 million years ago may look no different than one deposited 50 million years ago. The principles and processes of shale deposition may not have changed, but populations of organisms always change through time. If the shales contain different fossil populations, these can be used to determine that the two shale layers are not time equivalents. Figure 3-4 is an example of the *correlation* of strata by comparison of their fossils.

Implicit in the consideration of correlation of sedimentary rock layers is the problem of *rock units* versus *time*. Rock layers that are relatively constant in their composition and are widespread are often given **formation** names. For example, a sandstone that is exposed in several midwestern states is known as the St. Peter Formation. It can be distinguished from other sandstones by its physical nature and by its relationship with formations above and below it. When a new exposure of sandstone is found, if it meets the physical criteria and relative time relationships with adjacent formations that are characteristic of the St. Peter, it is said to correlate with the St. Peter Formation and is assigned the formation name.

The formation is a rock unit but, usually, does not everywhere

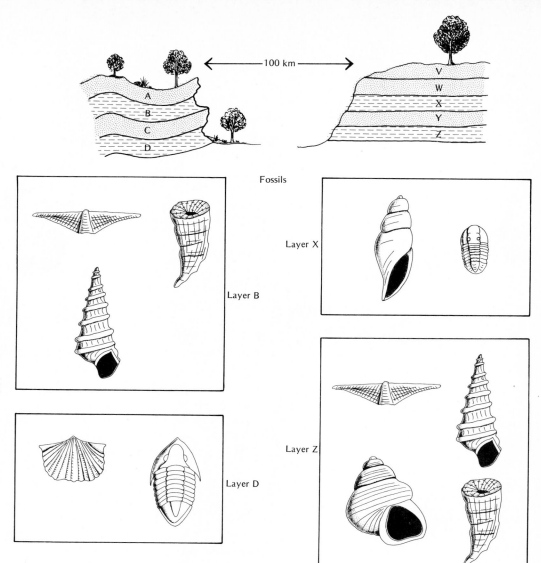

3-4. *Correlation of rock strata using fossils. The principles of faunal succession and superposition indicate that layer D is older than layer B and that layer Z, 100 km away, is a near time-equivalent of B. V, W, and X are therefore younger than any strata at the outcrop on the left. Without fossils, the strata could be correlated erroneously: D = Z, C = Y, and so on.*

Fossils

Layer B

Layer D

Layer X

Layer Z

represent the same period in time. Figure 3-5 illustrates this idea by showing the deposits associated with a sea that had been rising and advancing westward onto a continent. The figure indicates that separate exposures of a formation, easily correlated, may be equal in relative time but may not represent identical periods of real time. As Alfonso the Wise said in the thirteenth century: "Had I been present at the creation, things would have been a lot simpler."

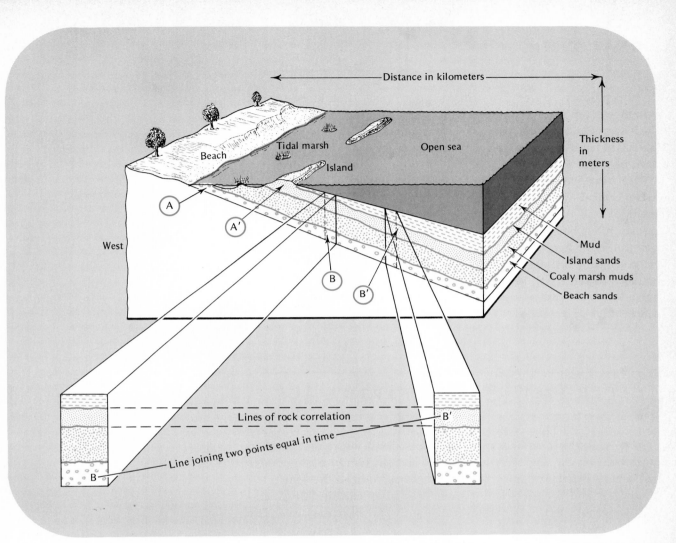

3-5. *Rock units versus time. The extended view on the lower left illustrates four rock formations; each is defined on the basis of its unique composition. To the right, kilometers away, the same four formations are shown. Note that in the main illustration each formation is a particular rock type deposited in a particular environment: beach, tidal marsh, barrier island, and open sea. The vertical sequence at any one point must represent a change of environments through time. In this case a rising sea advanced westward, and the sediments characteristic of environments further offshore were deposited, in sequence, over sediments characteristic of environments closer to shore.*

At the present time beach sand (A) is being deposited and so is island sand (A'); the different sediment types are time equivalents. The same situation existed in the past at points (B) and (B'). Once uplifted and exposed, as diagrammatically shown in the extended views, the time difference is not evident. The four formations will be correlated on the basis of their rock types (rock units), but, if the time factor is disregarded, subtle geologic confusions (and confused geologists) can result.

REAL TIME

As previously stated, we will consider "real" time to be time to which we can apply some measurement—seconds, years, or some such segmentation of events. These measurements are not absolutely invariable and they are only a classification system to make the passage of time recordable.

The year, probably because of readily observable seasonal and astronomical events, has been a fundamental unit for the measurement of real time since recorded history began. Undoubtedly the question has been pondered for thousands of years: How many years has the earth been here? The answer demands "real" time and can be given only if there is an earth "clock." The search for such a clock actually began in all seriousness a couple of centuries ago. Before considering the failures and successes of the searchers, let us examine a few older opinions about earth time.

OLDER VIEWS OF THE EARTH'S AGE

Some cultures have had much less constricted views of time than those held by the Christian world. The ancient Hindu philosophers had a very expansive idea of time. They had noticed the erosion of mountains and the general impermanence of all earthly things, and they generated the idea that everything is part of some cycle. The grandest cycle of all was the creation, the life, and the death of the universe. The span of one cycle of the universe was equivalent to one day of Brahma's life, or about 4.3 billion years. We are, according to Hindu scripture, about 2 billion years into one cycle.

This early concept of a creation billions of years ago was far removed from some of the Christian thought of later centuries. If geology, as a science, had evolved among the Hindu peoples, many of the concepts of time that we have only lately developed would have appeared in our thought long ago. But the science of geology is a product of the western world, and in some of the older Christian concepts concerning the birth of the earth, time was enormously contracted.

Men have long studied the *Book of Genesis,* and, although such notables as Plato considered its "6 days" of earth creation to be an allegory rather than 144 hr of time, some Christian factions later accepted the "day" in a more literal sense. One biblical clue that gave rise to ideas that the day in Genesis was not 24 hr was found in a psalm: "For a thousand years in thy sight are as but yesterday . . ." So 6000 years represented the 6 "days" of creation and was often accepted as the approximate time span from creation until the birth of Christ.

The idea of an approximately 6000 year old earth lasted for over 1500 years, until the middle ages. Shakespeare wrote in *As You Like It:* "The poor world is almost 6000 years old." We don't like it, and neither did many geologists in the past. Further study of the genealogies given in the Bible further reduced, instead of expanding, earlier thoughts on time, and in the mid-1600s Archbishop Ussher made the pronouncement: "The earth was created in October, 4004 B.C."

Although we now realize that ideas such as these may have retarded both science and theology, as a whole Christianity has supported the theory that there is some order in the universe and that this order can be interpreted by rational minds. Thus in the seventeenth century the observations of certain things by some rational people initiated a demand for more time. Layers of rock that were known to have been deposited beneath the waters of an ancient sea because they contain fossilized marine organisms were found, folded and contorted, high in mountain ranges. With only a few thousand years at their disposal, early geologists were forced down some very inelegant mental paths to explain such facts. The recognition that an immense variety of organisms once lived, became preserved in rock, and are now extinct required either a great deal of time or a multitude of creations and catastrophies. Soon it was found that, while some species became extinct others, which coexisted with them, did not and could be found in younger rocks. How did some species survive and the others die if catastrophe was the cause of extinction? A large amount of time seemed necessary to explain both the physical and biological phenomena of the earth.

The rates of erosion of a high mountain range could be estimated, at least roughly, and the observation that there were old, worn mountain ranges demanded either time or vastly increased rates of erosion in the past. It takes very long periods of time for the slow-acting processes of biological evolution and geologic change to achieve significant results. The observations of the modern world and the observations of rocks representing the ancient world left no doubt in the minds of some early geologists that the earth was a very old place.

But no matter how well constructed the arguments, no matter how obvious the requirements, if they could not find a "clock" to measure geologic time, there was no proof.

EARLY ATTEMPTS AT MEASURING EARTH TIME

In the nineteenth century several attempts were made to measure geologic time. All were thoughtfully conceived and all were in error. Nonetheless they did change our ideas, expand our concepts of time, and give us a much better understanding of earth history. Through the work of many men, the rock strata of the world had already been divided into units on the basis of relative age. Geologists knew, by principles discussed earlier in this chapter, that some units of rocks were older or younger than others. The rocks, by virtue of their fossil contents and relations with strata above and below, were placed into relative time periods that were given names, although their boundaries could not be dated. This system, over a long period of time and through many changes, resulted in a geologic time scale, which we will discuss later.

It was reasoned that, if the thickest pile of sediments deposited during each period of time were measured, totaled, and divided by some average rate of deposition, the age in years of the whole could be calculated. This formidable task was carried out, not once but many times. Most of the estimates of the total thickness of sedimentary rocks were in the range of 30–100 km. The rates of deposition used varied more widely, as would be expected. Depending upon the environment, sediment may accumulate at 1 cm or less per 1000 years or thousands of times as fast.

Most of the estimates of real time from all these efforts were 100 million years or less. Erroneous perhaps, but, for the geologist trying to explain how mountains were formed and destroyed, for Charles Darwin and other scientists who were trying to understand evolution, 100 million years was a far more comfortable framework than the 6000 years previously available. Some of the investigators of depositional rates had concluded, as we pointed out, that, in a particular area, large amounts of time had no record in the form of rock. So a factor to account for nondeposition and erosion was used, and some estimates of real time were "corrected" to several billion years.

This approach to the problem is quite logical but still verges on guesswork rather than estimate.

Early in the eighteenth century a proposal was made to investigate another type of earth "clock," the salinity of the oceans. The reasoning was that the sodium portion of the principal salt of the ocean (sodium chloride) was introduced to the sea by rivers. Knowing this, if one knew the total quantity of sodium in the sea and could make a reasonable estimate of the amount entering each year via the world's rivers, simple division would result in the age of the oceans.

When this idea was first proposed, no significant oceanographic expeditions had set forth, and the volume of the seas was unknown so that the total amount of sodium was unknown. By the latter part of the nineteenth century, however, enough data were available to make the estimate

$$\text{age of the oceans} = \frac{1.6 \times 10^{16} \text{ tons of sodium in the ocean}}{1.6 \times 10^{8} \text{ tons of sodium added each year}}$$

$$= 1 \times 10^{8} \text{ years or } 100,000,000 \text{ years}$$

This date concurred with several of the dates based on depositional rates and it certainly must have appeared to be true that the first sedimentary rocks and the first seas formed about 100 million years ago.

Unfortunately several factors make the salinity clock erroneous. All sodium that enters the sea does not stay there. Some is retained in salt deposits, some is in sea water trapped between grains of marine sediments on the ocean bottom or now on land, and there are several other ways by which sodium leaves the oceans. Also, the estimated rates of sodium entry into the sea were probably much too high. The continents now stand relatively exposed, and the amount of chemical weathering of rock and stream transport of sodium to the sea is high. Compared to much of the past we now have very little continental mass covered by the shallow seas that were so common in earth history.

The methods of age-dating the sedimentary rocks of earth or its oceans involve some very broad estimates that were open to serious questioning at the time. The geologists and biologists of the nineteenth century were not only driven by curiosity in their quest for a more precise, less questionable earth clock, they were driven by need as well. They could hardly devote years to building scientific theories on the basis of a time reference that might prove false. Uniformitarianism and evolution were two spurs for the search.

Nearly all of the life-supporting heat at the earth's surface comes from the sun, but the earth does have its own heat. The rate of heat flow from the earth is minute enough that it is usually not noticeable to our senses, but it has been measured in many places. On the average, there is an escape of about 40 calories each year from each square centimeter of surface. The earth's heat is easily detected if one descends into a deep mine or lowers a thermometer down an oil well. Although differing from place to place and nonlinear with great depth, a typical increase is about 2°C/100 m.

In the nineteenth century William Thompson, Lord Kelvin, one of the leading physicists of the day, made some startling calculations. Kelvin began with the assumptions that the earth originated as a mass of molten material and that the heat escaping from its surface was residual from the original heat. Although evidence has now induced most astronomers and geoscientists to reject the idea of an originally molten earth, Kelvin's assumptions were very sensible. The earth was viewed in the same light one might view a cannon ball heated almost to its melting point and allowed to cool. After a time the surface would be cool enough to touch, analogous to earth's surface, but inside it would still be very hot. Kelvin's idea was to calculate how long it took the earth to cool to its present state, based on the current rate of heat loss. Not only were Kelvin's assumptions sensible, but they had the luster of uniformitarianism. What was occurring now on earth was a means towards understanding its past.

However, the geologists were about to be hoist on their own petard. Kelvin and his colleagues coupled the calculations of the earth's heat flow with others related to the idea that the sun must be cooling off. The sum of this was that the earth would have been too hot for life some 20 to 40 million years ago. The proponents of geological uniformitarianism and an evolutionary faunal succession saw many of their concepts dashed on the firm rocky shore of mathematics. But if the physicists could take time away, physicists could give it back. For within the rock there is a clock whose tick finally was heard about the same time Kelvin made his summation.

RADIOMETRIC AGE DATING

In 1896, with biologists and geologists in full retreat, Becquerel, a French physicist, revealed the results of an interesting experiment. A photographic plate, analogous to modern film, became exposed in the dark if uranium ore was placed near it. Instead of visible light exposing the film, some other type of radiation, invisible to our eyes, was responsible. At this time Marie Sklodowska, a student revolutionary who had left her native Poland and studied physics in Paris, married the physicist Pierre Curie. The two studied the phenomenon discovered by Becquerel and named it **radioactivity.** One of their most significant discoveries was that in addition to the emission of rays, a radioactive substance gives off heat!

Here then was Kelvin's unknown source of error: most of the heat given off by the earth is produced by radioactive minerals in the earth; it is not heat left over from a molten stage in the earth's history. Within a few years it was calculated that there is sufficient radioactive material in the common rocks of the first tens of kilometers of the earth's crust to account for the heat emission. Nuclear processes are now known to be the source of the sun's energy as well, and it has not been decreasing its heat production at a rate anything like Kelvin's estimates.

The Nature of Radioactive Decay

As you may remember from the previous chapter, the number of neutrons in the nucleus of an atom of an element determines the isotope of that element. For example, for an element to be oxygen it *must* have eight protons; any more or any less and it is not oxygen. Over 99% of the oxygen you are breathing has eight neutrons as well, and, because its mass number is simply the sum of its protons and neutrons, it is known as oxygen-16 (^{16}O). However, about 0.2% of the oxygen you breath has ten neutrons instead of eight and is designated oxygen-18 (^{18}O). Although the two isotopes obviously have slightly different weights, their chemical nature does not seem to differ. Neither isotope is radioactive, and they are said to be "stable" isotopes.

In nature, two other situations occur. In the case of some elements, some isotopes are stable but others are unstable, and with other elements all isotopes are unstable. The unstable isotopes are radioactive. This means that they change, or **decay,** spontaneously, into other elements. To understand how they change it is necessary to know something about the emissions of radioactive isotopes.

Not many years after the discovery of radioactivity the emissions were found to be of three types:

1. An alpha particle consisting of two protons and two neutrons.

2. A beta particle consisting of an electron.
3. Gamma rays, similar to x rays but of very short wave length and with high powers of penetration.

Alpha Emission. It is quite evident that if an atom of an element spontaneously emits an alpha particle it can then no longer *be* that element, for it has lost two protons. For example, a small percentage of naturally occurring platinum is an unstable (radioactive) isotope, platinum-190 (^{190}Pt). All platinum has 78 protons, so ^{190}Pt has 112 neutrons. This isotope emits an alpha particle and so loses two protons and two neutrons; what remains are 76 protons and 110 neutrons. An element with 76 protons is osmium, and that is what the platinum has become! 76 + 110 tells us that it has changed to that isotope of osmium known as ^{186}Os (Figure 3-6).

Beta Emission. When a radioactive isotope emits a beta particle, it is a nuclear event; the particle comes from the nucleus. Yet, a beta particle is an electron, and anyone who has had modern general science or physics in high school knows that there are no electrons in the nucleus of an atom. We know that the emitted beta particle is not an electron that comes from the "cloud" of electrons surrounding the nucleus; that would simply be a common chemical event and an ion would be formed. The beta electron is envisioned as one of the products of the decay of a neutron.

If a neutron "breaks up" and emits a negative electron, it is left

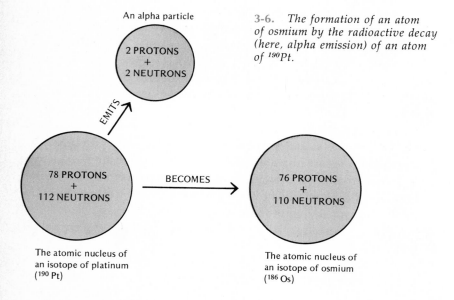

An alpha particle

2 PROTONS
+
2 NEUTRONS

EMITS

78 PROTONS
+
112 NEUTRONS

BECOMES

76 PROTONS
+
110 NEUTRONS

The atomic nucleus of an isotope of platinum (^{190}Pt)

The atomic nucleus of an isotope of osmium (^{186}Os)

3-6. *The formation of an atom of osmium by the radioactive decay (here, alpha emission) of an atom of ^{190}Pt.*

with a net positive charge of one; that is, the neutron has become a proton. And, of course, when the number of protons changes, a new element is born. For example, the element rubidium has 37 protons, and an isotope, ^{87}Rb (obviously with 50 neutrons), makes up over 25% of natural rubidium. This isotope is unstable and gives off a beta particle; thus, one neutron has converted to a proton. The new element formed should have 38 protons and 49 neutrons. Strontium is the element defined by 38 protons, and what has been formed is, indeed, ^{87}Sr.

The Concept of a Half-Life

Any one or more of these reactions in the nucleus of a given atom occur almost instantaneously. How, then, can radioactive decay be used as a "clock" for measuring time? There is no way to predict which particular atom of a radioactive substance will decay to another element or, for that matter, when. Experiments, however, have indicated how long it takes for a particular *quantity* of atoms of a radioactive substance to decay. This "rate of decay" is measurable and fairly accurately known for most isotopes. An apt analogy is an hourglass. If a particular hourglass filled with sand is inverted, no one can accurately predict *which* grain will flow through first, which will be second, and so on. Nonetheless, by experiment, it may be accurately predictable that one half of the total will have flowed through after 1/2 hr.

The rates of decay for most unstable isotopes have been experimentally determined, and these rates are defined in terms of the **half-life.** One half-life is the time required for one half of an original quantity of atoms of a radioactive substance to decay. Since a measurable amount of the radioactive substance is composed of an extremely large number of atoms, statistical predictions of the rate of decay are reasonably accurate. The rates of chemical reactions are governed by certain external factors, such as temperature and pressure, but nuclear reactions are not affected by temperature and pressure and continue at their self-determined rate independently of imposed physical factors. Therefore the half-life of a particular isotope is considered to be a time-value that does not vary.

Let us consider a theoretical unstable isotope of element A that decays to element B and has a half-life of 1000 years. A is the parent and B is its daughter. If there were 500 atoms of A at the start, after 1000 years had passed, 250 atoms of A would remain and 250 atoms of B would have been produced. However it would be erroneous to

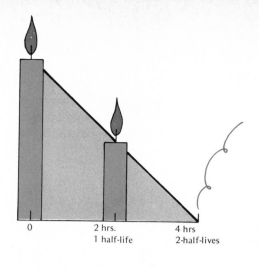

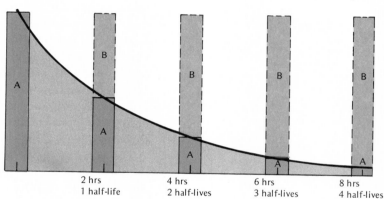

3-7. *Half-life. In the upper diagram a candle with a "half-life" of 2 hr is decaying to heat, light, and gases at a linear rate. The lower diagram illustrates the exponential nature of radioactive decay. The unstable parent isotope A, with the same half-life as that of the candle, is decaying to the daughter product B. After two half-lives the candle has decayed to zero, whereas one quarter of the parent isotope A still remains after the same period of time.*

assume that after two half-lives there would be no A remaining, only 500 atoms of B. The 250 atoms of A that remain after one half-life had passed, will continue to decay into B at the same half-life rate; that is, after the second 1000 years had passed, there would be 125 atoms of parent A and 375 atoms of daughter B.

Figure 3-7 illustrates two different types of decay; one type (illustrated by a candle) is linear and easily understood since it is familiar to us. Isotopic decay, however, is nonlinear and requires a bit of thought. If you consider the curve of the pattern of radioactive decay, you might think the process continues to infinity and the parent never completely disappears. For example, if one rolls an egg across the floor half way to a wall each time, will the egg ever get to the wall and break? That philosophical teaser is not exactly comparable because there is a finite number of atoms of any parent. Someday the last atom will decay to its daughter element.

Most types of radioactive decay are somewhat more complex than the theoretical example used above. These complexities and the computations used to obtain the date when decay began, although not difficult, are beyond our purpose. The assumptions upon which the method is based and the concept of how the process can be used to tell time are the important things.

The simplest conceptual model would be to know the quantity of the parent element and daughter element in an igneous rock. If we know the time required for one half of the parent to change to the daughter, a decay diagram could be constructed, such as that shown in Figure 3-8. Addition of the measured quantities of parent and daughter together results in the quantity of parent at time zero, when the rock crystallized. If we then divide that number into the amount of parent present we will obtain the fraction of parent remaining. For example, if there were 25 atoms of parent and 75 atoms of daughter, there were 100 atoms of parent at time zero. Dividing 25/100 indicates only one quarter of the original amount of parent is still there; by the graph, two half-lives have passed. Multiplying the half-life time for the particular parent isotope by the number of half-lives that have passed gives the time that has passed.

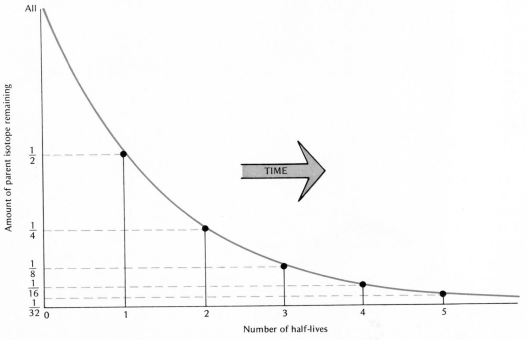

3-8. *Graph of time in half-lives versus amount of remaining parent isotope.*

To achieve any reasonable accuracy in dating, the quantities of parent and daughter must remain in a **closed system.** A closed system means that neither parent nor daughter has been added or removed by any process other than radioactive decay. Such a system might be a crystal that, after it formed at time zero, was never unduly heated or affected by chemical solutions such that isotopes escaped or entered the system.

Half-lives vary from one unstable isotope to another; some are seconds long whereas others are billions of years. For example, an unstable isotope of the gas radon is an intermediate product in the decay of uranium. Its half-life is only 3.8 days. Yet enough radon isotope is produced in one European uranium mine so that its radiation caused 50 times the normal rate of lung cancer in the miners. Most unstable isotopes with short half-lives are not of much use to geologists; for even those with long half-lives to be of much use, they must be relatively common to crustal rocks.

The first attempts to date rocks radiometrically in terms of real time were carried out in England by Rutherford and in the United States by Boltwood, and both attempts were made less than 20 years after the discovery of radioactivity. Today the practice is common, although some very sophisticated and expensive laboratory equipment is required. The dates are not precise in terms of human lifetimes, but the margins of error have been reduced to small percentages of the actual time elapsed in many cases. The history of concepts, discoveries, and applications in radiometric dating is a striking example of physicists, chemists, and geologists working together to produce a major advance in science. Some of the systems commonly used to date rocks are described below.

Age Dating Schemes

Uranium-Lead. All isotopes of uranium (there are about 15 known) are unstable. Although in their process of decay, many intermediate elements are formed, their final daughter products are stable isotopes of lead (Pb). Two uranium (U) isotopes are used for dating: ^{235}U, which changes to ^{207}Pb at a half-life rate of 713 million years, and ^{238}U, which changes to ^{206}Pb at a half-life rate of 4.5 billion years. In minerals naturally occurring lead is a mixture of isotopes, nearly half of which were produced by the decay of uranium.

Other than uranium ore, which is not very common, one of the best minerals to use for uranium-lead dating is zircon. The formula of zircon is $ZrSiO_4$, and one would not expect uranium to be found in it. Yet in the crystal lattice, enough uranium often substitutes for zirconium (Zr) to make measurement and age-dating possible. Zircons are typically found in small quantities in such common rocks as granite, and the computed date indicates the time when the granite first crystallized. One of the features of uranium-lead dates is that two different isotopes are decaying and two different daughters are produced; thus, there is a built in cross check to help eliminate error. If the ages calculated from each parent-daughter combination are the same, one can be confident of a closed system and an accurate date.

Potassium-Argon. There are three natural isotopes of potassium (K) and two of them, composing over 99% of the world's potassium, are stable. However, ^{40}K is not stable and decays in two ways. About 89% of ^{40}K decays by emitting a beta particle, thus converting a neutron to a proton. The potassium then becomes calcium (Ca), but this daughter, ^{40}Ca, is no different from the most common form of calcium, which was not produced radioactively. Thus, the probability that some of the ^{40}Ca measured in a mineral was already present at time zero makes the system of limited use.

The remaining 11% of ^{40}K decays by capturing one of its own orbital electrons into its nucleus rather than by emitting a particle. The addition of a negative charge to the nucleus converts a proton to a neutron, and the element with one less proton than potassium is argon (Ar). Argon is the third most abundant gas in our atmosphere (after nitrogen and oxygen), and much of it is the product of radioactivity. Potassium is a constituent of very common minerals, micas for example, and the ^{40}K —^{40}Ar method has been a favorite for age dating.

The problem here is that the daughter product, argon, is a relatively inert gas and can easily escape the system; therefore, the computed date may be too young. Fortunately some minerals tend to retain the argon better than others. If reheating is suspected to have occurred, some careful choices of which minerals to date must be made. The abundance of potassium-bearing minerals and a half-life for ^{40}K of 1.3 billion years allows a wide range of rocks to be dated. Under ideal conditions, rocks that formed only 50,000 years ago can be accurately dated. However, the small quantity of ^{40}Ar produced during this short time makes it necessary to achieve very precise measurements, and this limits the use of the method for such young rocks.

Rubidium-Strontium. The last system commonly in use for dating rock is rubidium-87 (^{87}Rb) to strontium-87 (^{87}Sr). There are some uncertainties as to the precise half-life of ^{87}Rb, but it is very long, about 47 billion years. This method is most often used to date certain micas and feldspars of either igneous or metamorphic rock.

A PROBLEM IN TELLING REAL TIME

Let us examine how an actual problem was attacked using radiometric dating techniques. This particular problem was chosen because it is of geological importance, it allows the formulation of some ideas about relative age relations, and because, in spite of good geologic work, the rocks did not yield their answers easily.

The Problem. To assign a date to the onset of the present continental glaciation of Antarctica.

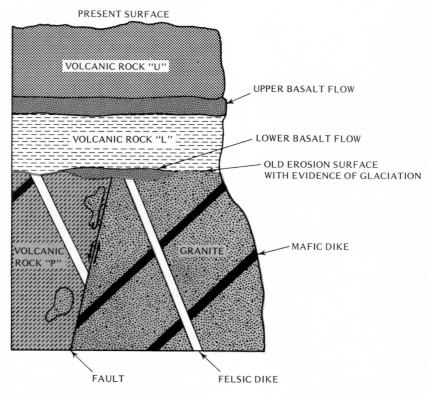

3-9. *A generalized cross section of rocks in the Jones Mountains, Antarctica.* [*Permission courtesy of R. Rutford, C. Craddock, C. White, and R. Armstrong.*]

PRESENT SURFACE

VOLCANIC ROCK "U"

UPPER BASALT FLOW

VOLCANIC ROCK "L"

LOWER BASALT FLOW

OLD EROSION SURFACE
WITH EVIDENCE OF GLACIATION

VOLCANIC ROCK "P"

GRANITE

MAFIC DIKE

FAULT

FELSIC DIKE

The Situation. (Refer to Figure 3-9.) A lower complex of granite, igneous dikes and volcanic rocks has an ancient erosional surface at the top. This old erosion surface, where exposed, shows scratches and grooves characteristic of glacial erosion (see Figures 11-39 and 11-40). In some places glacial rock deposits were found lying on the eroded surface. The surface shows no signs of having been exposed to weathering for any appreciable length of time; that is, it was glaciated then covered relatively quickly by the series of rocks overlying it.

With the exception of the glacial deposits of rock, the old glaciated surface is covered by a sequence of lava flows and other volcanic rocks. Near the erosion surface there is evidence of a complex interaction between the "lower basalt flow," ice, glacial deposits, and water. Therefore, it seems probable that, if the volcanic rock atop the old erosion surface can be radiometrically dated, one may assume that glaciation in this area was occurring at that time. The onset of glaciation would have occurred prior to the date of the lava flow.

The Observations. Based on the field relationships visible in Figure 3-9, all the rocks and events can be placed in a time sequence of relative age. This is for you to do. Below the "present surface" there are labeled: eight rocks, a fault, and an ancient surface of erosion. See if you can put all ten items into a series from oldest to youngest. The most probable age sequence is provided at the end of this chapter.

The Data. The granite was dated using both potassium-argon and rubidium-strontium methods, and the apparent age is approximately 200 million years. One of the felsic dikes yielded an apparent age of 104±4 million years. (If this sequence is not in accord with your observations on the relative ages of the two rocks, we suggest you revise your conclusions.) If a series of samples from the "lower basalt flow" had yielded apparent ages within a reasonably narrow range, this would have permitted a close estimation of the time of glaciation. Unfortunately, this was not the case.

The apparent potassium-argon ages of several samples from basalt flows within 3 m above the glaciated surface varied from 9 million to 300 million years! Hardly a narrow age range. Any of these ages over 104 million years (the age of the youngest rocks beneath the glaciated surface) seem geologically impossible. The anomalous ages were obtained in duplicate tests at two independent laboratories; hence, the erroneous data are due to something other than poor techniques of sampling or analysis. Samples from the "upper basalt flow" more than 100 m above the erosion surface yielded apparent ages of 7, 7, 9, 10, 10 and 24±12 million years. These ages seem more reasonable. Extremely wide variations in apparent ages, such as those from the lower lava flows, are not normal. In many instances very close ages can be obtained from different samples of the same body of rock and a high degree of confidence in the age is often possible.

Conclusions and Comments. In this area of Antarctica, glaciation began at least 7 million years ago and, perhaps, well over 9 million years ago.

The daughter product argon is the only isotope commonly used in age dating that is a gas. It is known that potassium-argon systems in various minerals may not remain closed due to the relative ease of escape or addition of argon. Therefore, the application of other radiometric dating methods, rubidium-strontium perhaps, may allow the problem to be resolved more accurately. In the case we have examined some of the apparent ages were obviously too great. For this to occur argon must have been added to the system. Since the basalt in the molten state may have incorporated fragments of other rock as the magma came through the crust to the surface, perhaps the extraneous ^{40}Ar was inherited from older rocks below.

This case study well illustrates that good geologic field work is a necessary requirement for the interpretation of radiometric ages.

THE GEOLOGIC TIME SCALE

With the advent of methods of radiometric dating, geologists were at last able to assign dates of real time to their previous scale of relative time. Figure 3-10 shows the **geologic time scale** now in use. The names were assigned prior to the twentieth century, and vast amounts of labor went into the present construction. Most of the names have some historical meaning, usually related to the area where rocks of that age were studied. For example, Cambrian comes from "Cambria," the Roman name for Wales. Rocks worldwide can be assigned to the Cambrian period by comparison with those in Wales. To assign strata to the Cambrian Period does not mean that the rock types must be the same as those in Wales, or even that any direct compari-

Era	Period	Epoch	Millions of years ago (approximate)	
CENOZOIC	QUATERNARY	Recent Pleistocene		
			2	Man as a user of fire and tools
	TERTIARY	Pliocene Miocene Oligocene Eocene Paleocene	25	First man-like primates
			65	
MESOZOIC	CRETACEOUS		135	First flowering plants First birds
	JURASSIC		190	
	TRIASSIC		225	First mammals
PALEOZOIC	PERMIAN		280	
	PENNSYLVANIAN } CARBONIFEROUS		325	First reptiles
	MISSISSIPPIAN		350	First land animals (amphibians)
	DEVONIAN		400	First land plants
	SILURIAN		430	
	ORDOVICIAN		500	First vertebrates (fish)
	CAMBRIAN		600	Invertebrate fossils common
PRECAMBRIAN				Rare invertebrate fossils (670) Well-preserved unicellular organisms (2000) Probable unicellular organisms (3000)
			4600	

Duration of Eras

Cenozoic:	65 million years
Mesozoic:	160 million years
Paleozoic:	375 million years
Precambrian:	4000 million years
Total.......	4,600,000,000 years-approx. age of earth

3-10. *Geologic time scale.*

son need be made with strata in Wales. Rock types may vary within a few kilometers and one would not expect that the same species of organisms lived in ancient seas which covered Wales as in seas of the same age which covered portions of the western United States, for example. Age assignments are usually made according to the nature of the fossil populations within the rock and upon the spatial relations of the strata in question with relatively younger and older rocks. The fossils and stratigraphic relationships of well exposed beds of Cambrian age in North America have been described, and a North American geologist normally uses this information to determine if strata he or she is concerned with are of Cambrian age.

Fossils, not the reading of radiometric clocks, are the most typical criterion for age assignment. The dependence on fossils is due to two related facts: most rocks exposed at the surface are sedimentary, and most sedimentary rocks cannot be dated by radiometric means. The latter is true because the mineral grains making up a sediment may have been derived from the weathering of very old rocks then transported to the site of deposition. Thus, because the fossils were formed at the approximate time of sedimentation, they date it with fair accuracy, whereas the particles may have been formed a billion years before. Generally, igneous rocks that can be dated and are associated with sedimentary rocks have provided the means for tying the whole relative time scale to a real time scale.

Since the beginning of the Cambrian Period some 600 million years ago, fossils have been very common in sedimentary rock. Prior to that they are extremely rare because, apparently, animals lacked shells or hardened exoskeletons (such as a modern lobster has) that could easily become fossils. Therefore we know far more about the life of the last 600 million years than about that of the 4000 million years of Precambrian time which preceded.

THE AGE OF THE EARTH

The oldest rocks of the earth's crust that have yet been dated occur in western Greenland and appear to have been formed about 3.8 or more billion years ago. Intuitively, it would seem that our chances of finding a rock whose date would coincide with the development of rock at the earth's birth are small. On a dynamic ever-changing planet such as ours, the older the rock, the greater the chance that some event, such as metamorphism by heat, has allowed the escape of the daughter product and reset the clock to zero.

Many meteorites have been dated, and very commonly seem to have crystallized about 4.5 billion years ago. In addition, careful measurements of all the lead isotopes in crustal rocks have permitted comparisons with the leads in meteorites. If we assume that the isotopic composition of nonradioactively formed lead in meteorites is indicative of the lead present as the earth formed, the present day quantities of nonradioactively formed lead isotopes in crustal rocks indicate formation at about 4.5 billion years ago. Moon rocks have also yielded dates in the same range, and it now seems evident that, whatever form the earth was in originally, it became a planet at least 4.5 or 4.6 billion years ago.

HALF-LIFE AND UNIFORMITY

If the half-life decay time for radioactive parent isotopes has changed through time, our accepted dates are in error. Is the assumption of uniformity a valid one for this case? As previously stated, temperature and pressure, at least those conceivable after the formation of earth, have little or no effect on nuclear reaction rates. However, if the speed of light has varied with time, or some other "universal constants" have altered, our ideas on time would need to be changed. Without specific discussions, there are methods by which half-life rates in the past can be investigated and no differences have been detected.

Just after the turn of the century Einstein put forth the idea that time slows as the speed of light is approached. Humans have no experiences at such speeds, so we are not aware of the phenomenon. We all still tend to deal with time as an invariable "flow" in the Newtonian sense, but under some circumstances, Newtonian time must be abandoned. Physicists have been able to create unstable isotopes in sufficient quantities for experimental purposes in particle accelerators. The half-life of such a mass of an unstable isotope can be measured as the atoms travel in the accelerators at velocities approaching the speed of light. While traveling, the half-life is significantly less than that of a mass of the same isotope at rest—time, therefore, is slowed as speed is increased, as Einstein predicted. This truth has not caused geologists to alter the dating assumptions because the isotopes, the minerals that contain them, and we, have all been moving in the same space-time continuum and, for us and for earth, our dates seem valid.

OTHER METHODS OF TELLING TIME

The isotopic "clocks" we have discussed are used mostly to assign dates to events quite remote from the present. For more recent events of geological and archeological interest other means have become available. One of the most valuable is known as **radiocarbon** dating.

Nitrogen, with seven protons, is the most common gas in our atmosphere, and the isotope of nitrogen with seven neutrons (^{14}N) comprises over 99% of the total. The earth is constantly bombarded by high speed cosmic "rays," mostly protons, and a relatively constant number of these strike the nuclei of nitrogen or oxygen in the atmosphere. When this occurs a neutron is emitted, then quickly

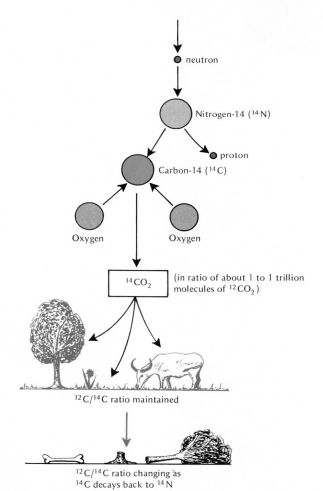

3-11. *An illustration of the production of ^{14}C in the atmosphere and of its uptake by organisms. ^{14}C continually decays to ^{14}N with a half-life of about 5730 years, but a fairly constant ratio with the stable ^{12}C is maintained in living organisms. Upon death, the ratio changes as ^{14}C continues to decay while none is being taken in. Measurement of the $^{14}C/^{12}C$ ratio in organic remains provides a means of calculating the time that has passed since the death of the organism.*

absorbed by a nearby unaffected atom of nitrogen. This nitrogen gives off a proton, and at that point is no longer nitrogen; it has transmuted into carbon with six protons and eight neutrons, ^{14}C. This newly created ^{14}C is quickly oxidized to carbon dioxide-14 ($^{14}CO_2$) and rapidly becomes distributed in the air and water of the world. Plants take up the $^{14}CO_2$ in their process of photosynthesis, animals eat the plants or eat animals that ate plants, and the ^{14}C is eventually taken in by all living organisms (Figure 3-11).

Cellular material is constantly replaced in living systems. The ^{14}C isotopes are decaying, but new ^{14}C is taken in along with the more abundant stable isotope, ^{12}C, as long as the organism lives. Thus a constant ratio of ^{14}C to ^{12}C is reached and maintained in life. However, when a tree fell or was burned in an ancient fire, when a snail died and left its shell, or when the bones of some mammoth were tossed aside after a meal, the radiocarbon clock began to tick. After the death of the organism no new carbon was taken in for cellular renewal, and the ratio of $^{14}C/^{12}C$ changed because the ^{14}C, with a half-life of about 5730 years, decays back to nitrogen.

Because it has a short half-life in geological terms, the ^{14}C quantity decreases rapidly with respect to the quantity of stable ^{12}C and the method is fairly accurate to about 50,000 years ago. Although the production of ^{14}C in the atmosphere has not been precisely constant through time, the introduced error is quite small. Known dates from historical records allow the system to be checked. Cloth wrapping from mummies of persons known to have lived, say, 4000 years ago, can be dated to within about 150 years.

This method has provided a means for understanding certain more recent geological events and has wide application in the study of man prior to written history. However, man, as a user of fire and tools, is thought to have appeared close to 2 million years ago, so the method cannot be used to study our origins.

There are a few other ways to tell real time. The growth rings of trees or the thin layers of certain lake sediments with seasonal variations can be counted. At the most these provide data for only a few thousand years. One other interesting method has given us insight into the variation of one of our most-used time units, the day.

The moon and, to a lesser degree, the sun cause tidal forces on earth. A certain amount of friction from these forces is causing the earth's rate of rotation on its axis to decrease. At the present time tidal forces appear to reduce earth's rotation about 2 sec every 100,000 years, but the moon is slowly getting farther away and, when it was closer in the past, this rate may have been slightly more. On the other hand, the rate of the earth's revolution around the sun is not thought

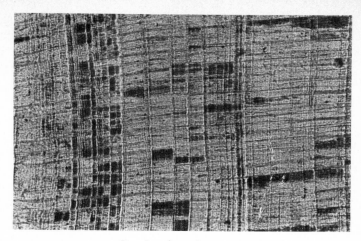

3-12. *Growth patterns of a clam shell. This photo-micrograph of a section of shell from* **Mercenaria mercenaria,** *a clam from the intertidal zone of the Atlantic coast of the United States, illustrates patterns of growth reflecting physiological responses to environmental changes. Daily, tidal, and seasonal patterns, such as these, from fossil organisms have allowed estimates to be made concerning the number of days per year in the remote past.* [*Courtesy G. Pannella.*]

Direction of growth ⟶

to have varied much. So, although the day, the time from sunrise to sunrise, has changed, the year itself has not.

If we assume that the present slowdown of the earth's rotation has been continuous, 600 million years ago, at the beginning of the Cambrian Period, the rotational rate was more rapid. The time from dawn to dawn of a Cambrian day was over 3 hr less than now. If the Cambrian day was less than 21 hr and the time of one year was about the same as now, there must have been about 424 days/year.

There is a means to check this. Several organisms, corals and clams, for example, add a small amount of material to their shells each day (Figure 3-12). By counting the daily growth additions and noting differences in thickness caused by variable growth rates according to monthly tidal or seasonal temperature changes, the number of days in the year can be calculated. Fossils from various geologic periods have been studied, and a steady lengthening of the day has been demonstrated at about the predicted rate at least since the Cambrian.

CONCLUSIONS ABOUT REAL TIME

The clocks of radioactive decay have given us two things: a tool to solve specific problems and a concept. Geologists can now provide reasonably definite answers as to when some ancient volcano spewed lava onto the surface or when the Appalachian or Himalayan mountains actively began to rise. It is usually impossible to assign a birthday to such events, but they can be placed into a period of real time.

Geology has used the dating methods to gain a clearer knowledge of many events in earth history. Evidence of life has been found in rocks over 3 billion years old; cells of organisms that differ little from living organisms have been found in rocks nearly 2 billion years old. These things have not only provided geologists with insights into the life, surface environments, and atmosphere of the primitive earth but, more important, they have expanded our concept of time. Whatever changes in dating methods the future holds, it is extremely unlikely that it will ever again be concluded that the earth is a very young planet. Geology has given mankind a realization of vast amounts of time and has provided a framework for theories of the physical and biological processes that have affected and developed our earth throughout its history.

It is now clear that the rocks that exist to tell earth's tale have only recorded fragments of the story. Science now sees itself trying to interpret a very thick history book with only scattered pages preserved. The startling thing is not how many mistakes can be found in studying the history of the geological sciences but how scientists have done so well with what is available.

In reference to Figure 3-9: The most probable relative age sequence, oldest to youngest is:

1. granite
2. volcanic rock "P" (note the inclusions of pieces of the pre-existing granite in the volcanic rock)
3. mafic dikes (they cut through both the pre-existing volcanic rock "P" and the granite
4. felsic dikes (they cut across the mafic dikes)
5. the fault (it cuts across the felsic dikes as well as all other pre-existing rocks and has brought the granite into fault contact with the volcanic rock)
6. erosion
7. lower basalt flow
8. volcanic rock "L" (by the principle of superposition the lower basalt flow must pre-date "L")
9. upper basalt flow
10. volcanic rock "U"

1. Why, even with the advent of accurate radiometric age dating methods, are most rocks placed into the relative time relations of the Geologic Time Scale rather than being assigned a definite numerical age?
2. Why is it usually easier to assign a marine sedimentary rock to a period of geologic time than it is for a rock deposited in a terrestrial environment?

four

The Earth in Space

There are two very important aspects to understanding the earth as a planetary body. One problem concerns the earth's composition and structure; the other has to do with how the earth came to be that way. In attempting to solve these problems we are faced with severe difficulties. Only the outermost portion of the earth, a tiny fraction of the whole, is available for direct observation, so the nature of the deeper parts must be deduced from indirect measurements. Unfortunately, these measurements may be interpreted differently, depending on our concepts about the early history of the earth. These concepts, in turn, depend on our conception of the origin of the solar system.

The solar system and the earth originated a very long time ago. To the best of our knowledge, the materials that formed the earth first came together about 4.5 billion years in the past. The long history of geological activity since that time has thoroughly obscured many of the characteristics of the early earth. This means that we never will know precisely all the physical and chemical conditions that attended its formation. The best that can be done is to form hypotheses as to the early history of the earth and solar system and test their consequences against those facts that we can determine today.

THE SOLAR SYSTEM: SUN AND PLANETS

To an astronomer, our sun is a thoroughly ordinary yellow star of smaller size and lesser energy output than most of the stars we can see in the night sky. However, within the bounds of the solar system (Figure 4-1) the sun is by far the largest, most massive, and most

Table 4-1   Physical data on the solar system

	Average Distance from sun, $\times 10^8$ km	Orbital Period, years	Rotation Period, days	Radius, $\times 10^3$ km	Mass, $\times 10^{27}$ g	Mean density, g/cm³	Number of Satellites
Sun			25.4	694	1.99×10^6	1.4	
Mercury	0.579	0.241	58	2.42	0.355	5.50	0
Venus	1.08	0.616	−243*	6.05	4.90	5.27	0
Earth	1.50	1.00	1.00	6.37	5.98	5.52	1
Moon	1.50	0.075	27.3	1.74	0.073	3.33	—
Mars	2.28	1.88	1.03	3.37	0.645	3.95	2
Asteroids	4	—	—	—	0.003(?)	3.5	—
Jupiter	7.78	11.9	0.41	71.4	1900	1.33	12
Saturn	14.2	29.5	0.43	60.5	568	0.70	10
Uranus	28.7	84.0	−0.45*	23.6	87.3	1.7	5
Neptune	44.9	164	0.66	24.8	103	1.6	2
Pluto†	59.8	247	6.4	3.2(?)	0.6(?)	4(?)	0

*The minus (−) sign on the rotation period of Venus and Uranus means that these planets rotate in an opposite direction from the other planets and from the direction of motion in their orbits.
†The radius, mass, and average density of Pluto are quite uncertain.

Table 4-2 Estimated Average Composition of the Sun*

Element	Atomic Mass	Per Cent of Atoms of Sun	Per Cent of Mass of Sun
Hydrogen (H)	1.008	92.3	74.3
Helium (He)	4.003	7.5	24.0
Oxygen (O)	16.000	0.08	1.1
Carbon (C)	12.011	0.05	0.46
Nitrogen (N)	14.008	0.009	0.10

*The sun is even more rich in hydrogen and helium than the estimated average composition of the universe. Most of the naturally occurring elements have been detected in the sun in at least trace amounts.

energetic body (see Table 4-1). The sun also differs from the planets in that its material is entirely in the gaseous state. Hydrogen is the dominant element in the sun, helium is next most abundant, then oxygen (Table 4-2). All the other elements form less than 0.1% of the sun's mass.

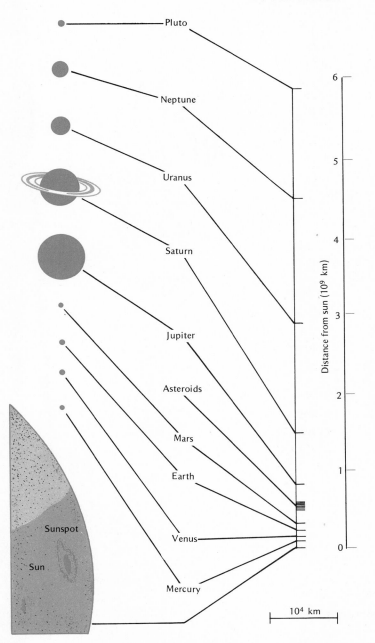

4-1. *Relative sizes and distance from sun of the planets of the solar system. Note the much greater size of the four major planets (Jupiter, Saturn, Uranus, and Neptune) compared to the other five. The distance shown for Pluto is its average distance. The orbit of Pluto is so eccentric that at times Pluto is closer to the sun than Neptune.*

Energy

The energy that the sun radiates into space, part of which we see as sunlight, comes from nuclear reactions deep within its central core. In the core the pressure is hundreds of millions times that of the earth's atmosphere, and the temperature is many millions of degrees. These conditions are sufficiently extreme to allow hydrogen nuclei to be fused into nuclei of greater atomic mass. There are two processes that operate within the sun, but both result in four hydrogen nuclei being combined to form one helium nucleus. Each hydrogen nucleus has a mass of 1.008 atomic mass units, whereas the helium end product has a mass of 4.003. This means that 0.03 atomic mass units have disappeared, turned into an amount of energy that can be calculated from Einstein's famous equation, $E = mc^2$. In this equation, m is the mass converted and c is the speed of light. This energy from the nuclear reactions is converted into heat, causing the sun's gases to glow and, at the surface, to radiate the energy into space as sunlight. Thus, the sun is consuming itself and slowly radiating away its substance as energy.

The concept of **gravitational energy** is fundamental to understanding the dynamics of the sun and of the planetary bodies that accompany it in the solar system. The fact that any two masses attract each other gravitationally means that energy will be released if they approach each other. If they do not approach, the energy remains stored in the system. Let us use some down-to-earth examples of the direct effects of gravitational energy for man, both desirable and undesirable. Water falling towards the earth over a waterfall releases energy, and this energy can be turned into hydroelectric power. On the other hand, if man himself falls, he also releases energy, which leads to any number of unpleasant consequences when the fall is stopped.

The sun consists of a great number of particles of mass, and a great deal of energy would be released if they could be packed into a smaller space. The ability of the sun to maintain itself as a gaseous body with a density of only 1.4 times that of water (Table 4-1) despite the tremendous gravitational pull of its mass is a result of the equally tremendous temperatures created by the nuclear reactions going on within.

The orbits of the planets about the sun and of satellites about the planets can also be understood in terms of energy. In the simplest case, one body moving around the other in a circular orbit, the distance between the two bodies always remains the same. Since they neither approach nor recede, the amount of gravitational energy in

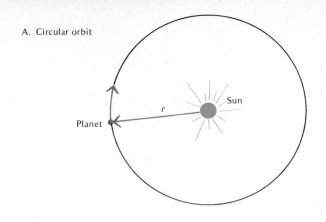

A. Circular orbit

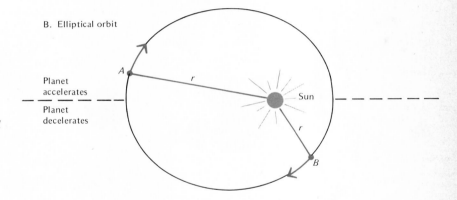

B. Elliptical orbit

4-2. *Orbits and energy. A. In a circular orbit, the planet is always at the same distance* r *from the sun and has constant gravitational and kinetic energy. B. If a planet moves in an elliptical orbit, its distance* r *from the sun is constantly changing. At point A, the planet is approaching the sun,* r *becomes smaller, and gravitational energy is converted into kinetic energy—the planet moves faster. At B, where* r *is increasing, the planet slows down, losing kinetic energy and storing it as gravitational energy.*

the system remains constant. The **kinetic energy,** the energy of motion of the planet or satellite, also remains constant, and the smaller body proceeds in its orbit around the larger one at constant speed (Figure 4-2A). The more general kind of orbit is **elliptical** (Figure 4-2B), where the distance between the two bodies is not constant. As the orbiting body approaches its **primary,** the larger body about which it orbits, gravitational energy is released and turned into kinetic energy—the orbiting body speeds up. As it swings about its nearest point to the primary body and begins to recede, the kinetic energy is tapped off as the body slows down and is stored as gravitational energy. Thus the two types of energy are traded back and forth, without any net gain or loss. Both circular and elliptical orbits can go on forever, as long as no external forces disturb the system.

The "Geography" of the Planets

The orbits of the major planets are almost circular and lie in very nearly the same plane as the earth's orbit. There is a very strong consistency in the motions of the planets. Seen from far above the North Pole of the earth, all the planets revolve counterclockwise in their orbits. The closer the planet is to the sun, the faster its orbital motion but the direction of all is the same. The sun and most of the planets also rotate counterclockwise about their axes. For a body as large as a planet, the gravitational energy to be gained by rearranging itself into the most compact form, that is, a sphere, is greater than the force with which its material can resist. Thus, the planets are all approximately spherical in shape.

As we go outward from the sun, the nature of the planets changes. The inner planets, Mercury, Venus, Earth, and Mars, are all relatively small, have high densities (Table 4-1), and only a small fraction, if any, of the mass is in the form of gaseous atmosphere. They are in essence rocky bodies, very unlike the sun in overall composition. The earth is unique among the inner planets, indeed among all the planets, in having a satellite of such large relative size. The Moon is not unusually large as satellites go, but the ratio of its size to that of its primary body, the earth, is much greater than for any other satellite in the solar system. The asteroid belt, between Mars and Jupiter, is made up of a very large number of small bodies. The largest has a radius of less than 400 km, and many are just irregular chunks of rock. The orbits of some of the asteroids are quite elliptical rather than circular in form.

The outer planets, Jupiter, Saturn, Uranus, and Neptune, have extensive atmospheres and low average densities (Table 4-1). In fact, Saturn is less dense than water. These four outer planets are also very large. Jupiter is the second most massive body in the solar system and may not be too much smaller than the minimum size necessary to become a star. The outer planets, with their low densities, must include a large amount of the light gases and are much nearer to solar composition than the inner planets. The outermost planet, Pluto, is quite anomalous. As far as we can tell, it is a rocky planet similar to the inner planets.

There is another change as we go outward from the sun that is purely related to distance. Because the sun's radiant energy spreads out uniformly into space, the amount of energy falling on any unit area, say 1 square centimeter (cm^2), diminishes with the square of the distance from the sun (Figure 4-3). Thus Venus receives almost twice as much solar energy per unit area than the earth does, and

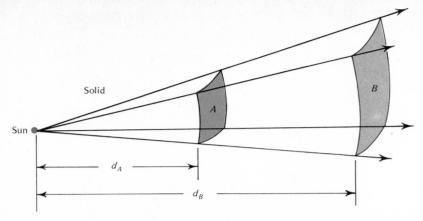

4-3. *The spreading of the sun's radiant energy. Because the rays of energy travel in straight lines, the same amount passes through surface B as through surface A. Since the area of the surfaces is proportional to the square of their distance from the sun, the energy per unit area is inversely proportional to the distance squared.*

Mars less than half as much. This means that the surfaces of the outer planets, and in particular of Pluto, are very cold places indeed, whereas the surface of Mercury is hot enough to melt lead. As will be discussed later, the earth is at a singularly fortunate distance from the sun, which allows the flourishing life of which we are part.

The Origin of the Solar System

The present physical state of the solar system is one kind of information that can be used to test ideas about its origin. The compositions, masses, and motions of the sun and planets are a result of their history, in particular their early history. Any theories about that history should be able to explain the configuration we see now. However, the present state of the system does not tell us much about its age. An additional problem is that often several differing theories can explain equally well the way the system looks now, although each theory invokes different early conditions. To help resolve these difficulties, we need samples of material left over, unchanged, from the time of origin of the solar system.

The earth is not a good place to find such materials because its surface has been so extensively reworked by geological processes that there is very little information left as to its original state. Fortunately, meteorites and the moon do provide us with accessible samples of really old material.

Meteorites are objects from space that have been trapped by the earth's gravitational pull and have fallen to its surface. They are composed of three general kinds of material—silicate minerals, the iron-sulfur compound that forms the mineral **troilite** (FeS), and metallic alloys of iron and nickel. Meteorites composed mostly of

silicate minerals are called **stony** meteorites, those of roughly equal parts silicate mineral and iron alloys are **stony-irons,** and those that are mainly iron-nickel alloy are termed **iron** meteorites. All three kinds contain troilite. There is evidence that none of these kinds of meteorites were ever incorporated in a body of planetary size, so they can be taken as left over samples of the kind of raw materials from which the planets were formed. The silicate minerals of meteorites include radioactive elements, making possible radiometric determination of their age. The ages lie in the range from 4.5 to 4.6 billion years, and this is generally accepted as representing the age of the solar system.

The moon also provides samples of very old material. Although its surface is scarred by its long exposure to meteorite impact (gravitational energy again) and has been modified somewhat by outpourings of molten rock (Figure 4-4), samples of rocks and soil returned by the lunar expeditions include materials with an age of 4.6 billion years. Thus the moon, like the meteorites, provides clues to the conditions that existed at the formation of the solar system. Moon rocks have obviously been incorporated in a large body with a subsequent history, so they may not be as representative of the solid raw materials of the solar system as meteorites.

What, then, is the most reasonable picture of the origin of the solar system? The hypothesis most generally accepted now is also the earliest one—the **nebular hypothesis.** In its modern version, it holds that just before 4.6 billion years ago, the material of the solar system was in the form of a **nebula,** or cloud of dust particles and gas.

Studies of meteorites have given us clues as to where this dispersed dust and gas cloud may have come from. There are certain elements that are so nuclearly unstable and have such short half-lives for radioactive decay that they are essentially **extinct** (have all decayed) in the solar system today. They do, however, leave behind distinctive decay end products, and some of these end products can be found in meteorites. This means that the meteorites must have been formed very soon after the synthesis of the now extinct elements. Because it takes a very high energy environment, some kind of violent stellar event, to create these unstable nuclei, the solar system must have come into existence in the neighborhood of, and soon after, such an event. It is now thought that a **super nova** explosion may have been responsible for synthesis of the extinct atomic nuclei. A super nova occurs when a particular kind of star burns up its nuclear fuel so that its gravitational pull is no longer balanced by nuclear energy. If the consequent gravitational collapse is fast enough, the release of gravitational energy results in an extremely violent ex-

4-4. *The near side of the moon. The large dark areas are maria, thought to be extensive lava flows. The smaller, circular craters are probably mostly caused by meteorite impacts. (Courtesy Lick Observatory, University of California.)*

plosion. Thus, the birth of the solar system may have taken place from the materials left over from the death of a star.

However the nebula may have originated, it would begin to contract as a result of the gravitational pull of its particles on one another (Figure 4-5A). As the particles spiraled inwards, most would end up in a central mass concentration; however, the accelerated motions would lead to turbulence in the cloud, and eddies and local mass concentrations might result (Figure 4-5B). The central concentration, as it contracted, would release tremendous energy, leading ultimately to the initiation of nuclear reactions and the beginning of the sun as a star (Figure 4-5C). The subsidiary mass concentrations would also collapse gravitationally, forming the planets, their satellites, and the material of asteroids and meteorites.

At this point in the evolution, two characteristics of the system must have been established. One is the lack of gases where the inner planets formed, shown by their meager atmospheres, and the retention of gases where the outer planets formed. The other is the odd distribution of kinetic energy in the system. If most of the mass ended up in the sun, most of the kinetic energy should have been

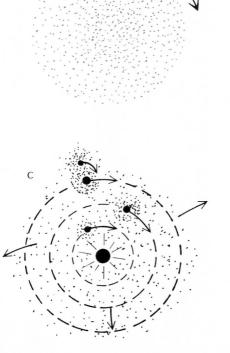

A

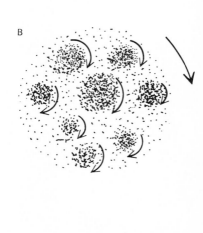

B

C

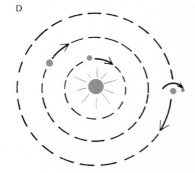

D

4-5. *The nebular hypothesis of origin of the solar system. (A) The nebula, a slowly rotating cloud of gas and dust, (B) begins to contract under its own gravitational pull, forming a disk with a large central mass concentration and several outer eddies ("whirlpools"). (C) These condense to form planets and satellites, while the central mass collapses and releases enough gravitational energy to start nuclear reactions. Meanwhile, the central mass transfers angular kinetic energy outward and sweeps the inner region relatively free of gases, and (D) leads to the solar system we see today.*

there also. In fact, the sun rotates so slowly that most of the angular kinetic energy is in the motion of the planets. Jupiter has well over half that energy now. Thus there must have been some way of moving the gases and the rotational energy outward. One possibility is the **solar wind,** an outward streaming of protons (hydrogen ions) and electrons from the sun. Pressure of the solar wind and the sun's early radiation may have pushed the gases away. It has been suggested that a rotating magnetic field of the early sun helped in this process, and also transferred kinetic energy outwards. As the magnetic field rotated, it would speed up hydrogen (H^+) ions in the vicinity of the sun and move them up into more distant orbits. At the same time, the rotation of the sun would be slowed, providing the energy passed out into the outer part of the solar system.

The exact details of how the planets in our solar system formed are still a matter of much debate. It is agreed that **accretion,** that is, gravitational accumulation of originally dispersed material, was the basic mechanism, but the question of how fast the material accreted is a subject of controversy. The rate of accretion is important because it controls the early state of the planets. If a planet, say the earth, accreted slowly enough, the gravitational energy released by the accretion would heat the growing earth sufficiently slowly that the heat energy could be radiated into space about as fast as infalling particles brought more heat in. A slowly accreted earth could have been, relatively speaking, a cold earth in the beginning. The slow decay of the important heat producing elements, uranium, thorium, and the radioactive isotope of potassium, would then warm the earth slowly until melting of some of the material became possible. In this model the gases now found in the ocean and atmosphere would originally have been included in and attached to the infalling particles; the gases would have been only slowly released to the surface of the earth as it warmed up.

On the other hand, if the rate of accretion was fast, gravitational energy would be released much faster than it could be radiated away, and an initially molten earth would result. In addition, if enough of the short-lived radioactive elements were incorporated into a quickly accreted earth, the heat generated by their decay would contribute to an early melting of the entire earth. This would have enabled the materials of the earth to have sorted themselves into something resembling the present structure very early in its history.

One question that arises when we consider the history of our solar system is: How many other planetary systems might there be in the universe? If the processes that made the sun also led naturally to an accompanying planetary system, other stars that formed in a similar way might also have planets in orbit about them. Even if the origin of our solar system was not typical, there is abundant evidence of other multiple systems in the nearby part of the universe. The frequent occurrence of **binary stars,** two stars orbiting around each other in a common system, shows that multiple stellar sized objects are easy to form. There is no reason to believe that smaller nonluminous planets, although difficult for us to detect, are not also common. If there is a multitude of planets in the universe, at least some of them should be favorable to the development of life. Earth might well not be alone in supporting life in the universe.

THE EARTH

Whatever the possibilities for other life in the universe, within our own solar system the earth is peculiarly suitable for the flourishing of the familiar forms of life. One of the most important characteristics of the earth in this respect is that it is just the right distance from the sun. The amount of heat it receives from sunlight balances the heat it radiates into space in such a way that the surface temperatures allow the existence of liquid water. Water in the liquid state is essential for the survival of most of the life forms on earth, and probably was necessary for the origin and evolution of organized life. The nature of the earth's early atmosphere was probably very favorable for the development of life. The composition of the atmosphere appears to have changed with time, and biological processes have almost certainly dominated the changes since life became common on the earth. In particular, the process of **photosynthesis,** by which plants use solar energy to convert carbon dioxide and water into organic carbon compounds and oxygen gas, has provided the oxygen in the air that we need to survive.

In addition to its central part in biological processes, solar radiant energy also drives many important physical geological processes. Sunlight provides the energy that makes possible the **surficial processes,** the work of wind, water, ice, and organisms. If it were not for the proper supply of solar energy to maintain water in its liquid state, the surface of the earth would look very different. Of course, in addition to the surficial processes, there are what we might call **internal processes,** driven by the internal heat of the earth. Some of that heat is derived from decay of long lived radioactive elements, and some may remain from the original gravitational energy of accretion. Plate tectonics, volcanism, and mountain building are examples of internally driven processes.

Bulk Characteristics of the Earth

By B.C. 250 Eratosthenes had arrived at a reasonable estimate of the size and shape of the earth (Chapter 1). Since then more sophisticated astronomical observations and, recently, studies of the orbits of artificial satellites have refined our knowledge of these properties of the earth. The planet is almost a sphere, with an average radius of 6371 km. There is a slight bulge about the earth's equator, and the polar regions are slightly flattened; therefore, the polar and equatorial radii differ by about 1 part in 300 or 22 km. This bulge is a result of the rotation of the earth.

As the earth rotates, the particles within it move in a curved path. The farther a particle is from the axis of rotation, the faster it moves. At the equator, a particle of mass completes a circle of more than 40,000 km circumference in 1 day. However, the law of inertia states that a mass will travel in a straight line unless a force is applied to it. In this case, the force of gravitation, due to the whole mass of the earth, keeps particles from flying off in all directions. Near the equator, part of the earth's gravitational force is devoted to just holding it together, and an equatorial bulge is raised in response to the tendency to fly apart. The fact that the earth has very nearly the shape of a fluid body rotating with the same speed shows that it cannot be completely rigid.

One might imagine that the mass of the earth would be very hard to measure. In fact, one theoretical law and one laboratory measurement proved sufficient. **Newton's principle of universal gravitation** states that any two particles of mass M_1 and M_2, separated by a distance r, attract one another with a force F given by

$$F = G \frac{M_1 M_2}{r^2}$$

where G is a universal gravitational constant for all particles. The constant G must be quite small, because gravitational forces become large only when the masses are very large.

It was not difficult to measure the force F that the earth exerted at its surface on a mass M_1 of 1 gram (g). In addition, Newton had shown that any body with spherically symmetrical mass could be considered, for gravitational purposes, to have all its mass concentrated at its center. This meant that the quantity r was just the radius of the earth. Three of the five quantities in Newton's equation applied to the gravity of the earth were thus known in Newton's time. All that was needed to find M_2, the mass of the earth, was the gravitational constant G. Around the year 1800, Cavendish succeeded in measuring the value of G in the laboratory by determining the weak gravitational attraction of large balls of lead. Thus Cavendish was able to "weigh" the earth.

The best value we have now for the mass of the earth is 5.976×10^{27} g. This number does not mean much to us until we divide by the volume of the earth to obtain its average density, 5.52 grams per cubic centimeter (g/cm^3). Since the average density of surface rocks is less than 3 g/cm^3, we can conclude immediately that much of the mass of the earth is concentrated downwards and that material of higher density must be within.

The Problem of Seeing Deep

Just where this high density material might be located is a more difficult question. Only the very outermost part of the earth is accessible for direct sampling, and we must infer the structure and composition of what lies below from what we can observe at the surface. Some sort of remote sensing techniques, some kind of indirect observation must be used. For this we need forms of energy that are capable of penetrating earth materials and returning to the surface some record of what they have been through.

One of the most useful ways of studying the interior of the earth is to use wave energy to "illuminate" its structure. **Seismology** is the branch of geophysics that uses vibrational waves to study earth structure, and much of what we know about the earth's interior comes from seismology. For problems of the deeper portions of the earth, earthquakes are the most important source of wave energy for seismic studies.

Earthquakes and Seismic Waves. Earthquakes are the result of fracturing of large masses of rock. If forces within the earth apply stress to a rock mass faster than it can deform by flowage, **strain energy** is stored up within the rock. When the rock cannot accommodate any more strain, it breaks suddenly and the stored strain energy is released violently. Some of this energy is released as **seismic waves**, and it is the energy of seismic waves that gives earthquakes their great destructive potential. This seismic energy also enables us to study the interior of the earth.

Vibrational energy can be transmitted through the earth as several kinds of seismic waves. **Surface waves** travel along the surface of the earth or along boundaries within it, whereas the **body waves** pass

4-6. *Compressional waves. As a P (compressional) wave passes through a material in the direction of the heavy colored arrow, the particles in the material vibrate back and forth along the same direction. The colored arrows show the motion of the particles at one instant of time. After the wave has passed, the particles end up in the places they started from. The gray arrows at the right show the total motion the particles went through as the wave passed. Thus energy is transmitted through the material without any permanent displacement being involved.*

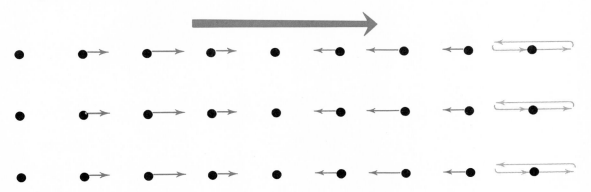

4-7. *Shear waves. As an S (shear) wave passes in the direction of the heavy colored arrow, the particles in the material it is traversing move at right angles to the direction the wave is traveling. The colored arrows, show the motions of the particles at one instant of time. The gray arrows at the right show the complete motions the particles go through as the wave passes. Shear waves will propagate only through solid materials and have a slower velocity than compressional waves in the same material.*

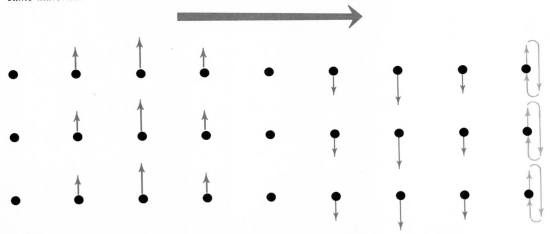

through the interior. The body waves are of two types: *P*, or **compressional waves,** and *S*, or **shear waves.** During passage of a *P* wave, the particles in the material through which it is moving vibrate back and forth in the same direction the wave is traveling (Figure 4-6). The familiar sound waves in air, by which we hear, are of this type. In the case of an *S* wave, the particles move at right angles to the direction in which the wave is transmitting energy (Figure 4-7). Shear waves are analogous to the vibrations of a plucked guitar string.

There are two important differences in the behavior of *P* and *S* waves. One is that *P* waves can travel through any kind of material, whether solid, liquid, or gas, whereas *S* waves are transmitted only by solids. Thus, if a region of the earth's interior allows *S* waves through, we have adequate reason to believe that it is solid. The other important difference is in **seismic velocity.** *P* waves *always* travel faster than *S* waves in any particular solid material, but the velocity of each may be different in different solids.

This difference in seismic velocity makes possible one method of locating where an earthquake occurred. For a simplified example, let us suppose that the velocity of *P* waves (v_P) is 8 km/sec and that of shear waves (v_S) is 5 km/sec. Then, if the seismic waves travel 1000 km from the earthquake to a **seismograph** (seismic wave recording instrument), the *P* waves will arrive in 125 sec and the *S* waves in 200 sec. There will be a time gap of 75 sec between the *P* and *S* arrivals. For a seismograph station to which the seismic waves travel twice as far, that is, 2000 km, the gap will be $400 - 250 = 150$ sec.

Thus, if one measures the time between arrival of *P* and *S* waves at a single seismograph station, one can calculate how far away the earthquake occurred. It must have happened somewhere on a circle, with the station as center and a radius that can be determined from the delay between *P* and *S* arrivals. If we have measured the delays at three seismograph stations, we know that the earthquake was located somewhere near where the three circles of distance intersect (Figure 4-8). This point on the earth's surface directly above the earthquake's **focus,** or area of origin, is called the **epicenter.** In practice, a more complicated method is used to locate earthquakes, which determines not only the epicenter but also the depth of the focus.

Information on the earth from seismology. The more immediate problem, however, is not where earthquakes occur, but what can be learned about earth structure from the seismic energy they emit. Let us suppose that we know where an earthquake occurred from some surface expression such as fracturing or faulting or from the damage

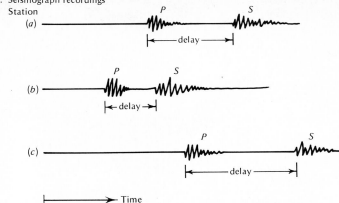

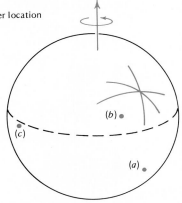

B. Earthquake epicenter location

4-8.   *Location of earthquake epicenters. A. Sketch of the waves from an earthquake recorded at each of three stations. The length of time between arrival of the P and S waves for each station is proportional to the distance from the earthquake. B. Circles of distance corresponding to the P-S delays are drawn around each of the stations. The earthquake must have occurred where the three circles cross.*

done to human habitations. The time at which the earthquake happened will then also be known. From how long it takes the seismic waves to reach seismograph stations scattered over the globe, we can tell how fast these waves travel in various parts of the earth. Figure 4-9 shows some of the paths various seismic waves follow through the earth, and the time it takes them to travel along those paths. By studying the **travel times** for many large, energetic earthquakes, seismologists have built up a picture of the interior of the earth.

This picture, is, of course, initially in terms of seismic velocity. One of its most striking features is that the earth is made up of concentric layers, with different characteristic seismic velocities. These layers are indicated in Figure 4-9, and Figure 4-10 illustrates how the *P* and *S* wave velocities vary with depth within and between the layers. The general increase of seismic velocities with depth within each layer is what causes the seismic wave paths of Figure 4-9 to be curved upwards. We will examine these layers—core, mantle, crust, and their subdivisions—in more detail.

A great deal more than the speed of transmission of seismic energy can be learned from the seismic studies if they are combined with

4-9. *Seismic wave travel paths through the earth. The time in minutes it took each wave to travel from the earthquake is shown where the paths surface. Over a little less than half the earth, directly opposite the earthquake focus, there are no direct, unreflected S wave arrivals. Note the reflected P wave (15.2 min) and S wave (27.3 min), in which the seismic energy bounced from the surface back into the interior. The paths are curved because of the increase of seismic velocity with depth within the earth.*

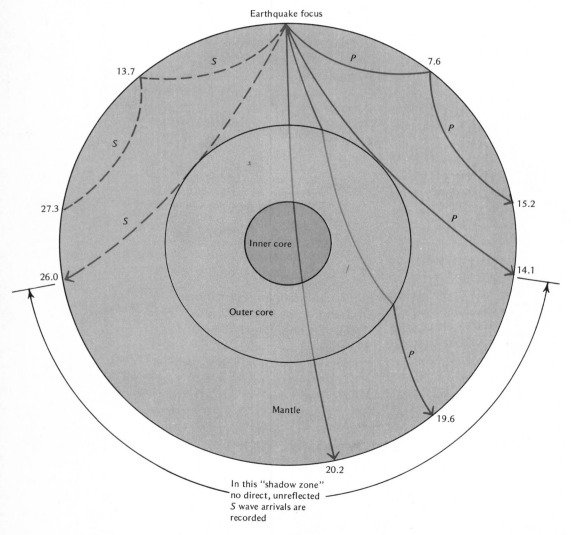

other sources of information. For example, the speed of sound in a material is determined by the chemical composition of the substance, its physical structure, and the temperature and pressure to which it is subjected. By placing limits on some of these variables, the seismic information can tell us about the possible values of the other variables. A very careful combination of theoretical work, on how materials ought to behave within the earth, and laboratory work, to measure how their properties actually do change when high pressures

4-10. *The variation of seismic velocity with depth within the earth; v_P is compressional (P wave) velocity, and v_S is the shear (S) wave velocity. Note that the S wave velocity drops to zero in the outer core. The velocities for the inner core are somewhat speculative. The region of low velocities, near 100-km depth, is called the seismic low velocity zone (colored arrow).*

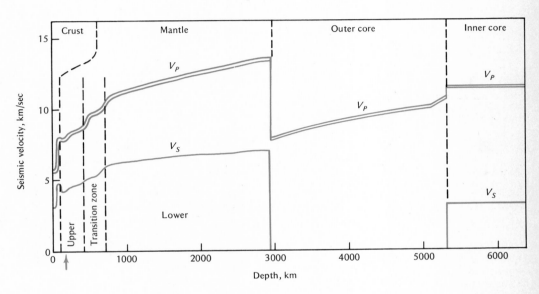

4-11. *Approximate variation of density with depth within the earth. The density of the inner core is not well known. Note the inversion of density—heavier material on top of lighter—at about 100-km depth.*

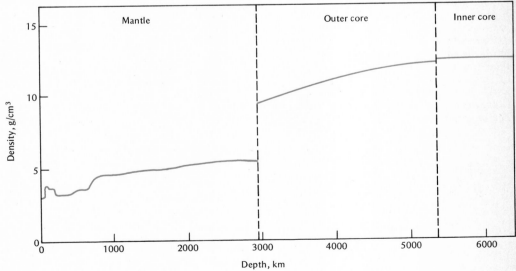

4-12. *One model of the variation of temperature with depth within the earth. In its deeper portions, the curve could be in error by as much as 1000° C.*

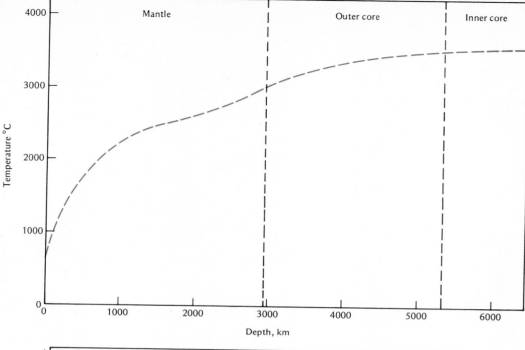

4-13. *The variation of pressure with depth within the earth. Because the calculated pressure is not very sensitive to errors in assumptions and observations used to deduce it, the curve is not likely to be seriously wrong. The pressure at the center of the earth is more than 3 million times as great as that of the atmosphere; 1 megabar is approximately 1 million atmospheres.*

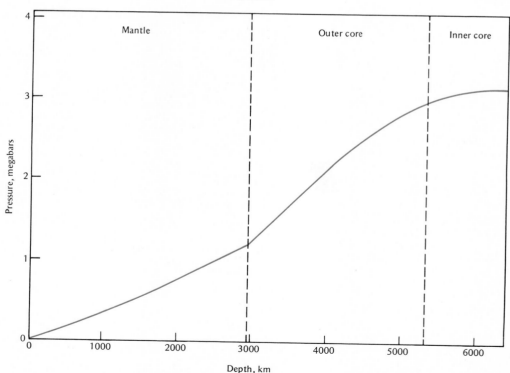

and temperatures are artificially applied, is necessary to interpret the seismic data.

Other important information sources are the gravity field of the earth, from which we can find the earth's total mass, the rotational characteristics of the earth, magnetic and electric fields, and **free oscillations.** The latter result when a particularly large earthquake sets the entire earth vibrating, "ringing like a bell."

When all these carefully measured phenomena are combined with educated guesses as to the earth's composition, mainly influenced by analogies to meteorites, theoretical models of the conditions within the earth can be calculated. Examples of these for density (Figure 4-11), temperature (Figure 4-12) and pressure (Figure 4-13) show the general kind of variation of these properties with depth within the earth. Of these curves, pressure is probably the best determined, the density is less well known, and the exact details of how the temperature changes with depth are quite speculative.

THE STRUCTURE OF THE EARTH

The Core

The large scale interior structure of the earth, a series of concentric shells, is summarized in Figure 4-14. The two innermost regions together are called the core and have by far the greatest average density of any of the shells, well over 10 g/cm³. Despite its great density the core includes less of the total mass of the earth than the mantle, which has considerably greater volume (see Appendix 3). Even at the extremely high pressures near the center of the earth, the most reasonable material to form a core of such great density is a metal. Analyses of meteorites suggest that iron is the most common metallic form occurring in the raw material of the planets. In meteorites, the iron is found alloyed with a lesser amount of nickel. Thus it is fairly well accepted by earth scientists today that the most likely composition for the earth's core is an iron-nickel alloy.

One problem with a purely iron-nickel core is that it would be too dense to match the observed characteristics of the earth's core. It is therefore usual to postulate that there are also some lighter elements mixed with the iron and nickel. This question of light elements in the core is a good example of how problems of the early history of the

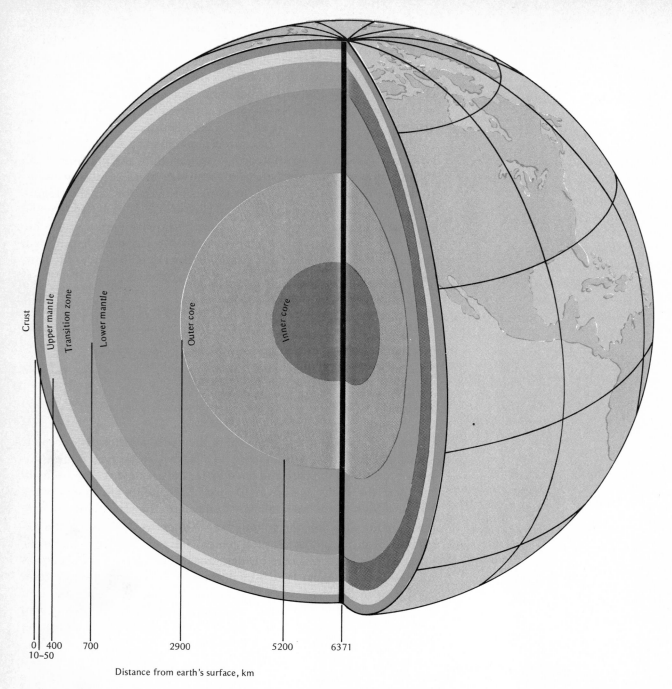

Crust

Upper mantle

Transition zone

Lower mantle

Outer core

Inner core

0 | 400   700   2900   5200   6371
10–50

Distance from earth's surface, km

4-14. *Regions of the earth's interior. The depth shown for the crust, 17 km,
is the weighted average of oceanic crust (7.5 km) and continental crust (35 km).*

earth interact with interpretation of its present condition. Silicon, an abundant light element, has been proposed as an alloying element in the core. To be dissolved in the core material, silicon would have to be in its reduced or elemental state. However, the silicon observed in surficial earth materials and meteorites is oxidized, in the form of silicon-oxygen tetrahedra. Thus, to maintain that silicon is the light element in the core, one must postulate that early in earth formation conditions must have been proper to reduce great quantities of silicon, stripping away the oxygen atoms and adding electrons. It must also be assumed that this process did not affect meteorites.

Recently it has been suggested that sulfur is the light element that lowers the density of the core. This possibility leads to quite a different picture of early earth history. As mentioned before, the three main components of meteorites are silicate minerals, iron-nickel alloys, and the iron-sulfur compound troilite. Certain mixtures of iron and iron sulfide have a lower melting point than most silicate minerals. Thus, if the earth were formed of material like meteorites the mixture of iron and iron sulfide would be the first to melt as the earth heated up by radioactivity or by accretion energy. The melt would be very dense and move downward, releasing gravitational energy and causing further melting. The process would accelerate rapidly, leading to formation of the core very early in earth history and probably to the release of sufficient energy to melt the entire earth.

Whatever the origin of the core, certain aspects of its present structure are clear. One is that it acts as an effective block to the transmission of seismic S waves within the earth. As shown in Figure 4-9, there is a "shadow zone" on the far side of the earth from an earthquake in which the only direct, unreflected seismic waves recorded are P waves, with no direct S wave arrivals. Remember that liquids will transmit P waves but not S waves; therefore the immediate conclusion must be that at least the outer part of the core is molten. However, study of P waves that travel through the inner part of the core show that the seismic velocity is higher there, leading to the suggestion that the inner core is solid. This idea is confirmed by study of free oscillations and the detection of seismic wave arrivals that are thought to have travelled through the inner core as S waves (Figure 4-10).

Magnetic field. The magnetic field of the earth (Figure 4-15) is also thought to be related to the liquid nature of the outer core. Very roughly, the earth's magnetic field resembles that of an enormous bar magnet placed near the center of the earth and pointing toward the poles. However, permanent magnetism of earth materials cannot be the cause of the magnetic field. Temperatures within the earth are far too high to allow permanent magnetism of any material. In addition, the earth's field changes too rapidly in direction and strength to be the result of any static magnetization. Motions of the molten iron in the outer core, acting like a gigantic dynamo, must be the source of the magnetic field. No one has succeeded in proving, however, just how this dynamo might work or from where the energy to maintain the magnetic field comes.

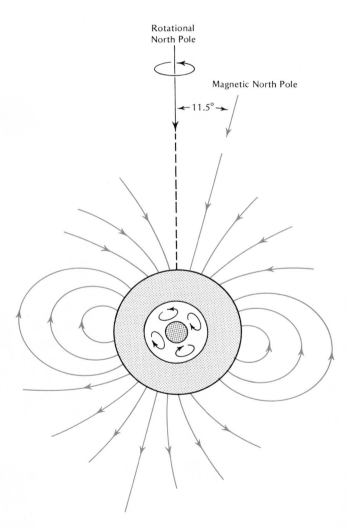

4-15. *The magnetic field of the earth is thought to originate in the motion of the liquid iron in the outer core. The core motions are shown schematically, because nobody really knows what they are. At present the magnetic north pole is inclined by about 11° to the rotational North Pole. The colored lines represent lines of force, which show the direction of the field, the direction that a compass needle free to move in a vertical plane would be pointed by the field. On the surface of the earth, near the Pole, the field is near vertical, whereas near the equator the field becomes horizontal. The closer the lines of force are, the stronger is the field. It reaches its greatest strength at the magnetic poles, and is weakest at the magnetic equator.*

The Mantle

The next layer outward, the mantle, includes most of the earth's volume and mass (Appendix 3). Its average density, 4.5 g/cm^3, indicates that the mantle is composed of rocky rather than metallic material. The exact composition of the mantle is in doubt, but oxygen and silicon must predominate, with the most abundant metallic ions being magnesium and iron. This conclusion is based on the mantle's density and seismic velocity, on the composition of rocks thought to have come from the mantle, and, by analogy, on meteorite compositions. Like the core, the mantle can be divided into several concentric shells on the basis of seismic properties.

The Lower Mantle. The **lower mantle** extends from about 700 km deep within the earth to the top of the core at around 2900 km. Although the temperatures are very high (Figure 4-12), the fact that the lower mantle transmits S waves (Figure 4-10) indicates that its material is essentially solid. The steady increase of density and seismic velocity downwards within the lower mantle is probably the result of pressure. The ever higher pressures (Figure 4-13) will compress the crystal structures of the minerals present there more and more, leading to a greater density at greater depths. In materials of constant composition and crystal structure, greater density under pressure will result in higher seismic velocities. It has been suggested by some workers that the amount of iron in the silicate minerals increases with depth in the lower mantle. This possibility would also increase the density and seismic velocity.

In contrast, the relatively sudden jumps in seismic velocity at the top of the lower mantle and within the **transition zone** (400−700 km) are probably not related to changes in composition but to alterations in crystal structure. These **phase changes** are also caused by the increase of pressure at greater depths but they are discontinuous rather than gradual changes. For example, the mineral olivine has a structure that is stable under the conditions present in the upper mantle. When the pressure reaches a certain value, the olivine structure becomes unstable and the atoms of an olivine crystal are rearranged into a more compact form that resembles the structure of the mineral spinel (Figure 4-16). This particular phase change probably occurs at a depth of about 400 km. An increase of seismic velocity occurs with the increase in density due to each phase change in the transition zone. It is interesting to note that the deepest known earthquakes, at about 700 km, coincide with the seismic velocity in-

A. The low-pressure
form—olivine

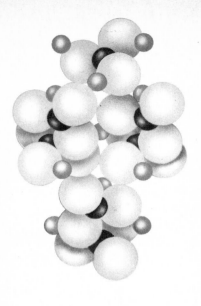

Oxygen ions

Magnesium ions

B. The high-pressure
form—spinel
structure

Silicon ions

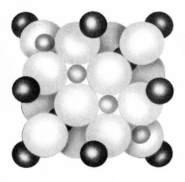

4-16. *A typical phase change—low and high pressure forms of olivine.* **A.** *At shallow depths within the mantle, olivine is stable in the same crystal structure it has at the surface.* **B.** *At greater depths and higher pressures, the olivine is stable in a crystal configuration similar to spinel. The high pressure form is slightly denser.*

crease and presumed phase change that marks the top of the lower mantle.

The Upper Mantle. The deepest layer of the earth that we have any hope of sampling, although indirectly, is the **upper mantle.** This layer extends from the base of the crust to a depth of about 400 km. The upper mantle is important to our study of geology at the surface of the earth because its motions and history are coupled to those of the crust. The most obvious and visible example of communication between upper mantle and crust is in the matter of diamonds, the dense crystalline form of carbon. As we saw in Chapter 2, diamonds can form only under very high pressures. These pressures, about 50 kilobars (kbar) or 50,000 atmospheres (atm), indicate a minimum depth of about 150 km, since pressures in the mantle increase about 1 kbar for every 3 km depth (Figure 4-13). A depth of 150 km is well within the upper mantle. The rocks that accompany diamonds in the **kimberlite pipes** (Chapter 8) and other material suspected to be of upper mantle origin lead us to believe that the upper mantle is composed mainly of olivine, pyroxene, and garnet. At upper mantle pressures, these minerals would be stable in the same crystalline structures that they have at the surface.

Seismically, the most remarkable feature of the upper mantle is the **low velocity zone.** This region has lower S and P wave velocities than the material above it and below it (Figure 4-10). Its top is at a depth of from 80 to 120 km in various places, and its thickness is approximately 100 km. The low velocity zone is particularly well defined for S waves and is most prominent in the upper mantle underlying the ocean basins. Under some continental areas, the low velocity zone is either missing or is very difficult to detect.

The most likely explanation for this zone of relatively low seismic velocities is melting in the upper mantle. Figure 4-17 illustrates the relationship between actual temperature in the upper mantle and the melting temperature of mantle materials that can lead to a confined zone in which melting takes place. Because the melt formed from most materials is less dense than the solid, an increase of pres-

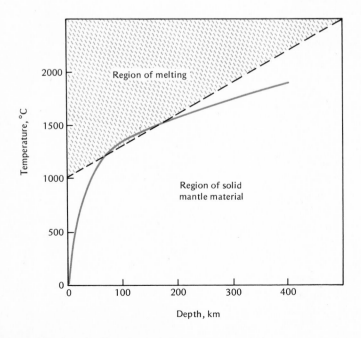

4-17. *A possible explanation for the seismic low velocity zone. Within the temperature-pressure conditions represented by the colored area, mantle material will be partly molten. In this illustration, depth is equivalent to pressure. The solid line is one model of temperature variation within the upper mantle. Over a restricted depth range, the temperature passes into the zone of partial melting.*

sure favors the solid phase. The only important exception to this is the ice-water system. Thus, as the pressure increases, the temperature at which part of the mantle minerals begins to melt also increases (Figure 4-17). The actual temperature within the mantle, which increases steeply downwards from the surface, must level off at some depth. Thus, the temperature can rise into the melting zone and then fall below the melting temperature again at a greater depth (Figure 4-17).

The situation, then, is probably something like this: small pockets of molten material scattered throughout a mass of solid minerals, mostly near their melting point. Both the presence of the liquid and the nearness to melting of the solids will decrease the seismic velocities markedly, especially the shear velocity. Another very important consequence of the semimolten nature postulated for the seismic low velocity zone lies in the mechanical strength of the material. Any substance that is shot through with pockets of melt and close to melting itself is not likely to be very strong. As we shall see later, this zone of the upper mantle is one in which considerable flow and motion takes place.

Moho. The next seismic boundary towards the surface is one discovered early in the twentieth century by a seismologist named Mohorovičić. In his honor the boundary is called the Mohorovičić discontinuity, or Moho for short. The Moho consists of a sudden increase of seismic P wave velocity from around 6.5 km/sec above the discontinuity to about 8 km/sec below. *By definition,* the Moho is the boundary between the crust and upper mantle. The Moho may in places be a compositional boundary, may in places represent a phase change, or could be both.

The Crust

This seismically defined crust, upon whose outer surface we live, comes in two distinct types. The presence of oceans and land areas on the earth's surface is not accidental—each is underlain by a distinctive kind of crust.

Oceanic crust. About 65% by area of the earth's crust is oceanic in nature. This oceanic crust is covered, on the average, by about 4 km of water. Below this rests an average of 0.5 km of sediments. The basement rocks of the oceanic crust consist of about 1.5 km of basaltic volcanic rocks, called seismic layer 2, and approximately 6 km of seismic layer 3 above the Moho. This lower layer, distinguished from layer 2 by its greater seismic velocity, is not very well known. It most likely consists of metamorphic equivalents of the basaltic volcanic rocks of layer 2, together with more iron- and magnesium-rich igneous rocks. All in all, the oceanic crust is remarkably uniform in its structure wherever it is studied.

Continental Crust. The same can by no means be said of the continental crust, which varies in thickness from less than 25 km (compare the oceanic crust's thickness of 8 km) to more than 50 km, with an average thickness of 35 km. The average elevation of the continental surface above sea level is only about 800 m, although Mt. Everest, the highest point, reaches up more than 8.8 km. The structure of the continental crust is definitely not simple and uniform. The upper part, which we can observe directly, consists of a complicated jumble of sedimentary, igneous, and metamorphic rocks. The lower part of the continental crust, though poorly known, is observed to have higher seismic velocities than the upper part. This could reflect a change in composition, or might be the result of phase changes in a crust of approximately constant composition. The assemblages of minerals stable in the upper crust can react among themselves and form new, denser assemblages under the higher temperatures and pressures in the lower crust. A higher seismic velocity would result.

Oceanic and continental crust differ in more than elevation, thickness, and structure. Their overall compositions, as far as they can be estimated, also differ considerably. Table 4-3 shows that continental

Table 4-3 Estimates of the Average Composition of Oceanic and Continental Crust

*Elements arranged by abundance by weight in oceanic crust.

Element*	Ionic radius, Å	Oceanic Crust Weight %	Oceanic Crust Atomic %	Oceanic Crust Volume %	Continental Crust Weight %	Continental Crust Atomic %	Continental Crust Volume %
O^{2-}	1.40	44.45	60.63	93.70	46.63	62.09	94.03
Si^{4+}	0.42	23.31	18.11	0.76	28.93	21.95	0.90
Al^{3+}	0.51	8.46	6.85	0.51	8.26	6.52	0.48
Ca^{2+}	0.99	8.04	4.38	2.39	4.07	2.16	1.16
Fe^{2+}	0.74	5.67	2.22	0.51	3.02	1.15	0.26
Mg^{2+}	0.66	5.20	4.67	0.76	1.87	1.64	0.26
Na^{+}	0.97	2.03	1.93	0.99	2.30	2.13	1.07
Fe^{3+}	0.64	1.56	0.61	0.09	1.82	0.69	0.10
Ti^{4+}	0.68	0.85	0.39	0.07	0.48	0.20	0.03
K^{+}	1.33	0.22	0.12	0.16	2.41	1.31	1.70
Mn^{2+}	0.80	0.12	0.05	0.02	0.08	0.03	0.01
P^{5+}	0.34	0.07	0.05	0.001	0.13	0.09	0.002

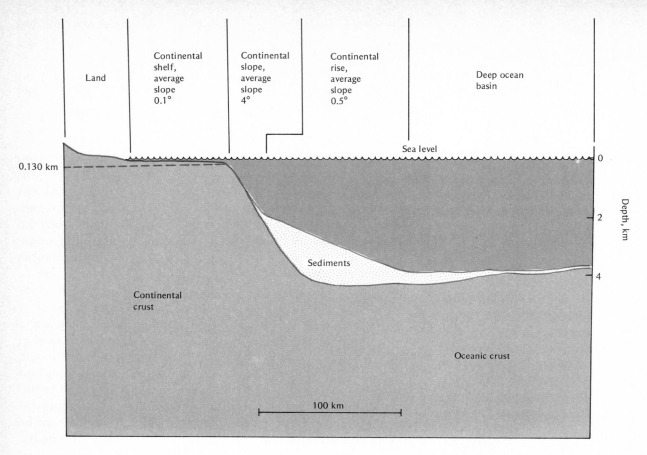

4-18. *One type of continental margin, typical of eastern North America. The slopes in this diagram are highly exaggerated because vertical distances are much expanded relative to horizontal distances.*

crust has more silicon and potassium and less iron, magnesium, and calcium than oceanic crust. The net result of these compositional differences is that oceanic crust is somewhat denser than continental crust.

The transition between oceanic and continental crust is also of interest. In general, the continents do not end at the shore line but extend under water some distance as shallow continental shelves (Figure 4-18). Thus, although water covers 71% of the globe's surface, about 6% of the total area consists of continental shelves. This leaves 65% as oceanic crust. Beyond the edge of the continental shelf is the continental slope, leading down into deeper water, and at the base of some continental slopes a gentle continental rise lies on the edge of the deep ocean basin. Both the slope and rise most likely consist of sediments derived from the continents and carried over the edge of the shelf into deeper water. It is very difficult to tell exactly where the continental crust leaves off and the oceanic crust begins.

Isostasy

The difference of elevation and thickness between oceanic and continental crust serves as a nice example of the very important role that gravity plays in determining the surface form of the earth. Let us suppose that the crust of the earth is floating on a denser layer below that is mobile and flows easily. If we wish, we can identify this layer with the seismic low velocity zone. Let us suppose further that the crust is not stiff enough to support by its own strength any great weight placed upon it. Under these conditions, various parts of the crust will behave in a way analogous to separate ice cubes of different thicknesses floating in water (Figure 4-19). Buoyancy forces caused by the earth's gravity and the differences in density between crust and underlying mobile material will ensure that each crustal block will seek its own level. The idea that each column of crustal material is floating in gravitational equilibrium with other crustal columns is called the theory of **isostasy.** Isostasy was first suggested and has been largely confirmed by studies of the earth's gravity field over crustal structures.

We can make the concept of isostasy more quantitative. At any level within the mobile layer and below any horizontal variations in density, the pressure will be everywhere the same. If it were not, the mobile material would flow sideways from under areas of high pressure and the overlying columns would sink until the pressures became equal. This means that any column of crust and upper mantle

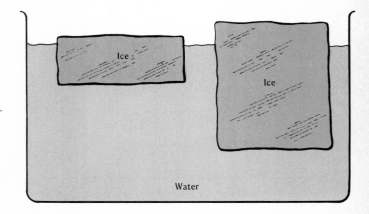

4-19. Ice cubes floating in water in isostatic equilibrium. The thicker ice cube not only sticks up out of the water further but also extends deeper into the water. Each ice cube will seek a level such that the weight of the ice above the water line is just balanced by the buoyancy force due to displacement of water. That buoyancy force is equal to the volume of water displaced times the difference in density between ice and water.

down to some level, say 100 km, within the mobile layer will contain the same total mass and, thus, total weight as any other column. This enables us to make calculations about the effects of isostasy.

One important example lies in the difference in elevation and thickness of oceanic and continental crust. Neglecting the minor difference in density between oceanic and continental crust, Figure 4-20 illustrates in a simplified way how the two float in isostatic equilibrium, by balancing the weights of columns down to 100 km. Sea water overlies oceanic crust because oceanic crust floats lower in the mantle, leaving a hole for the water to run into. If the earth had no surficial water, the ocean basins would not be quite as deep rela-

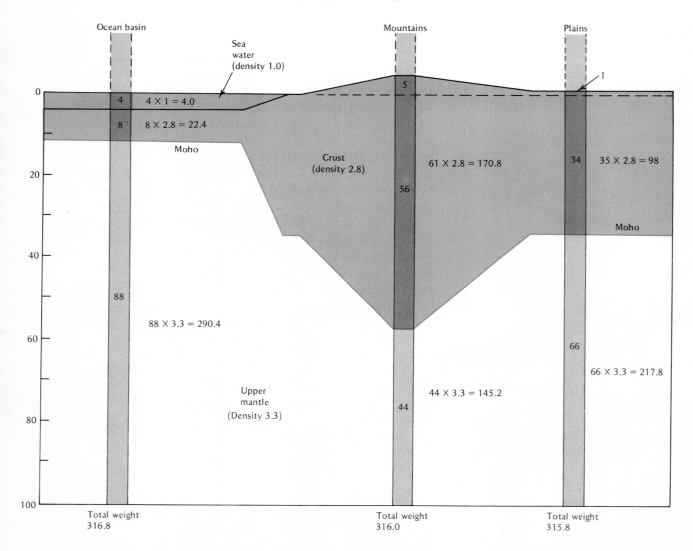

tive to the continents because the oceanic column would not bear the weight of the water. The basins would still exist, however.

A second important example is within the continents. Mountain ranges represent a considerable amount of extra mass above sea level. If isostasy is to be maintained within the continents, the crust must be thicker or less dense under mountains than under low-lying plains (Figure 4-20). If we assume that the crust below mountains has the same density as the crust below plains, 1 km of extra elevation must be compensated by about 6 km greater depth to the Moho. In a 1-cm² column, that 1 km represents 2.8×10^5 g extra, which must be made up for by displacing mantle material (density 3.3 g/cm³) below with crustal material of density 2.8. Every kilometer greater depth of the Moho compensates for 0.5×10^5 g, so we need $(2.8 \times 10^5)/(0.5 \times 10^5) =$ 5.6 km greater depth. Thus mountain elevations must have crustal roots if they are to remain in isostatic equilibrium. In general, the existence of these roots has been confirmed by seismic studies.

Isostatic Uplift or Rebound. As the name suggests, isostasy is a matter of static balance of masses, but it has dynamic effects also. Consider what happens when the top of a mountain range is eroded away. Mass has been removed from one crustal column and added to others. Thus, the mountain range will rise and the area where erosion products come to rest will sink to maintain isostatic balance. The mountains can continue to rise and be eroded away until their roots are raised to the surrounding level of the Moho. Another example of isostatic uplift is going on today in Scandinavia, where the land is rising in response to removal by melting of the great weight imposed on it by the glaciers of the last ice age.

In conclusion, it is not hard to believe that the two distinctive kinds of crust, with their differing compositions, structures, thicknesses, and isostatic levels also have very different histories and origins. How oceans and continents come to have the characteristics we observe today will be the subject of several of the following chapters.

4-20. *An example of isostatic balance among ocean basins, low lying continental crust, and mountains. The vertical rulings represent 1-cm² columns extending down to the arbitrary depth of 100 km below sea level. The numbers within the columns are the thickness of each layer in kilometers. To the left of each column is a calculation of the weight of the parts of the column. The weight for each part is shown in units of 10^5g. The total weights of the three columns are very close to being equal, 3.168×10^7g for the oceanic column, 3.160×10^7g for the mountains, and 3.158×10^7g for the plains. Thus the crustal structure shown is in isostatic balance. The vertical scale is depth in km; the horizontal scale is schematic.*

QUESTIONS

Starred (★) questions in this and the following chapters are somewhat more difficult.

1. How is the energy that makes the sun shine produced? What two kinds of energy keep the planets in their orbits?

2. Explain how the major differences between the inner and outer planets might have been caused.

3. How can the mass and density of the earth be measured?

★4. The headwaters of the Mississippi River are 6 km closer to the center of the earth than the mouth of the river at New Orleans. How can we still say that water flows downhill?

5. Why can the seismic energy radiated by earthquakes tell us something about the interior of the earth?

6. Explain why geologists believe that the core of the earth is composed mostly of iron and nickel, and why it is believed that the outer core is liquid.

★7. What is the likely explanation for the seismic low velocity zone, and why is the S wave velocity within this zone reduced by more than the P wave velocity?

8. Discuss the major differences between oceanic crust and continental crust.

9. What are the basic assumptions involved in the theory of isostasy?

★10. The theory of isostasy predicts that if a layer of material one km thick is eroded from a mountain range, the average elevation of the mountains will be reduced by only 0.15 km. Why is this so?

Plate Tectonics

As we saw in Chapter 1, men have studied the earth for a long time; until very recently, however, they were limited to observing the geology of dry land and very shallow water. The more than 70% of the earth's surface covered by the deep ocean remained unknown. Much of what was seen on land also seemed mysterious in origin. With the development of techniques to study the ocean bottom, there has come not only some understanding of the geology of oceanic crust but a set of ideas that has revolutionized and unified our knowledge of the large scale geologic processes affecting both continents and oceans. Mysteries remain, but there is now a framework in which our understanding can be expanded. This framework is known as the theory of plate tectonics. In the next few chapters, plate tectonics will be used as the setting in which we discuss the geology of the outer part of the earth.

THE THEORY OF PLATE TECTONICS

Tectonics is the study of deformation of earth materials and the structures that result from deformation. The "tectonics" in plate tectonics refers to deformation on a global scale. The matters of concern are where such deformation is taking place, the particular kinds of deformation, their consequences, and how these consequences are expressed in the geologic record. We will first discuss plate tectonics as a model of the behavior of the surficial portions of the earth and then give a historical sketch of how the model evolved from crucial observations by workers in geology and geophysics.

The basic assumptions of plate tectonics are amazingly simple, considering their far reaching consequences. The first assumption is that large areas of the earth's outer portion act as rigid caps on a sphere (the **plates** of plate tectonics) that undergo no significant internal deformation. That is to say, if London, Paris, and Moscow are all located on the same plate and we measure the distances between them this year, a hundred or a million years from now those distances would be the same. Thus, the plates are just those areas within which there are now no horizontally directed tectonic processes of significance taking place. There can be small vertical movements within the plates, but they will be of much smaller magnitude than the horizontal movements between plates we are about to discuss and do not alter the application of the ideas of plate tectonics.

The second assumption is that each plate is in relative motion with respect to the other plates on the surface of the earth, and these plates undergo significant deformation only at their boundaries with each other. If New York is situated on a different plate from London, Paris, and Moscow, its distance from those cities will change from year to year, and a zone of deformation must lie in between. Except for surficial geologic processes—the work of wind, water, ice, and gravity—and the slow vertical movements, geologic activity is concentrated along the plate boundaries. As we discussed in Chapter 4, earthquakes represent the release of tectonic energy by fracturing of rocks. Thus, the boundaries of the plates are marked by long narrow belts of seismic activity. As can be seen by comparing Figures 5-1 and 5-2, the location of the earthquake belts is one of the most important tools for deciding where the plate boundaries are. If this concept is correct, Figure 5-1 shows that London, Paris, and Moscow do seem to lie on one plate, and New York on another.

Plate Geography

The plates outlined in Figure 5-2 are the larger ones that have been identified, and together they cover most of the earth's surface. As can be seen easily, their edges do not always correspond exactly to geographical boundaries. The floor of the Pacific Ocean is shared between a number of separate plates. The largest, labelled Pacific in

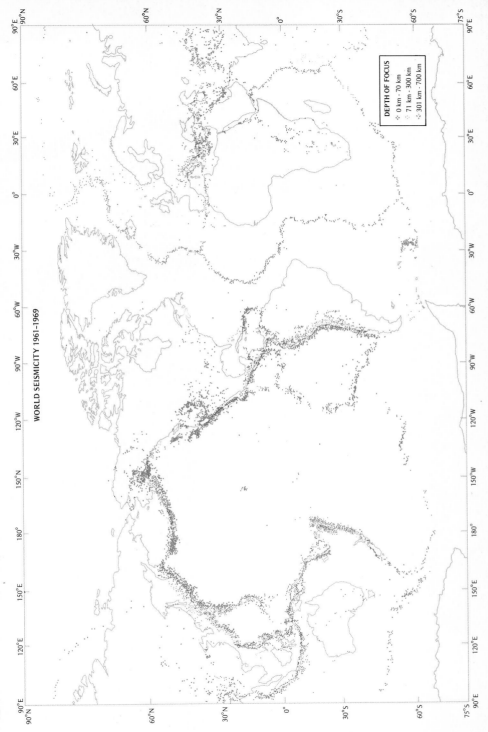

5-1. *World map of earthquake epicenters. The solid colored dots represent shallow earthquakes originating at depths of less than 70 km. The long linear belts of shallow earthquakes mark the boundaries of the plates. The light colored dots mark earthquakes with focal depths between 70 and 300 km. The black dots represent earthquakes whose foci lay deeper than 300 km. Most of these deeper earthquakes occur within a down-going plate rather than on a boundary between plates. [Data from Environmental Science Services Administration (ESSA).]*

WORLD SEISMICITY 1961-1969

DEPTH OF FOCUS

0 km - 70 km
71 km - 300 km
301 km - 700 km

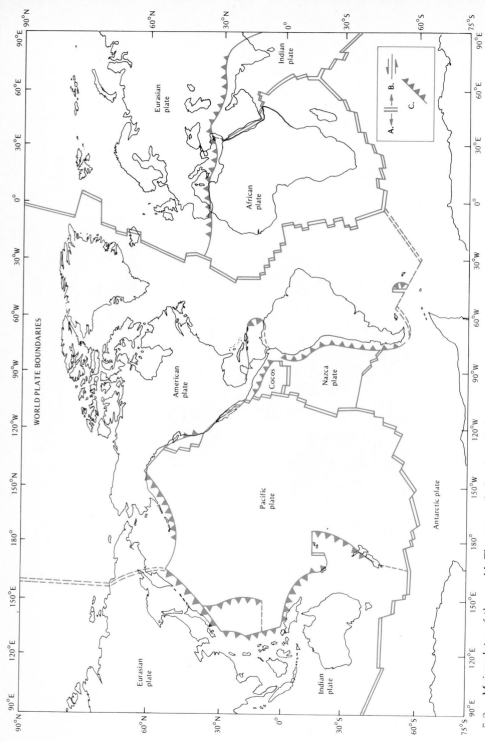

WORLD PLATE BOUNDARIES

5-2. *Major plates of the world. The types of plate boundaries, marked by heavy lines, are discussed later. A. Midoceanic ridges at which the plates move apart are represented by double lines. B. Transform fault boundaries are shown by single lines. C. Trenches and other subduction zones are marked by lines with teeth on one side. The teeth point down the descending slab. Dotted lines are used where the exact location or nature of the boundary is uncertain.*

Figure 5-2, contains about a fifth of the world's surface area. It also happens to include part of continental North America, in the form of a thin sliver of California and Baja California. The Atlantic Ocean is divided between three major plates that also contain the continental masses surrounding it. These plates are of mixed nature, with both continental and oceanic regions. In fact, there are no major plates composed of continental crust alone, although several are purely oceanic. The reason for this will become apparent soon. At the other end of the size scale, geophysicists have identified some very small areas, extending only a few degrees in latitude and longitude, that seem to act as separate rigid plates.

With the help of the narrow belts of earthquake activity, we have been able to map the surficial extent of most of the plates. Measuring their thickness is a more difficult problem. However, it seems reasonable to conclude that the material in the zone of low seismic shear wave velocities, discussed in Chapter 4, is too warm and soft to behave as part of a rigid plate. The top of the low velocity zone is also the bottom of the lithosphere. This would make the oceanic parts of the plates about 80 km thick, and the thickness of the plates under the continents from 100 to 150 km. Other ways of estimating the thickness give comparable results.

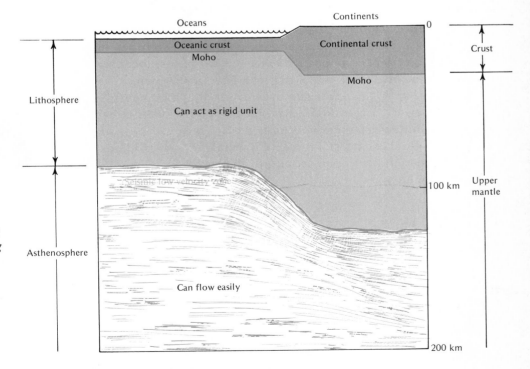

5-3. Cross section of part of a single plate containing both oceanic and continental lithosphere. Note that each kind of lithosphere is capped by a distinctive kind of crust and that the lithosphere includes both the crust and some of the upper mantle.

Lithosphere and Asthenosphere

The rigid outermost portion of the earth, located above the low velocity zone, and divided into plates, is called the **lithosphere** (*litho*, from the Greek word for rocks). It includes both the crust and some of the upper mantle. The material below the lithosphere is termed the **asthenosphere** (*astheno*, Greek for weak) and is presumed to consist of rock material hot enough to deform easily and be capable of internal flow. The asthenosphere might extend as deep as 700 km. This is the depth of the deepest known earthquakes. Figure 5-3 illustrates the relationships of crust, upper mantle, lithosphere, and asthenosphere. The high mobility of the material in the asthenosphere is what allows the plates of the lithosphere to move about on the surface of the earth. The asthenosphere is in fact an effective lubricating layer.

PLATE BOUNDARIES AND RELATIVE MOTIONS

One of the most important aspects of plate tectonic theory is the nature of the relative motions between the plates. If we consider for a moment a point on the boundary between two plates, there are three kinds of relative motion that can take place at that point. The plates can move apart, leaving a gap in between; they can move toward each other, making it necessary that one or the other either buckles and shortens or otherwise loses surface area; or they can slide past each other in contact. Figure 5-4 illustrates the possibilities. Each kind of motion does occur in nature, and each has its definite geological consequences.

Divergent Plate Boundaries

Sea Floor Spreading. One can imagine many possibilities of what can occur when two plates move apart at a divergent plate boundary, as in Figure 5-4A. What actually occurs at the divergent plate boundaries is a set of phenomena known as **sea floor spreading.** As the plates recede from one another, hot and partially molten material rises from the asthenosphere to fill the gap (Figure 5-5). The pressure

Two plates in contact

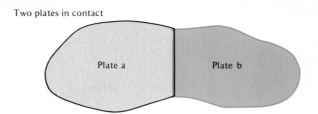

A. Can move away from each other

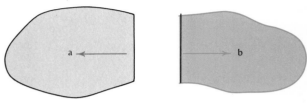

B. Can move toward each other

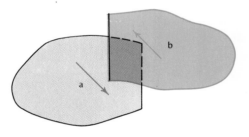

5-4. *The three possible kinds of relative plate motion. Possibility A, where the plates move away from each other, is called a divergent plate boundary. The process of sea floor spreading creates new oceanic lithosphere at divergent plate boundaries. Possibility B, where the plates approach each other, is called a convergent plate boundary. The process of subduction destroys lithosphere at convergent plate boundaries. Possibility C, where the plates slide past each other in contact without either approaching or diverging, is called a transform fault boundary. Transform faulting is the corresponding process.*

C. Can slide past each other in contact

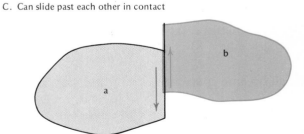

of the lithospheric plates resting on the asthenosphere can supply the energy to move the material upward. When this material cools and becomes more rigid by losing heat through its surface, it becomes part of the plates on either side of the divergence, or spreading zone. The structure and composition of the new material is the same as that of older oceanic crust and upper mantle. In fact, as far as we

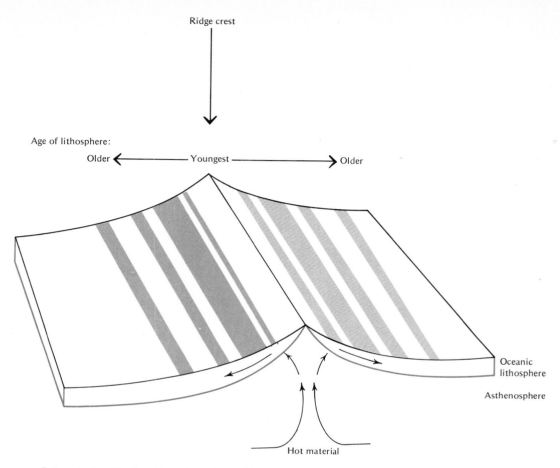

5-5. *A simplified section through a midoceanic ridge, showing the process of sea floor spreading. Hot material rises from the asthenosphere under the ridge crest, cools, and is added to the lithosphere on either side. The meaning of the stripes will be explained later in this section.*

know, all oceanic crust that exists today was generated by this same process.

In Chapter 6 the process is discussed in more detail. The new material usually seems to be added in equal amounts to both the diverging plates, so that the divergence zone remains in the middle. Thus, sea floor spreading is usually symmetrical.

The reason that the major continental plates now existing also include oceanic lithosphere and crust is that they were formed by fragmentation of larger continental masses, and the sea floor spreading that accompanied their motion away from each other has added oceanic lithosphere to their edges. Sea floor spreading thus provides a means for plates to grow in size by increasing the area attached to their trailing edges.

Oceanic Ridges. Because most materials contract in volume when cooled, the initially hot lithosphere added to the diverging edges of the plates shrinks as its heat is lost through the surface to the overlying sea water. It takes a long time for the excess initial heat to work its way up through the roughly 80 km of oceanic lithosphere, so the cooling and shrinkage is gradual as the new lithosphere moves away from the zone of spreading. This greater volume of the new, hot lithosphere is the major reason why the divergence zone stands elevated above the surrounding older sea floor. The depth of water over the oceanic lithosphere increases as the lithosphere ages.

The long linear elevations along the zones of spreading and the belt of shallow earthquakes along the ridge crest had been discovered before anyone had thought of sea floor spreading. These crests were

5-6. *Profiles of water depth across midoceanic ridges. A. A section across the Mid-Atlantic Ridge of the Atlantic Ocean. This is a typical slow spreading ridge. B. A section across the Pacific-Antarctic Rise of the Pacific Ocean. This is an example of a fast spreading ridge. Note that a high vertical exaggeration is used, so that 5 km vertically is the same as 500 km horizontally. This emphasizes slopes and greatly sharpens gradual features. [Data from B. C. Heezen: The deep sea floor. In S. K. Runcorn (ed.): Continental Drift. Academic Press, New York, 1962, pp. 235–288.]*

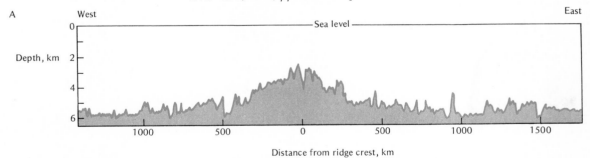

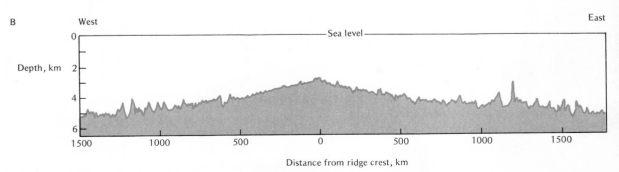

called **midoceanic ridges** because they occupied central positions in the Atlantic and Indian Oceans (see Figure 5-2). They are located in the middle of those oceans that were formed by approximately symmetrical spreading following the dispersion of previously adjacent continental masses. In the Pacific, which also had symmetrical spreading but was not formed by the breaking up of a continent, the ridges are not even near the center of the ocean basin.

Figure 5-6A shows a profile of depth across the Atlantic Ocean, with the central elevation of the Mid-Atlantic Ridge standing about 2.5 km above the surrounding ocean bottom. Slow spreading, such as has occurred along the Mid-Atlantic Ridge, tends to create rough bottom topography, and the zone of divergence or **rifting** is often marked by a deep central valley on the crest of the ridge. Faster spreading has created the topography shown in Figure 5-6B, a profile across the Pacific-Antarctic Rise. The bottom is smoother near the ridge crest and no central rift valley is seen.

Geophysics of Spreading Centers. In addition to the characteristic topographic elevations produced by sea floor spreading, there are typical geophysical patterns observed at divergent plate boundaries. We would expect that the cooling of newly created hot lithosphere would be reflected by a high flow of heat through the earth's surface in the vicinity of the **spreading centers** or divergence zones. In general, this high heat flow is actually observed when measured near the ridge crests. Since the heat produced by radioactive decay of unstable elements is continually escaping from the interior of the earth, even the deep and ancient ocean basins are marked by some outward flow of heat. However, the heat flow caused by creation and cooling of young lithosphere has been estimated to be as much as 45% of the total heat lost by the earth.

There are also typical seismic patterns associated with spreading centers. The earthquakes occur in quite narrow bands centered on the ridge crests, and they are all shallow, usually much less than 100 km in depth of origin. These earthquakes tend to be small in magnitude; the great destructive earthquakes occur on the other kinds of plate boundaries. From the nature of the seismic waves generated by an earthquake, seismologists can tell not only where it occurred but also what kind of faulting produced it. The studies of seismic wave patterns that can tell what type of faulting produced an earthquake are called **focal mechanism solutions.**

Spreading center earthquakes show focal mechanism solutions typical of **normal faulting.** Normal faulting occurs when the fault or fracture plane is not vertical, and the side overlying the fault moves

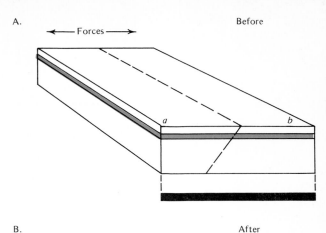

A.
← Forces → Before

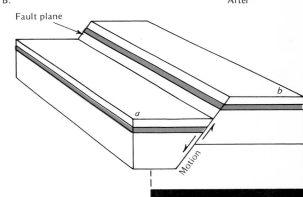

B. After

Fault plane

Motion

5-7. *A simplified block model showing the effects of normal faulting, in which the block overlying the fault plane moves downward. A. The blocks before the fault develops, showing the extensional forces acting to create a normal fault. B. The blocks after the faulting occurs. Note that the points a and b have moved away from each other, as well as to different levels. The solid bar scale has the same length in both A and B and thus shows the amount of extension. Normal faults usually have steep fault planes, sometimes close to vertical.*

downward relative to the side under it. As we can see in Figure 5-7 this kind of faulting lengthens or extends the material in which it takes place. This is just what we would expect if the forces causing a fault are extensional, directed toward pulling the material apart, and extensional forces are just what we would expect near a divergent plate boundary. This normal faulting is one factor that helps create the rough topography of the ocean bottom. The faults seem to occur within the very young lithosphere on either side of the spreading center, rather than at the contact between the two plates.

Magnetic Evidence. The most important single geophysical aspect of sea floor spreading was discovered by analysis of the strength of the earth's magnetic field over oceanic crust. Most of the magnetic field originates at great depth within the core of the earth, but there are also small local fluctuations in the field at the surface that come from variations in the magnetic properties of surficial rocks. If we average the observed magnetic field over a large enough area, the

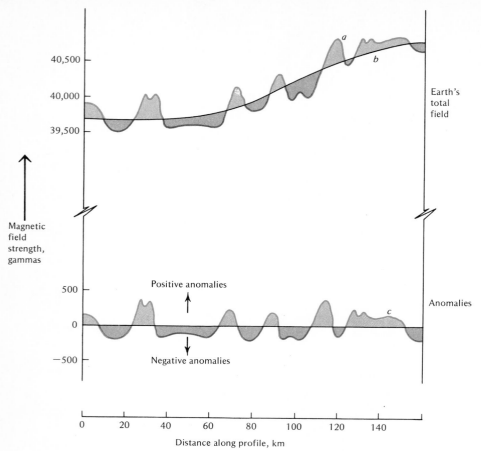

5-8.   *Magnetic anomalies. The wiggly line* a *is the actual observed total field strength along the profile. The smoother line* b *represents the slowly varying part of the earth's field that originates within the core. Subtracting* b *from* a, *we are left with line* c *below, the magnetic anomalies due to crustal variations in magnetic properties. Geophysicists commonly use the units of* **gammas** *for magnetic field strength. The main field of the earth has a strength of about 60,000 gammas at the pole and about 30,000 gammas near the equator.*

local variations will have little effect and we can find what the smoothly varying regional field of deep origin looks like. If this regional field is then subtracted from the observed field at each point, what is left are the local variations, the so-called **magnetic anomalies.** It is these anomalies that reflect crustal magnetic structures (Figure 5-8). Mantle material does not contribute to the magnetic anomalies, because temperatures in the mantle are above the Curie point (Chapter 2) of natural magnetic minerals.

The most striking thing about oceanic magnetic anomalies, quite unexpected at the time of discovery, is that they are strongly *linear* in many places. This means that areas of positive and negative magnetic anomaly are not randomly distributed but tend to form narrow parallel bands, as shown in Figure 5-9A. To understand these anomalies, one needs to know that the main magnetic field of the earth occasionally and unpredictably reverses its polarity. At the time of

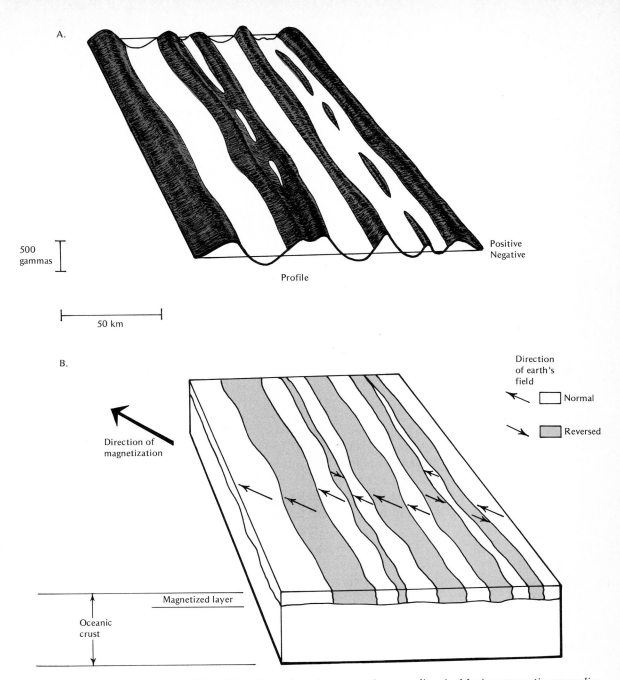

A.

500
gammas

50 km

Profile

Positive
Negative

B.

Direction
of earth's
field

Normal

Reversed

Direction of
magnetization

Magnetized layer

Oceanic
crust

5-9. *The origin of marine magnetic anomalies. A. Marine magnetic anomalies, with typical height and width shown by the bar scales. The linear nature of the anomalies is evident. B. An interpretation of the underlying crystal structure in terms of blocks of normally and reversely magnetized sea floor.*

one of these **magnetic reversals,** the magnetic north pole of the earth becomes instead the magnetic south pole and vice versa.

The core dynamo that produces the geomagnetic field seems equally happy in either the **normal mode,** yielding the present day field direction, or the **reversed mode.** During the Tertiary, the field has tended to stay in one mode for an average time of one half million years before switching to the other; however, the period between reversals has varied widely. During the Late Cretaceous, the field remained normal for as long as 25 million years, but before that another period of frequent reversals occurred. For much of the Permian, the field remained reversed.

Now, if the sea floor is spreading, new hot material is being added to the oceanic crust. The new surficial material cools through the Curie point of its magnetic minerals, and they can freeze in the magnetic field direction present when they cooled, either normal or reversed. These mineral grains, oxides of iron and titanium, act as tiny magnets aligned along the field direction, and once cooled they can retain their original orientation of magnetization for great lengths of time. Recent work has shown that only as much as the upper 400 meters (m) of the oceanic crust probably contains most of the crustal magnetization. This thin skin of the crust acts as a tape recorder, preserving with more or less fidelity the history of geomagnetic reversals.

As the sea floor moves away from the spreading center, a series of alternating normally and reversely magnetized blocks are formed (Figures 5-5, 5-9B). For the same reversal history, a fast spreading ridge will produce wider blocks than slow spreading will. We can play back the recordings by using a ship or airplane to tow a **magnetometer,** a device that measures magnetic field strength, above the sea floor. In some places, the field due to magnetization locked into the thin upper layer will reinforce the earth's deep field, producing a positive magnetic anomaly. In other places a negative anomaly will result. From the anomalies, the underlying block structure of magnetization can be deduced. The anomalies are linear because sea floor spreading generates new oceanic crust along linear midocean ridges. Figure 5-9B shows a block model interpretation of the anomalies in Figure 5-9A. This idea relating marine magnetic anomalies and magnetic reversals is called the Vine-Matthews hypothesis, after the men who first published the idea.

As we will discuss later (page 000), the concept of a relationship between sea floor spreading, lineated magnetic anomalies, and geomagnetic reversals led eventually to the concept of plate tectonics.

Besides this, the importance of the theory lies in the power it gives to measure the age of oceanic crust and to determine the rate of sea floor movements. Because the reversals are worldwide, all actively spreading ridges record the same reversals at a given time. If a distinctive pattern of magnetic reversals preserved in a particular piece of oceanic crust can be recognized, we can tell when that crust was generated. How this time is known will be discussed later under "Geophysical Developments" in this chapter.

The vertically stacked stratigraphy of fossils on land is replaced under the oceans by a horizontally spread out sequence of identifiable magnetic anomalies. Fortunately, the geomagnetic reversal history is erratic enough that distinctive patterns of long and short reversals do exist. We can thus date vast areas of sea floor simply by towing a magnetometer over them and interpreting the magnetic anomalies. Furthermore, since all the ridges record the same sequence of reversals but spread at different speeds, their rate can be measured by comparing the width of the anomalies with the time between the reversals that created them.

Spreading Rates. In Figure 5-10 an actual observed magnetic anomaly profile and a theoretical profile calculated from a block model are compared. The agreement of the profiles is clearly excellent. Spreading rates are usually quoted as half-rates, the speed recorded on one side of a spreading center. If the spreading rates are the same, the plates are separating at twice the spreading half-rate. The width of the blocks on either side of the ridge crest will also be symmetrical. Because the exact form of the magnetic anomalies depends on the orientation of the earth's field, their shapes may be only approximately symmetrical.

Along the world-girdling system of spreading centers, typical separation rates are from 2 to 20 cm/year, with the average being about 6 cm/year. These rates may not seem very large, but geologically speaking they are very fast because the spreading goes on for many millions of years. With a separation rate of only 2 cm/year, 20 km of new lithosphere will be created every million years. The slow spreading ridges mentioned above, which have rough topography, are those spreading at less than about 4 cm/year separation rate. Once the rate of separation on divergent plate boundaries has been measured with the help of the magnetic anomalies, the simple geometric properties of rigid plates on the surface of a sphere make it possible to calculate the relative rates of motion on the other kinds of plate boundaries.

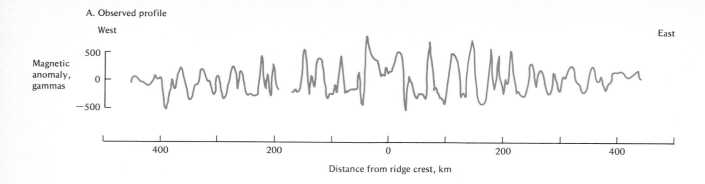

A. Observed profile

West   East

Magnetic anomaly, gammas

Distance from ridge crest, km

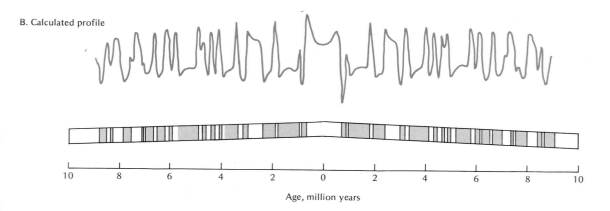

B. Calculated profile

Age, million years

5-10. *A magnetic anomaly profile from the Pacific-Antarctic Ridge. A. The actual observed magnetic anomalies. B. A theoretical profile calculated using the Vine-Matthews hypothesis. The block model beneath the calculated profile shows the reversal time scale used. Open blocks are normally magnetized, whereas the colored areas are reversely magnetized blocks. The age of the reversals in millions of years is given at the bottom. Note the excellent agreement between the observed and the calculated profiles. This particular observed profile shows very strong symmetry about the ridge crest. [Modified from J. R. Heirtzler: Evidence for ocean floor spreading across the ocean basins. In R. A. Phinney (ed.):* The History of the Earth's Crust. *Princeton University Press, Princeton, N. J. 1968, pp. 90–100.]*

The process of sea floor spreading that generates oceanic crust is much simpler than the complex series of events that create and later modify continental crust. We are indeed fortunate that the history of the ocean basins has proved to be relatively easy to unravel, since it has provided insight into the more complicated evolution of the continents.

Convergent Plate Boundaries

Subduction. If the surface area of the earth remains constant, as seems likely, the creation of new lithosphere at divergent plate boundaries must be balanced by destruction of lithosphere at convergent plate boundaries, where plates move toward each other as in Figure 5-4B. Although many ways can be imagined for the earth to accomplish this destruction, one process, **subduction,** actually takes place. When two plates converge, one or the other of them must be deflected down into the asthenosphere rather than both remaining on the surface to be crumpled and buckled.

Because oceanic lithosphere is directly derived from underlying mantle material during sea floor spreading, its overall composition must be similar to that of mantle material. Being cooler and therefore denser, it readily sinks back into the mantle and is eventually resorbed there. Thus, when two oceanic plates collide, one of them returns to the mantle from which it came. And, when an oceanic plate approaches a continental plate, it is the oceanic lithosphere that is consumed in the **subduction zone.**

The continental crust is not of the same composition as mantle material and is made of considerably less dense rocks; therefore, continental lithosphere is not easily forced down and consumed by subduction. When two continental plates are in collision, the effect is to throw up great mountain ranges, such as the Himalayas. Presumably, the work necessary to erect these ranges is so great that the approach of the continental masses eventually stops, and they remain welded together. Figure 5-11 summarizes the types of subduction zones.

Oceanic Trenches. As an oceanic plate enters a subduction zone, it bends down over a distance of about 100 km, then straightens out again and slides down into the asthenosphere at a steep angle. The contact between the descending and the overlying plate is usually some distance down the bend. The subduction of oceanic plates is therefore marked by a distinctive topographic feature, the great **oceanic trenches,** as sketched in Figure 5-11. The deepest known water in the world is in the Marianas Trench of the western Pacific, which has a depth of over 11 km.

Volcanos. It takes a long time for the initial heat of the oceanic lithosphere to escape. Once the lithosphere cools and then goes down into the asthenosphere, it takes a long time, about 10 million years, to warm up again until it is no longer distinguishable from mantle

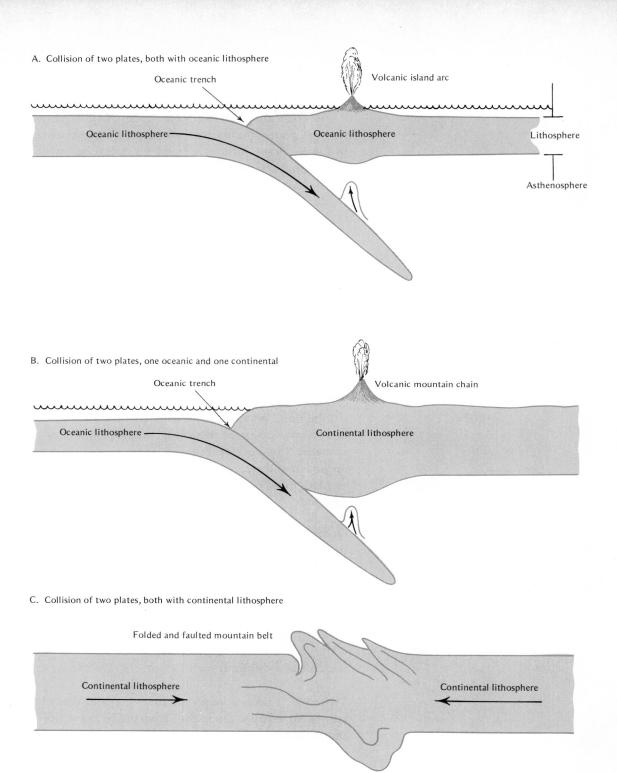

A. Collision of two plates, both with oceanic lithosphere

Oceanic trench

Volcanic island arc

Oceanic lithosphere

Oceanic lithosphere

Lithosphere

Asthenosphere

B. Collision of two plates, one oceanic and one continental

Oceanic trench

Volcanic mountain chain

Oceanic lithosphere

Continental lithosphere

C. Collision of two plates, both with continental lithosphere

Folded and faulted mountain belt

Continental lithosphere

Continental lithosphere

material. Thus, the descending lithosphere can retain its identity to depths as much as 700 km.

In the depth range of 150–300 km, either the upper part of the oceanic crust which is in contact with the mantle begins to melt, or additional melting in the mantle is triggered by the descending slab of oceanic lithosphere. The resulting molten rock rises into the overlying plate and forms volcanos when it breaks through to the surface. Most of the world's active subaerial volcanos are located above subduction zones. The volcanos build **island arcs** if the overlying lithosphere is oceanic (Figure 5-11A) or volcanic mountain chains if the lithosphere is continental (Figure 5-11B). Japan is a good example of an island arc, while the Andes mountains are the most extensive continental volcanic mountain chain. As will be seen in later chapters, we have strong reason to believe that it is this process that creates new continental crust.

Seismic Effects. The process of subduction also has its typical geophysical patterns, of which the seismic patterns are most important. Subduction zones are seismically very active. Most of the world's earthquakes, including some of the largest and most destructive, occur along subduction zones. For most subduction zones, four areas with distinctive kinds of earthquakes can be identified.

The **outer zone** lies on the descending plate near the outer wall of the trench, and most of its earthquakes originate within the oceanic crust or upper part of the lithosphere. The seismic wave pattern of these earthquakes is typical of normal faulting, and they are thought to be the result of stresses generated by the bending of the down-going slab.

The **shallow zone** of earthquakes is located along the boundary between the plates, at depths less than 100 km. The quakes there represent friction along the contact between the down-going lithosphere and the overlying plate. The focal mechanism solutions for earthquakes in the shallow zone indicate that **thrust faulting** is the dominant type there. Thrust faulting is a special case of reverse faulting (Figure 5-12), which happens when the block lying over a fault plane inclined at a low angle moves upward relative to the underlying block. This is just what we would expect along the upper contact of the descending slab. Thrusting is the result of compressive, or shortening, forces. Some of the shallow zone earthquakes have been among the most energetic ever recorded.

Intermediate depth earthquakes are those that originate between 100 and 300 km deep. They seem to occur within the down-going slab rather than on its contact with the asthenosphere. Their focal

5-11. *Types of convergent plate boundaries, depending on the kinds of lithosphere in collision. The drawings are schematic and not to scale. A. If two plates containing oceanic lithosphere collide, one of them is subduced into the asthenosphere. Note that the volcanism occurs over the descending slab at some distance from the trench. B. Convergence of oceanic and continental lithosphere results in subduction of the oceanic plate. As for the volcanic island arcs in A, the volcanic mountain chain that results in some distance from the trench. C. Convergence of two plates both bearing continental lithosphere. Continental lithosphere is not readily subduced, so great complicated mountain chains are thrown up. These three types of subduction zone illustrate that oceanic lithosphere returns to the mantle but continental lithosphere does not.*

Before   B   After

→ Forces ←

Fault plane

5-12. *A block diagram showing the effects of* **reverse** *faulting, in which the block overlying the fault plane moves upwards relative to the block beneath. If the fault plane is only slightly inclined to the horizontal, the term* **thrust** *faulting is used. The contact between the down-going and the overlying litho-sphere in a subduction zone lies at a shallow angle to horizontal, so subduction involves great thrust faults.* A. *The blocks before faulting, showing the direction of the forces involved.* B. *After faulting, points* a *and* b *have approached one another as well as moving to different levels. This is shown by the solid bar scales, which are the same length in both* A *and* B.

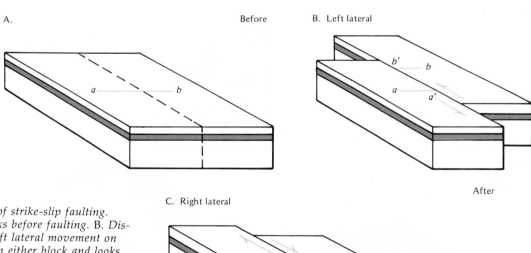

A.   Before   B. Left lateral

C. Right lateral

5-13. *Block diagram of strike-slip faulting.* A. *Position of the blocks before faulting.* B. *Displacement caused by left lateral movement on a fault. If one stands on either block and looks across the fault, the displacement of the other block is to the left.* C. *Displacement caused by right lateral faulting. Looking across the fault, the block on the other side is displaced to the right. Note how both senses of strike-slip faulting separate the segments of the once continuous line* a-b.

After

130   THE EVOLVING EARTH

mechanism solutions show, for different subduction zones, either extension or compression directed down the dip of the slab. Presumably, intermediate depth earthquakes reflect the stresses acting within the descending lithosphere. If the down-going slab is denser than the surrounding mantle, it will be sinking under its own weight and extensional stresses might result. Compressional forces could result from resistance of the mantle to the downward motion.

The **deep zone** of earthquakes include those of depths from 300 to 700 km. Deep zone earthquakes are not found under all subduction zones. These earthquakes, the deepest known and all related to past or present convergent plate boundaries, have focal mechanism solutions that indicate compression within the descending slab. This suggests that they are encountering material resistant to their movement.

Because all zones of earthquakes are related to the stiffness of the down-going lithosphere, they happen only within the lithosphere and enable us to follow its progress as it slides down into the mantle.

Transform Fault Boundaries

Boundaries where adjacent plates slide horizontally past each other in contact without loss or gain in area are called transform fault boundaries. They were given that name because their sliding motion can transform one kind of relative motion at one of their ends into another kind at the other. The type of faulting they represent is called **strike-slip** because the direction of movement is horizontal and parallel to the *strike* of the fault plane, the direction of its surface trace. Figure 5-13 illustrates the two possible senses of motion on strike-slip faults. **Left lateral** movement indicates that, if we stand on either block looking across the fault zone, we will see the other block move to the left. **Right lateral** movement is in the opposite sense.

Transform faults can join convergent and divergent plate boundaries together in various combinations. Figure 5-14 shows some of these. One common type is the **ridge-ridge** transform fault (Figure 5-14A). Note some of its peculiar properties. Although the *apparent offset* of the ridge crest is left lateral, the actual relative motion across the transform fault between the ridge segments is right lateral. Instead of an originally straight spreading center having been offset by motion on the transform fault, the ridge crest was offset from the beginning, and the fault is just a part of the plate boundary that happens to be parallel to the direction of relative movement. Notice

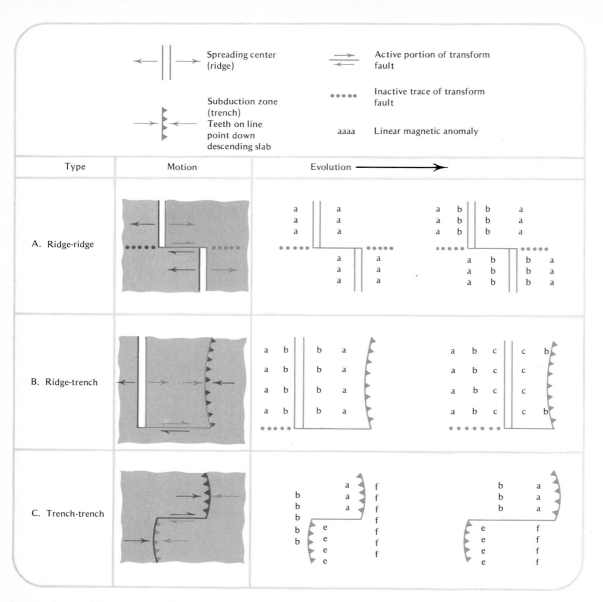

5-14. *Some of the possible types of transform fault. The first column names the features joined by the transform fault, the second column illustrates the relative motions that take place, and the third column shows how the system of fault and the features it connects would look if surveyed at two consecutive, well separated times. The magnetic anomalies are in alphabetical order of age, aaaa being oldest. Note that in A the length of the transform fault remains constant, in B the fault becomes shorter with time, and in C it lengthens.*

also that the only portion of the transform that is active is the part between the two ridge segments, the only place where the fault forms a boundary between the two plates. Beyond the ends of the ridges, the inactive trace of the fault lies entirely within one plate, and carries only slight vertical movement as the oceanic lithosphere cools and subsides.

Topographic Features. Transform faults also have topographic expressions, both on the active part and on the inactive trace. The shearing motion across the fault produces ridges and troughs that run roughly parallel to the motion. This characteristic topography is carried along as the plate on which the features lie moves beyond the active part of the transform. Beneath the oceans, the rough areas created by transform faulting are called **fracture zones.** On the continents, they form great **transcurrent fault zones** such as the San Andreas fault zone of California and the Anatolian fault of Turkey. Earthquakes on the transform faults occur on the active portions alone and have focal mechanism solutions that indicate the appropriate sense of strike-slip motion. These earthquakes are usually very shallow, originating at depths of less than 15 km. If they occur within cold, and therefore brittle, crust, they can be of great destructive force.

Direction of Relative Motion. One important geometric characteristic of transform faults is that their strike or trend is parallel to the direction of relative motion between the two plates. If the fault slipped in any direction except parallel to the motion, then the plates would be either diverging or converging along the boundary and it would no longer be a pure transform fault. This characteristic has proved very useful in deducing plate motions. The spreading center ridges usually are at approximately right angles to the direction of relative plate motion, but not necessarily exactly so. Thus, they are less reliable indicators of motion direction.

Relative Motions of Plates

Figure 5-15 is a summary of the effects of the various kinds of relative motion we have discussed. A comparison of this illustration with Figure 5-2 and 5-14 might help in understanding how plate interactions are actually expressed at present on the surface of the earth. Note also in Figure 5-2 that there are a number of places where three plates meet. These points are called **triple junctions,** and the

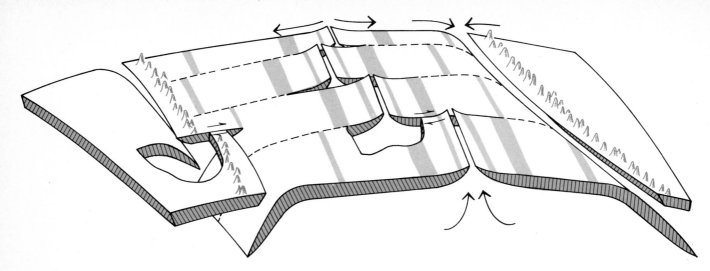

5-15. *Summary diagram of the types of relative plate motion and their geological consequences. The solid arrows show the directions of relative motion of the four plates depicted. The magnetic stripes shown do not correspond to the actual known reversal time scale.*

plate boundaries that join there can be any combination of the three kinds—divergent, convergent, and transform. Only certain configurations are likely to persist for any length of time without evolving into other forms. A ridge-ridge-ridge triple junction, such as that between the Pacific, Cocos, and Nazca Plates, is the only combination that is always stable.

At this time, it might be wise to emphasize the reason why the word "motion" has so often been preceeded by "relative" in this discussion. In local phenomena such as crustal faulting (Figures 5-7, 5-12, 5-13) it is impossible to tell whether one block was fixed and the other moved, whether both moved in opposite directions, or whether both moved in the same direction but one moved farther. If it is impossible to tell which of these possibilities actually happened, it is meaningless to define anything but the relative motion.

The same is true for the larger scale plate motions. One can arbitrarily fix a coordinate system to one plate, but then every other plate and every ridge, transform fault, and subduction zone will move about in that coordinate system during the course of their evolution. Somebody else can with equal justification choose to pin their coordinate system to some other plate. About the only place in plate tectonics where the absolute motion is relatively clear is in subduction—we do indeed know that one plate goes down and the other stays at the surface.

HISTORY OF THE IDEAS OF PLATE TECTONICS

Continental Drift

The theory of plate tectonics is a recent one, but more than a century of geological research, speculation, and controversy has gone into its formation. A good deal of the controversy has been over the question of continental drift—whether the continents have always been fixed in their present positions, or whether they have moved around on the surface of the earth. Continental drift is now seen as necessary and understandable from the plate tectonic viewpoint, but its general acceptance is quite recent.

The first observation leading to the idea of continental drift was the close match in shape of the east coast of South America and the west coast of Africa that faces it across the Atlantic. Francis Bacon in 1620 and Francois Placet in 1658 had noted the similarity, but Antonio Snider-Pelligrini in 1858 was the first to suggest that the similar coastlines were two sides of a break separating a formerly continuous land mass. However, it should be clear from Chapter 4 that the continents do not end where the waters begin. Thus, in more modern attempts to reassemble the continental jigsaw puzzle, the edge of the continental shelf or a line half-way down the continental slope is used to fit the pieces together. Such a reassembly of South America and Africa is shown in Figure 5-16.

To be relatively certain that the continents are displaced parts of a jigsaw puzzle, it is not enough that the shapes fit together nicely. The geological patterns, at least those that existed before breakup, must also match across the junctions. Snider-Pelligrini's assembly was in support of a similarity of fossil plants found in coal beds in both Europe and North America. His work seems to have been totally ignored. The subject was reopened early in the twentieth century by two American geologists who independently invoked continental drift to explain the distribution of modern mountain systems. However, it was the work of Alfred Wegener, a German meteorologist, astronomer, and geophysicist, that founded the modern theory of continental drift and precipitated controversy.

Wegener brought together many lines of evidence—not only the similarity in shapes, but the patterns of ancient climates, the continuity of geologic structures such as the roots of old mountain belts, the distribution of rock types and ages, similarities in the fossils of the various continents, and the pattern of Permian-Carboniferous glaciation in the Southern Hemisphere. He used all this evidence in support of the idea that the continents had previously been assembled into one, **Pangaea** (Greek, meaning all earth), and had then rifted and drifted apart at various times.

Unfortunately, some of the arguments Wegener used were not sound, and the mechanism he proposed was unsatisfactory. The continents were depicted as sailing through the oceanic crust like great rafts, driven by the earth's rotation. This mechanism could easily be shown to be completely inadequate to overcome the measured strength of rocks. Partly because of weaknesses in Wegener's arguments and partly because of mental inertia, the possibility of continental drift was rejected by most geologists and geophysicists in the Northern Hemisphere. In the Southern Hemisphere, where the geologists were living in the midst of the best evidence, the idea lived on. Alexander du Toit of South Africa did field work in both Africa and Brazil and uncovered much supporting geological evidence. Later Lester King in South Africa and Warren Carey of Australia carried on and developed the ideas.

Evidence of Continental Drift

Let us consider a small part of the geologic evidence for continental drift, concentrating on Gondwanaland. This assemblage of Africa, Antarctica, Australia, India, Madagascar, and South America may have existed in the Paleozoic era and drifted apart in the Mesozoic and Cenozoic eras. Wegener knew that the southern continents had a distinctive Permian **flora,** or association of fossil plant remains. This was called the **Glossopteris flora** for a typical fossil leaf (Figure 5-17) that occurs in it. At the time Wegener formulated his ideas on continental drift, the Glossopteris flora had been found in Africa, Australia, India, Madagascar, and South America. It has since also been discovered in Antarctica. It has not been found on any of the other continents. If the continents had always been in their present positions, why had this flora been able to spread among the southern continents and across the equator to India, but never anywhere else? It was more reasonable to suppose that the places on which it grew had formed a single continental mass in the Permian.

Similarly, the evidence of Upper Carboniferous and Permian glaciations that occurred in the Gondwanaland continents are scattered today at all latitudes; however only a single large ice cap on a

5-16. *The geometric and geologic fit of Africa and South America. The shaded areas are the continental shelves of the two continents. Areas where they overlap in the reconstruction are shown in black. The largest of these is the Niger delta of Africa, which has been deposited subsequent to the rifting. The colored areas are continental platforms or cratons that give radiometric ages of greater than 2 billion years. Note particularly how the craton in northern Africa is matched by a small part of one in Brazil. The colored lines show the trends of fold belts and other tectonic features that are younger than 2 billion years. The colored crosses represent some of the localities in which the Permian reptile* **Mesosaurus** *has been found.*

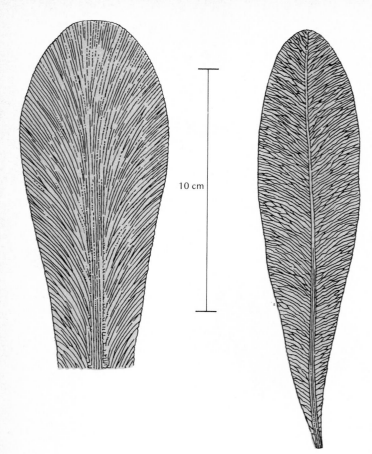

10 cm

5-17. *Drawing of leaves of plants from the* Glossopteris *flora. Left,* Glossopteris *itself; right,* Gangamopteris. *Remains of these and related plants have been found on all the Gondwanaland continents.*

reassembled Gondwanaland is required to explain them (Figure 5-18).

There is a small amphibious Permian reptile called **Mesosaurus** (Figure 5-19) that is found only in South Africa and Brazil, in very similar sedimentary beds laid down in shallow lakes (locations shown in Figure 5-16). If this reptile walked all the way around the Atlantic, why is it found in these two areas only? It also seems unlikely that a breeding pair of these small shallow water reptiles could swim more than 6000 km across the Atlantic. Again, it is more reasonable to suppose that the homelands of Mesosaurus were once one.

The trends of structural belts mapped by Du Toit show the same jigsaw puzzle fit across the South Atlantic (Figure 5-16). Also plotted in Figure 5-16 are the more recently determined **radiometric age provinces** of Africa and South America. Age provinces are areas within which the old basement rocks all have similar ages as found by

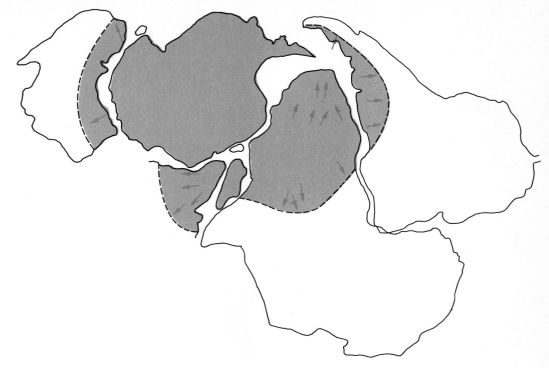

5-18. *The regions affected by the Permo-Carboniferous glaciation of the Southern Hemisphere are shown in color on one possible reconstruction of the original positions of the Gondwanaland continents. The colored arrows show the directions of ice movements deduced from striations left in underlying rocks.* [*Modified from A. G. Smith: Continental drift. In I. Gass, P. Smith, and R. Wilson (eds.):* Understanding the Earth. *Artemis Press, Sussex, 1971.*]

5-19. *A cast of the fossil remains of* **Mesosaurus,** *a small aquatic reptile of Permian age found both in South America and Africa.* [*Photograph by M. Kauper.*]

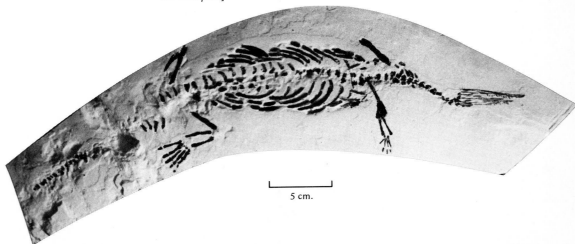

5 cm.

radioactive dating methods. Note especially how the northern region of very old rocks in Africa is matched precisely by a fragment left on the coast of Brazil. To these examples of similarities between the southern continents can be added many others.

Geophysical Developments

A more feasible mechanism for continental drift than Wegener's was proposed in 1927. It was suggested that radioactive heat generated within the earth caused material to expand and rise toward the surface, then spread out, cool off, and descend again in a closed pattern of **convection currents,** which carried the continents on their backs.

The results of **paleomagnetism,** developed in the 1950s led to a wider acceptance of the idea of continental drift. Some rocks on the continents, like the rocks formed during sea floor spreading, preserve the magnetic field present at their time of formation. By measuring the direction of this "fossil" magnetic field, it can be calculated where the magnetic poles were relative to the rocks' location when the rocks were formed. Pole positions deduced from older rocks on different continents did not agree. However, if the continents were put back together into the original positions indicated by their shapes and their paleomagnetic poles were moved with them, the agreement for certain time periods was very much improved. This suggested that the rocks had acquired their magnetism when the continents were together and had since drifted apart.

Meanwhile, the oceanographers were also developing improved instruments, replacing the lead weight and line previously used to measure depth by **echo sounders** that send out a sound pulse and measure the time it takes to reach the bottom of the ocean and return. Knowing the speed of sound in sea water, the depth can be calculated. By 1960 the existence of a world-encircling system of mid-oceanic ridges was known, and sea-going magnetometers had been used to discover the linear magnetic anomalies that lie off the west coast of North America. In 1962 Harry Hess suggested the idea of sea floor spreading as an explanation for the morphology, composition, and structure of the midoceanic ridges and as a mechanism for continental drift.

At the same time, progress was being made in determining the history of geomagnetic field reversals. In 1906, rocks were discovered that were clearly magnetized in a direction opposite to that the earth's field has now. Unfortunately it was not possible to prove whether

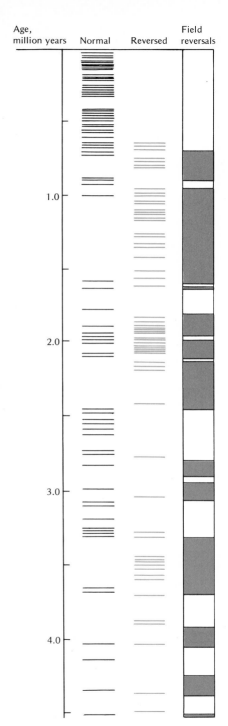

Age, million years	Normal	Reversed	Field reversals

the field had reversed or whether the rocks had perversely magnetized themselves (**self-reversing**) in a direction opposite to the field they experienced at formation. Combined geologic and paleomagnetic studies showed that the first alternative was more likely, but a few examples of self-reversing rocks have been found. The deciding evidence came from careful magnetic studies and equally careful radiometric age determinations for young volcanic rocks in California. It was shown in 1963 that, for the last 4 million years, rocks of the same age all had the same magnetization, normal or reversed. This was convincing evidence that the field had been reversing, rather than the measured reversals being due to the peculiar behavior of particular rocks. Furthermore, when the same reversal pattern was found in completely dissimilar rocks, sediments retrieved from the ocean bottom, the case for geomagnetic field reversals was very clear. Figure 5-20 shows the geomagnetic reversal history for the past 4.5 million years.

Hypotheses Concerning Crustal Motion

The next, very important step on the path to the theory of plate tectonics was to put together the concepts of sea floor spreading, geomagnetic field reversals, and magnetic anomaly lineations of the oceanic crust. In 1963 the Vine-Matthews hypothesis (page 000) was proposed. The basic tool for measuring rates of sea floor spreading was now available. Verification of the hypothesis did not take long, for within 5 years it had been successfully applied to magnetic anomalies in all the oceans.

In 1965 J. Tuzo Wilson proposed the transform fault hypothesis, and by 1967 it had been proven consistent with earthquake locations and focal mechanism solutions. As predicted by the hypothesis, earthquakes were limited to the parts of fracture zones between the adjoining ridge crests; moreover, the sense of motion was just that predicted for transform faults.

5-20. *The geomagnetic reversal time scale for the last 4.5 million years. The first column shows the radiometric ages of volcanic rocks, determined by the potassium-argon dating technique. The lines in the second and third columns represent the ages of particular rock samples; The sample is indicated in the second column if normal magnetization is preserved in the rock and in the third column if the rock was reversely magnetized. The fourth column displays the sequence of normal (open) and reversed (black) periods of the earth's magnetic field, as deduced from the rock measurements. [Modified from A. Cox: Geomagnetic reversals,* Science, *163(3684): pp. 237–245, (1970).]*

Progress was also made in extending the geomagnetic reversal time scale. The radiometric reversal time scale is only reliable for the last 5 million years of earth history. Farther back than this, the errors in the potassium-argon dates are larger than the average duration of a reversal period. In looking for a way of extending the reversal time scale, a group of marine geophysicists made a brilliant guess. They hypothesized that spreading rates in the South Atlantic had been constant for a very long period of time. They took the present spreading rate at the ridge crest in the South Atlantic and assumed that all the anomalies in that ocean had been formed at the same rate. On this basis, they constructed a time scale for reversals of the past 80 million years. This extrapolation, from 4 to 80 million years, had no right to be correct, but it was—within 4% in the older portion. Proof of its correctness came from the Deep Sea Drilling Program, which found that fossil ages from the oldest sediments lying on the oceanic crust in the South Atlantic matched the ages from the anomaly time scale almost exactly (Figure 5-21).

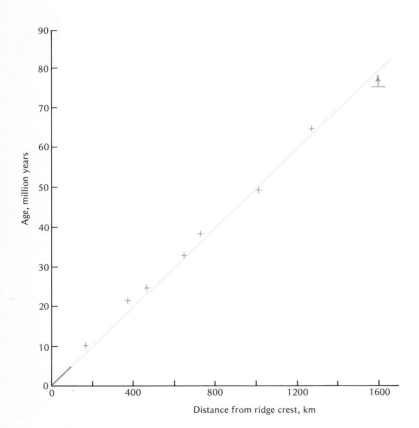

5-21. *Deep Sea Drilling Program results in the South Atlantic, compared to ages found from magnetic anomalies. The crosses show paleontological age and distance from ridge crest of the oldest sediments overlying oceanic crust at the bottom of the drill holes. The solid line is the age of the crust deduced from magnetic anomalies by assuming that the spreading rate has been constant for the last 80 million years and equal to that measured at the ridge crest using the radiometric reversal time scale. The increase in paleontologic age with distance from the ridge is itself a striking confirmation of sea floor spreading. The agreement with an assumed constant spreading rate is also remarkable.*

Concept of Plate Tectonics

The Vine-Matthews hypothesis is important because it is a quantitative, predictive model whose success was the most convincing evidence for sea floor spreading and because of the precise information on rates of crustal motion it gives when applied to the marine magnetic anomalies. There was great interest and excitement about it among oceanographers, but the revolution in the rest of the earth sciences did not get fully underway until the concepts of plate tectonics were advanced. Ironically, people who had been fitting coastlines together were implicitly using the concept of rigid continental plates. The transform fault hypothesis also required rigid behavior of the crust away from spreading centers and trenches. Finally, by 1968 the full implications of rigid plate behavior had been realized and published by Dan McKenzie and Robert Parker, and independently by W. Jason Morgan.

Later in 1968, determinations of present relative motions of six of the major plates of the world appeared. This set of motions was quickly demonstrated to be consistent with the results of earthquake seismology, especially focal mechanism solutions, and the revolution was launched. A fundamental rethinking of the concepts and data of geology had begun and continues to this day.

THE ENIGMATIC DRIVING FORCE

Plate tectonics is a **kinematic** theory; that is, it deals with motions and their observable effects. Fortunately its success does not rely upon knowledge of the reasons for those motions. Even now the **dynamics,** the forces causing the motions, are not at all clear. It is generally agreed that thermal energy from radioactive decay must be the main driving force, but just how it works is not known.

We can say that the outer part of the earth is a system involving the coupling of thermal and gravitational energy by hydraulic and mechanical means, but having said this we still do not know what controls the motions. Various possibilities have been proposed. Thermal convection in various forms, either in great closed cellular currents or as individual local thermal uprisings, is one possibility. It has been suggested that the oceanic lithosphere slides down the slope of the midoceanic ridges of its own weight, and that its density, greater than that below, causes it to be pulled down into the asthenosphere. All of these processes and more might contribute to driving the motions. The problem is that, although the effects of the motions can be seen at the surface, their causes lie deeper and are not as accessible to direct measurement. The rigid plates, so convenient for understanding surface motions, also obscure what is going on below.

CONCLUDING STATEMENT

We have already discussed many of the observations and ideas that went into formulation of the plate tectonic model. The fundamental differences between continental and oceanic crust have become understandable. In the guise of continental drift, plate tectonics can explain the fit of continental outlines, patterns of fossil organisms in the Southern Hemisphere, and the stratigraphy of the southern continents. The truncation of age provinces and structural trends at the continental edges is no longer mysterious.

As we will see in Chapters 7 and 8, the plate tectonics model also encompasses the origin of these structural trends and the reasons for the process of mountain building. It explains why earthquakes are concentrated in linear belts, and why active volcanos have a similar distribution. Most important of all, it has replaced previous views of the earth as a static collection of continents and oceans with a dynamic, mobile concept of the earth. In addition, plate tectonics is a precise model that can predict as well as explain. In the future, more sophisticated models may replace plate tectonics, as plate tectonics has replaced earlier models, but never again can the continents and oceans be regarded as isolated static phenomena rather than parts of a single, evolving earth system.

QUESTIONS

1. What are the basic assumptions of plate tectonics?
2. Why are the belts of shallow earthquake epicenters useful in locating plate boundaries?

3. Explain why the rigid plates of lithosphere are thought to contain both crust and part of the upper mantle.

4. Why are the areas where sea floor spreading is taking place marked by midoceanic ridges?

5. Explain why the major plates surrounding the Atlantic Ocean contain both oceanic and continental lithosphere.

6. How are the linear bands of marine magnetic anomalies created, and what specific information do they give about sea floor spreading?

7. Illustrate with a sketch how oceanic lithosphere is ultimately destroyed in a subduction zone. What is the reason for the existence of oceanic trenches?

8. Which of the earthquakes at a convergent plate boundary represent the two plates rubbing against each other, and which occur within one of the lithosphere plates?

9. Show, with a diagram, why a ridge-ridge transform fault has active horizontal motion only on the part between the two spreading centers.

★10. Why are transform faults always parallel to the local direction of relative plate motion?

11. Discuss briefly the kinds of geologic evidence from the continents that support the idea of continental drift.

★12. What is it about plate tectonics that makes it difficult to tell exactly what causes the plates to move?

The Life Cycle of Oceanic Crust

The plate tectonic model requires that new crust of oceanic type be generated at divergent plate boundaries and that the presently active divergent plate boundaries are marked by a worldwide system of midoceanic ridges. For the reader to understand the processes by which new oceanic crust is created, it is necessary to digress briefly from our central theme to discuss the origin of magma and the formation of igneous rocks.

IGNEOUS ACTIVITY

Magma

In broad terms igneous activity refers to the formation, movement and eventual solidification of magma, which, as described in Chapter 2, is naturally occurring molten rock material. Magma is best envisioned as a hot, viscous liquid, in which elements such as magnesium, iron, calcium, sodium, and potassium occur in solution as ions (cations) in electrostatic balance with silicon-oxygen tetrahedra (anions).

The important fact about magmas is that they are liquid, and until they solidify they retain the properties of a liquid, the most important of which is fluidity. In many cases magmas contain gases and particles of solid material such as suspended crystals or fragments of solid rock; however, provided the content of solid material is not too great, the fluidity or mobility of the magma is not seriously affected.

The ranges of temperatures over which rocks can melt to form magma, or below which magma crystallizes to form igneous rocks (600–1400°C) are known from laboratory studies and, to a lesser extent, from direct measurement of the temperature of magma in the craters of volcanos. The most important factors that influence the temperature at which specific magmas can exist (that is, be melts) are

1. The compositions of the melt material. Magmas that contain relatively large amounts of magnesium and iron can only exist at temperatures in excess of 1000°C. Magmas that contain small amounts of magnesium and iron, but relatively large amounts of silicon, sodium, and potassium can exist at temperatures much lower than 1000°C.
2. The depth (pressure) at which melting or crystallization takes place. At greater and greater depths within the earth the pressure increases. The temperature at which rock of a specific com-

6-1.   *Plot to illustrate effect of pressure on melting (and crystallization) temperature of dry basaltic and granitic assemblages. Note increase of melting temperature with pressure. This relationship applies only to essentially dry systems In water-saturated systems increase in pressure leads to a decrease of melting temperature.*

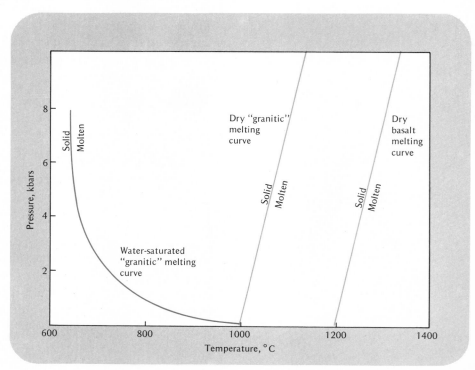

position is able to melt, or the equivalent magma crystallize, is influenced by pressure. In general, increase of confining pressure elevates melting and crystallization temperatures of magma (Figure 6-1).

3. The availability of water at the site of melting or crystallization. Water, even in small amounts, has the property of lowering the temperature at which melts can develop. Thus, a dry magma will begin to crystallize at higher temperatures than a magma of similar composition that contains several per cent water.

Movement of Magma

Magma is invariably less dense than the surrounding materials at the site of its generation. Provided a large enough volume of melt has been formed, it will tend to move upward in the same way a droplet of oil will rise through a denser medium such as water. As magma rises from its high temperature source area, it will encounter cooler overlying rocks but, provided the density contrast between the magma and the adjacent solid rock is maintained, it will continue to rise. The magma will either force the surrounding rocks aside and upward bodily or detach and incorporate blocks of the overlying rock into itself. Eventually the magma will either reach the earth's surface or solidify at some distance below the surface.

An interesting and well-documented case of the upward movement of low density material through the uppermost layers of the earth's crust (**diapiric** movement) is provided by salt domes (Figure 6-2). Before exploring this phenomenon further, a brief explanation of the concept of **viscosity** of liquids is needed. In general terms, viscosity refers to the flow properties of a liquid. Liquids that exhibit low viscosity flow readily (for example, water), whereas liquids of high viscosity are much "stiffer" and do not flow readily (for example, molasses). Although salt (NaCl) is a crystalline solid material under surface conditions, it behaves as a highly viscous liquid when subjected for long periods of time to the increased pressures and temperatures associated with deep burial. In this environment sedimentary salt locally tends to pierce its way upward gradually through denser overlying sedimentary rocks and, in some instances, will actually reach the earth's surface. Because economic concentrations of sulfur and petroleum can be associated with salt domes, much effort has been expended in the search for these structures and hundreds of salt domes have been discovered, especially in the Gulf Coast area of the United States and the Persian Gulf of Iraq and Iran.

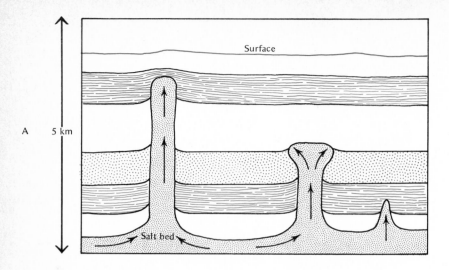

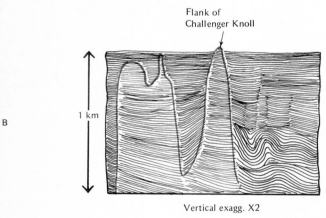

Flank of
Challenger Knoll

Vertical exagg. X2

6-2. A. *Digramatic cross section illustrating salt domes piercing their way toward the surface through overlying rock layers. The original salt bed was deposited as a result of the evaporation of sea water and subsequently buried by deposition of the overlying sedimentary units. B. Actual seismic reflection record of an area in the Gulf of Mexico showing penetration of salt domes through overlying marine sediments. [From Burk et al.: Deep sea drilling into the Challenger Knoll, central Gulf of Mexico. Amer. Assoc. Petroleum Geol. Bull., 53 (7): pp. 1338−1347 (July, 1969).]*

Eruptions of Magma

Most magma is much less viscous than deeply buried salt. Viscosity of magma decreases with increasing temperature, and thus high temperature magmas flow more readily than lower temperature magmas. Frequently magma reaches the earth's surface (or the ocean floor) through extensive elongate fissures (**fissure eruptions**) or through vents of more or less circular shape. Eruptions of magma from fissures are typically nonexplosive in character, whereas vent eruptions can be either nonexplosive or explosive. Some magmas contain several per cent of water and other gases, which are incorporated either at the site of their generation or have been absorbed from water-saturated rocks through which the magmas passed just prior to eruption. In such instances the gases expand violently during the eruption event and fragmental volcanic material is ejected from the vent into the air. Magmas of low water and gas content tend to erupt more quietly and generate **lava flows.**

Fissure eruptions generally result in the formation of flat-lying sequences of lava flows, whereas vent eruptions tend to build the distinctive forms known as volcanos. Volcanos can be built almost entirely of fragmental material, or lava flows, or mixtures of both types of material. Some volcanos, formed from relatively viscous magma form impressive steep sided cones around the central vent (Figure 6-3). The world's largest volcano, Mauna Loa, Hawaii, has

6-3. *Shishaldin Volcano, Aleutian Island. The volcano displays the typical conical shape of vent-type volcanic edifices.* [*Courtesy U.S. Department of the Interior Geological survey.*]

been built up from lavas of low viscosity and, as a consequence, exhibits gentle slopes. Despite this, it rises nearly 10 km above the surrounding ocean floor from a base that is 600 km in diameter.

COMPOSITION OF IGNEOUS ROCKS

Igneous rocks can have a wide range of compositions, and an alarming number of names has been used by geologists to denote igneous rocks of specific compositions and mineralogical and textural characteristics. It is unnecessary for the beginning student to know all but the most important igneous rock names; however, it is extremely important to be entirely familiar with these major types of igneous rocks, their mineralogical constituents and grain size, and some aspects of their chemical composition. This information is presented in graphic form in Figure 6-4 and, although this diagram may not convey much information to the reader upon brief perusal, it does in fact contain a great deal of information and merits careful study until fully understood.

In general terms, the diagram portrays the spectrum of important igneous rock types, from those with relatively low total silica (SiO_2) content and high magnesium oxide (MgO) and ferrous oxide (FeO) content (**ultramafic igneous rocks**) through transitional types (**mafic*** and **intermediate igneous rocks**), to those of high silica content and low magnesium oxide and ferrous oxide content (**felsic** igneous rocks**). Also indicated are the approximate melting or crystallization temperatures of the rock types within the compositional spectrum. Note that rocks rich in ferromagnesian (iron and magnesium bearing) minerals melt or begin to crystallize only at temperatures in excess of 1000°C, whereas rocks rich in quartz and alkali feldspars melt and crystallize at considerably lower temperatures.

The importance of a thorough understanding of Figure 6-4 cannot be stressed too strongly. In this and following chapters, a clear understanding of our discussion of the formation and occurrence of igneous rocks in various plate tectonic environments will depend on the reader's familiarity with this diagram.

*The term *mafic* is derived from the terms *ma*gnesium and *F*e (iron), and consequently is used to denote minerals or rocks rich in magnesium and iron (ferromagnesium).

**The term *felsic* is derived from the terms *fel*dspar and *si*lica, and consequently is used to denote rocks rich in feldspar and silica.

6-4. *Mineralogical grain size and chemical relationships of common igneous rock types. This diagram, which should be read from the bottom up, gives information on the mineralogical composition of common igneous rock types (center box) and the variation of ferromagnesian and alkali oxides in these rocks (top box). Also indicated are the percentage SiO_2 contents of the rocks (between the boxes) and data on the temperature of the magmas from which such rocks crystallize (above top box).*

For example, it can be read from the diagram that an average andesite (plutonic equivalent equals quartz diorite) contains 20% ferromagnesian minerals, 50% plagioclase, approximately 10% quartz, and 20% alkali feldspar. Furthermore such a rock would contain 57% SiO_2, 1% K_2O, 2% Na_2O, 6% FeO, and 3% MgO, and crystallize on the average of about 930°C.

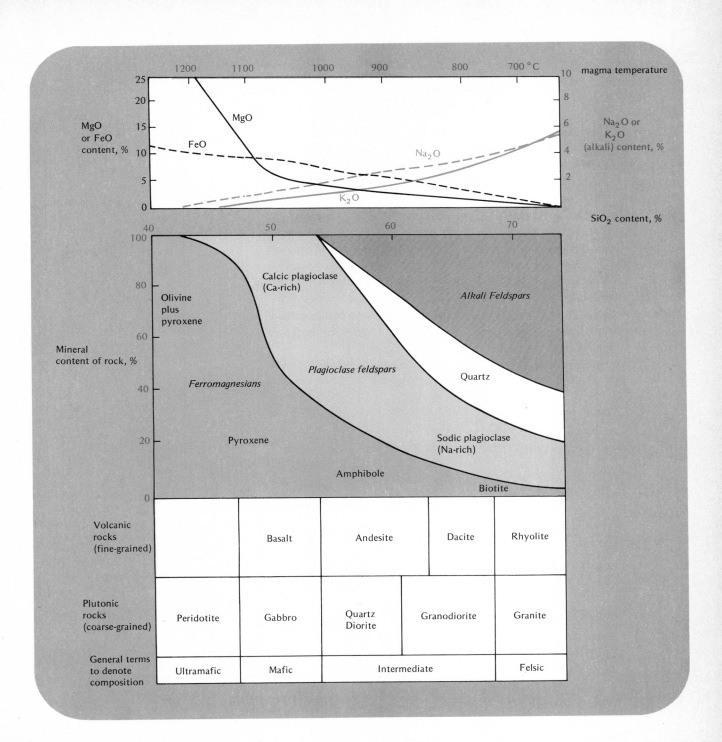

Causes of Diversity in Composition

By what general processes then are igneous rocks of diverse compositions created? The answer to this question lies in the manner in which magma crystallizes. N. L. Bowen, one of the great pioneers in laboratory experiments on silicate melts, proposed in 1922 that the differences in igneous rock compositions are related to the cooling history of their parent magma. This concept is now called Bowen's reaction principle and is diagrammatically illustrated in Figure 6-5.

In essence, Bowen demonstrated that as a magma cools certain minerals will crystallize from it before others. Once this occurs two things can happen. Either the crystals of the early formed minerals, which differ in their composition from that of the remaining cooling magma, react with the magma to form other mineral species stable at lower temperatures (reaction process), or the early formed crystals may settle out of the magma due to their higher density. The removal of early formed ferromagnesian crystals in this manner will affect the bulk composition of the melt that is left behind (fractional crystallization process) and, as this residual melt evolves in composition, becoming more felsic, and decreasing in volume, igneous rocks of diverse compositions can be created.

As the temperature of a cooling magma decreases, igneous min-

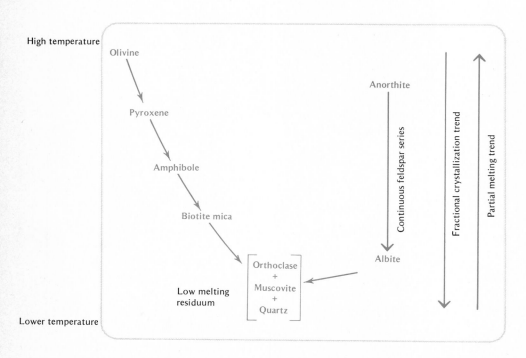

6-5.   *Bowen's Reaction Series, showing sequence of mineral phases involved in magmatic crystallization. Also indicated are fractional crystallization and partial melting trends. The figure should be studied in conjunction with Figure 6-4. Note that diagram refers essentially to the crystallization sequence of a magma of* **basaltic** *composition.*

erals that are stable at lower temperatures start to form, either by crystallization of new mineral species or by reaction of the melt with earlier formed crystals (see Figure 6-5). In general terms, early forming magmatic minerals are rich in magnesium, iron, or calcium, whereas late forming magmatic minerals are rich in silica, alumina, sodium, and potassium. A comparison of the data portrayed in Figure 6-4 and the Bowen reaction series (Figure 6-5) will serve to emphasize the close relationship between major igneous rock types, their composition and melting temperatures, and the mineralogical sequence of fractional crystallization. The central point we wish to make is that igneous rocks of diverse compositions can result from a single parent magma.

Partial melting is the reverse process of fractional crystallization. The process occurs when those minerals in a rock that have the lowest temperature stability melt first to form magma. The net result is a mixture of magma and solid material, the latter composed of minerals that have not reached their melting temperature. Obviously, if such a magma were squeezed out of the rock body in which it was formed, its composition would not be representative of the total rock composition at the site of its formation. It will in fact always be richer in the felsic (see Figure 6-5), lower melting components than the surrounding material. Thus, these two processes, partial melting and fractional crystallization, operating to varying degrees in different crustal and subcrustal environments, are the most important factors that determine the eventual composition of an igneous rock.

As we shall see in this and following chapters, the operation of these processes within the varied geological environments produced by plate interactions provides earth scientists with a partial explanation, not only for the diversity of igneous rock compositions but also for the relative volumes of igneous rock types found in distinctive geographical areas.

THE BIRTH OF OCEANIC CRUST

The creation of new oceanic crust at the spreading ridges that mark divergent plate boundaries is the result of igneous activity, and it now appears to be certain that the ~7-km thick oceanic crust underlying all the ocean basins of the world is composed predominantly of rocks of magmatic parentage. Spreading ridges therefore represent major zones of igneous activity, and large volumes of igneous rocks are continually being formed along them.

By what process are these large volumes of magmas generated below spreading ridges, and what controls their composition? It is now known that partial melting of mantle peridotite occurs below spreading ridge systems at depths of less than 100 km, with the production of basaltic magma.

The mechanism by which this magma is generated is related to the effect of pressure on the temperature at which melting can occur in the mantle. The tension that results when plates move away from one another causes a reduction in the pressure exerted on the underlying hot mantle material. This in turn lowers the temperature at which melting is possible to values below the actual temperature in the upper mantle, and partial melting of mantle peridotite results. The melt formed is of basaltic composition, and therefore more felsic than the ultramafic mantle material from which it was generated. This basaltic magma moves upward in sufficient volume to form the volcanic layer of oceanic crust and continually fills the uppermost part of the crack between the diverging plates. At depths greater than 10 km under ridge crests, upward and outward movement of mantle material probably takes place by solid state creep.

The total volume of basaltic magma that must be produced at present to create new oceanic crust along the spreading ridge systems of the world is estimated to be ~20 cubic kilometers per year (km^3/year). Although spreading rates vary from one spreading ridge system to the next, or even along the same ridge system, apparently oceanic crust of a more or less constant thickness is always generated. An overall balance between spreading rate of a ridge system and supply of magma along its axis is thus indicated. This is not as surprising as it might appear because, if the supply of magma is related to partial melting induced by a decrease in pressure, this effect will obviously be more marked along ridges that spread rapidly (and thus generate oceanic crust faster) than along ridges that spread slowly.

The reader, mindful of the fact that virtually the entire length of the world's spreading systems lies beneath the surface of the oceans, is probably wondering on what factual data the model we have been discussing is based. Supporting evidence has come from the following areas, which will be discussed in turn.

1. Seismic data from shallow earthquakes along ridges.
2. Heat flow profiles across spreading ridges.
3. The geology of Iceland, Surtsey, and other islands along spreading ridge systems.

4. Study of rock samples dredged from the vicinity of spreading ridges and undersea photographs of similar areas.
5. The study of slices of oceanic crust that have become elevated above sea level. These are called ophiolite complexes.

Seismic and Heat Flow Evidence

As we saw in Chapter 5, the present midoceanic ridge systems are the sites of numerous shallow (< 100-km depth) earthquakes. Analysis of the seismic data obtained from these earthquakes indicates that they are caused by failure of rock due to tensional forces. Thus the seismic data suggest that the crust and outermost layers of the mantle are undergoing extension along the oceanic ridge systems. Measurements of heat flow from the oceanic crust in the vicinity of ridge crests demonstrate that the amount of heat coming from the earth's interior is very much higher near oceanic ridge crests than away from them. These observations are entirely consistent with the idea that large volumes of hot basaltic magma are involved in the creation of new oceanic crust along spreading ridge systems.

The Spreading Ridge Systems

Further evidence for tension or pulling apart of crust and large scale basaltic magmatism comes from studies of the geology of Iceland, the only land area of any size that lies astride a midocean ridge system. In Iceland geologists have found evidence that tensional rifting (Figure 6-6) and fissure eruptions of basaltic magma are taking place along a northeast-southwest line approximately in the center of the island. Outwards from this central zone the basalt lava flows are of progressively greater age (Figure 6-7, map of Iceland).

In 1964 a series of spectacular volcanic eruptions occurred at the crest of the Mid-Atlantic Ridge southwest of Iceland (Figure 6-8). The building up of volcanic materials eventually led to the creation of a new volcanic island (Surtsey) that continued to produce lava and fragmental material for several months. In the early stages of the creation of Surtsey, large volumes of glassy volcanic ash and coarser fragments were ejected from the volcanic center in violent eruptions. This probably resulted from mixing of sea water and the hot basaltic magma prior to eruption. Upon consolidation, volcanic ash forms a rock called **tuff**, and coarser fragments form **agglomerate**.

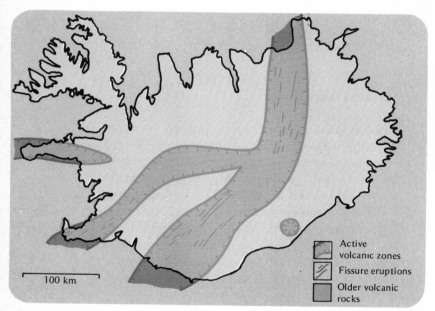

100 km

Active volcanic zones

Fissure eruptions

Older volcanic rocks

6-7. *Map of Iceland showing distribution of zones of Recent (<20,000 years) volcanic activity. The rocks outside of these zones are essentially similar but somewhat older.*

One aspect of spreading ridge systems that requires discussion is their topographic form. Why are the crests of spreading ridges elevated hundreds to thousands of meters above the floors of the ocean basins? Before dealing with this question it should be pointed out that slow spreading ridges (2–5 cm/year separation rate, Atlantic Ridge System) tend to have a strong topographic expression, whereas fast spreading ridges (>10 cm/year, East Pacific Rise) are marked by

6-8. *Surtsey Volcano in eruption. The island of Surtsey was 15 days old at the time this photograph was taken and had reached a height of about 100 m. It is now considerably larger. [Courtesy of S. Thorarinsson.]*

ridges with very gentle slopes (see Figure 5-6). The precise reason for the topographic form of spreading ridges is not fully understood, but as mentioned in Chapter 5, it is probably an expression of the thermal expansion of hot underlying lithosphere. Slow spreading ridges in most cases exhibit a well-defined trough or axial valley that runs along their crests. This feature is apparently formed as the consequence of the upfaulting of blocks of recently created oceanic crust some tens of kilometers away from the central line of tensional rifting. When this process occurs symmetrically on both sides of the central line of tensional rifting, a well-defined troughlike feature will mark the axis of the ridge.

Dredge Sampling and Photography

In an attempt to understand the origin of midoceanic ridges, marine geologists have made numerous dredge hauls at various points along them to investigate the nature and composition of the rocks present. Careful study of these has indicated that virtually all the samples obtained from ridge crests consist of rapidly chilled basalt. Samples dredged from the steep walls present in places where the ridges are offset by transform faults are also, in most cases, basaltic in composition. The size of the individual mineral grains in these rocks is much larger, indicating that they were produced by slow cooling of basaltic magma. Thus, rocks underlying the basaltic volcanic layer of oceanic crust are largely gabbros (Figure 6-9).

Undersea photographs of the material present at ridge crests in-

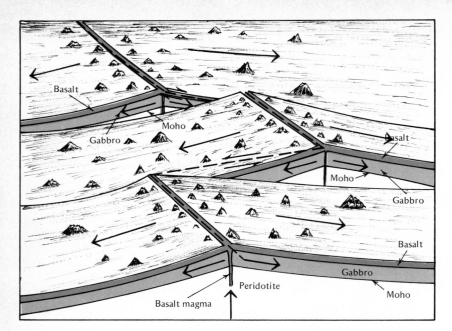

6-9. *Diagrammatic illustration of spreading ridge system cut into offset segments by transform faults. It is due to spreading configurations similar to this that the offset patterns of marine magnetic anomalies are generated.*

variably demonstrate the presence of pillowed basalts (Figure 6-10A). This term is applied to volcanic rocks that are extruded under water and chilled so rapidly that they tend to form discrete, elongate, pillowlike masses at the site of extrusion. The presence of pillow lavas in volcanic rocks of any age is interpreted by geologists to indicate that those rocks formed in a submarine environment (Figure 6-10B and C).

Exposed Oceanic Crust

Probably the most convincing evidence for the igneous character and the mechanism of formation of igneous crust has come from the study of slices of oceanic crust (ophiolite complexes) that, due to special circumstances, have been thrust above sea level. This, of course, makes them accessible for geologic study. Many examples of fragments of oceanic crust that have been exposed by dynamic processes related to plate tectonics are now known. They have been recognized and studied around the margins of the Pacific Ocean and in particular in a belt from the Mediterranean to the Himalayas.

One sequence of igneous rocks that is interpreted to represent a slice of oceanic crust and underlying mantle is well exposed on the island of Cyprus in the northeastern Mediterranean. Recent detailed

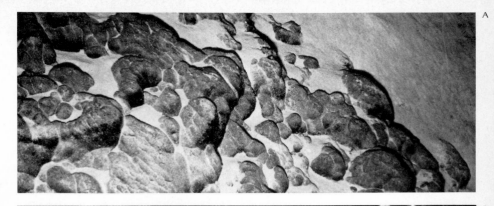

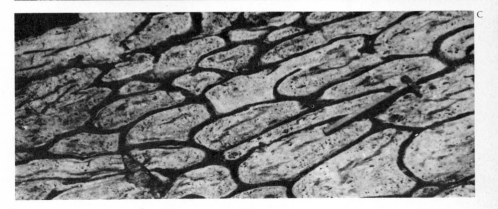

6-10. *Examples of pillow lavas.*
A. Undersea photograph of
modern basaltic pillows in the
Pacific Ocean. [Courtesy of
Scripps Institute of Oceanogra-
phy.] B. Pillow lava of Middle
Cambrian Age; Trinity Bay,
Newfoundland. Note elongate,
lightly flattened shape of indi-
vidual pillows. [Courtesy Geo-
logical Survey of Canada, (No.
152581).] C. Early Precambrian
basaltic lavas that exhibit well-
defined pillow structures, indi-
cating formation in a submarine
environment. Note that the pil-
lows have been somewhat flat-
tened and deformed by post
formational structural processes.
[Courtesy Geological Survey
of Canada, (No. 84169).]

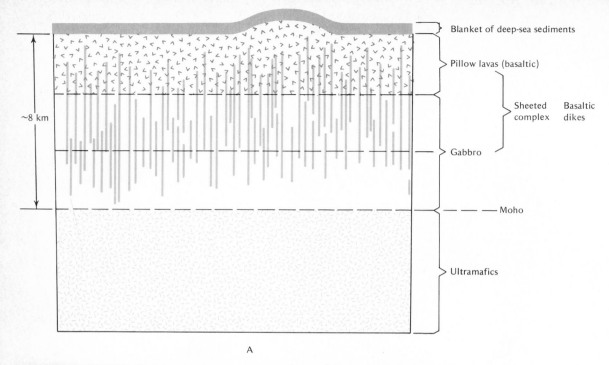

~8 km

Blanket of deep-sea sediments

Pillow lavas (basaltic)

Sheeted complex — Basaltic dikes

Gabbro

Moho

Ultramafics

A

B

6-11. A. *Interpretative cross section of a slice of oceanic crust on the island of Cyprus. This is probably one of the best models geologists have of the present day crust in the ocean basins of the world.* [*Modified after E. Moores and F. Vine: The Troodos Massif, Cyrpus, and other ophiolites as oceanic crust: evalutaion and implications. Phil. Trans. Roy. Soc. London, Sect. A, 268: pp. 443–466 (1971).*] B. *Pillow lavas of the uppermost part of the slice of oceanic crust exposed of the island of Cyprus. The geologist in the picture is Frederick Vine, one of the originators of the sea floor spreading concept.* [*Photograph courtesy E. Moores.*]

study of this area revealed the following relationships (Figure 6-11A). Overlying a sequence of pillowed basalts (Figure 6-11B) is a thin veneer of sedimentary rocks of typical oceanic character. Towards the base of the volcanic sequence, the pillow lavas are cut by a large number of parallel vertical sheets of intrusive basalt rock or **dikes.**

Below the base of the pillow lavas the rock consists almost entirely of this sheeted material. Careful study of sets of dikes within the sheeted zone indicated that many dikes intruded into the central portions of earlier dikes. Thus, continuing crustal extension must have been taking place locally during emplacement of the dikes, which now fill the fractures through which the magma that formed the overlying pillow lavas passed. Below the dikes, the rocks are coarser grained **gabbros,** which represent the more slowly cooled plutonic equivalents of the overlying dike rocks and pillow lavas. Underlying the gabbroic rocks are ultramafic rocks, composed almost entirely of pyroxene and olivine, called **peridotites.** These are considered to be representative of upper mantle material.

Such sequences of rock types, from deep sea sediments downwards through pillow lavas, dikes, gabbro, and peridotites are called **ophiolite complexes.** If ophiolite complexes represent normal oceanic crust and immediately underlying mantle, then, according to the plate tectonic model, they must have been created at a spreading ridge system.

The model presented earlier for the character of oceanic crust formed at spreading ridge systems is therefore in agreement with observations made on exposed sections of oceanic crustal rocks. In broad terms then, the hypothesis of sea floor spreading and the creation of oceanic lithosphere by magmatic processes is compatible with the available data from modern ridges and ophiolite complexes. Much work remains to be done, however, before the full spectrum of processes involved in the creation of oceanic lithosphere is understood.

LIFE OF OCEANIC CRUST

As newly formed oceanic lithosphere moves away from the ridge system it thickens, cools, and slowly subsides. The subsidence is extremely gradual and, for this reason, the new oceanic crust does not sink to the level of the deep ocean floors ($\sim$4 km) until it has moved over 1000 km away from the ridge crest. During its cooling history the uppermost layer of pillow lavas undergoes some alteration in its mineralogical character due to reaction with sea water, which apparently enters the pillow lavas readily. One effect of this interaction is to cause some mineralogical rearrangement of the iron oxide minerals in the lavas. It is iron oxide minerals that recorded the signature of the earth's magnetic field at the time the lavas first crystallized. This rearrangement of iron oxides fortunately does not change, or destroy, the magnetic anomaly patterns of the newborn crust; it does, however, lead to an observable decrease in the intensity of these anomalies.

Sedimentation

As it moves away from the ridge system, our newborn oceanic crust now enters a major, but relatively uneventful, part of its life cycle. This consists of a gradual recession from the spreading ridge and the slow accumulation on its surface of a covering blanket of sediment. Studies of the sedimentary material covering the volcanic crust in the present ocean basins indicate a progressive increase in sediment thickness further and further away from a spreading ridge system.

The marine fossils obtained in drill core samples of the sediments that immediately overly the volcanic rocks also provide support for the sea floor spreading concept. The geological age of these fossils increases with increasing distances from the spreading ridge (see Figure 5-21). Clearly the evidence supporting the hypothesis that oceanic lithosphere is created at spreading ridge systems and moves outward from them in symmetrical fashion is extremely broad based. The sea floor spreading model is one of the cornerstones of plate tectonic theory and represents a major advance in our understanding of the dynamic processes that shape the surface of our planet.

We must now answer the question, What is the source of the sedimentary material that is deposited on oceanic crust? Investigations of sediments deposited in the deep ocean basins of the world indicate that they are composed of variable amounts of organic and inorganic materials. In certain instances there are significant amounts of mineralogical components that have grown in place at, or just below, the boundary between sediment and overlying seawater.

Far from such sediment sources as continental masses and oceanic islands, most of the material that rains down through the deep ocean water represents the skeletal remains of tiny marine organisms. These skeletal remains are of two contrasting types, one type composed of calcium carbonate ($CaCO_3$) and the other of silica (SiO_2), depending on the species of microorganism they represent (Figure 6-12). Immense numbers of these organisms live in sea water, eventually die, and sink to the ocean floor. This fine material accumulates

6-12. *A mixture of foraminifera and radiolaria that accumulated on the floor of western Pacific. Open structured skeletons are radiolaria, the remainder are foraminifera (× 150). [Courtesy of S. Kling.]*

on the ocean floor to form what are called **pelagic** (Greek *pelagios,* meaning of the ocean) oozes.

Rates of accumulation of pelagic ooze are extremely slow and range from less than 1 mm per thousand years to ~10 mm per thousand years. Furthermore, when these water-rich oozes eventually become buried and compacted to form sedimentary rocks, they undergo a considerable reduction in volume. Not surprisingly, rates of accumulation of pelagic sediments depend to a large extent on the amount of marine life present in the uppermost layers of the ocean water in that area. In areas of high biological productivity, the downward rain of skeletal remains is more intense and sedimentation rates are at the high end of the spectrum.

Another factor that affects sedimentation rates is water depth. The solubility of calcium carbonate increases with depth in the oceans, and thus, in areas where the water depth is in excess of 5 km, most or all of the skeletal material composed of calcium carbonate is dissolved before it reaches the ocean floor. Skeletal remains composed of silica are not subject to this dissolution effect; consequently, in the deeper parts of the ocean basins, the pelagic sediment is composed essentially of siliceous ooze. Minor amounts of meteoric and continentally derived dust and fine volcanic ash particles also make up a fraction of oceanic sediments.

Closer to land areas, inorganic weathering products such as sands and clays derived from continents or islands, or ash derived from volcanic eruptions (both subaerial and submarine) provide much of the sediment that collects on the deep ocean floor. The supply of material from sources such as these tends to be intermittent and most of the weathering products of land areas do not reach the deep ocean floor. In some areas, however, the supply of land derived sedimentary material is sufficiently great to bury the original topography of the ocean floor completely and create flat **abyssal plains.** The mechanisms by which these sediments are deposited on the deep ocean floor involve lateral submarine transport and will be discussed in Chapter 8. Oceanic sediments tend to exhibit a layered character because during their deposition materials of different type and sporadic falls of volcanic ash accumulate on the ocean floor.

Of the materials that form in place on the ocean floor, the most interesting and potentially important to man are manganese nodules. Large areas of the surface of the ocean floor have been found to be covered by nodular masses of manganese oxides that range in size from less than 1 cm to greater than 1 m (Figure 6-13). Manganese nodules exhibit a concentric structure and apparently grow by the precipitation of successive layers of manganese and iron oxides.

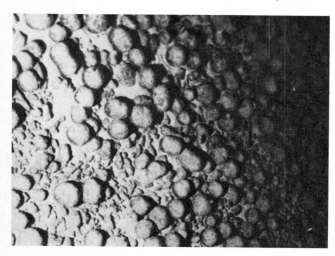

6-13. *Cannonball sized manganese nodules lying on the floor of the south eastern Pacific Ocean at a depth of 4559 m. [Courtesy of C. D. Hollister.]*

Their possible economic significance lies in the fact that in many areas the nodules contain potentially economic quantities of copper, nickel and other metals. Considerable attention is now being directed by some mining companies toward the possibility of mining manganese nodules from the ocean floor. It is only a matter of time before man attempts to exploit this ocean floor resource.

Igneous Activity

In the deep ocean basins the sea floor is highly irregular in most areas. This is due to the presence of numerous volcanic structures, some of which reach the ocean surface. Pelagic sediments inevitably blanket many of these volcanic structures, but in most cases this serves merely to subdue their topographic expression. In areas where sufficiently large volumes of continentally derived sediments are laid down on the sea floor, the irregular topography of ocean crust can become completely buried. As a result the sea floor in these areas is characterized by virtually featureless abyssal plains.

Many presently inactive oceanic volcanos have been found to be much younger than the surrounding oceanic crust; therefore, they could not have been formed at spreading ridges and merely conveyed to their present locations by sea floor spreading. Igneous activity in oceanic areas is thus not restricted to spreading ridges, and the generation and upward movement of the basaltic magma that forms these oceanic volcanos can take place away from plate boundaries (that is, in intraplate environments). The factors that control the generation of this magma in intraplate environments are not well understood at present, but its ultimate source is thought to be from depths of 100 km or more, within the upper mantle. As in the case of volcanism at spreading ridges, basalt is the predominant rock type, but careful studies by petrologists of the rock types on volcanic islands have indicated that the composition of some of the volcanics is distinctive from spreading ridge basalts.

The volume of magma involved in building volcanic structures in intraplate oceanic environments has been estimated to amount to 0.2 km³/year, only one hundredth of the 20 km³/year estimated to be produced annually along current spreading ridge systems. Nevertheless, until the advent of detailed studies of undersea topography after World War II, geologists had no realization of the widespread extent of oceanic volcanism. In some areas of the Pacific Ocean the supply of basaltic magma from the mantle to the ocean floor has been considerable. As a result of the Pacific Plate moving over these local

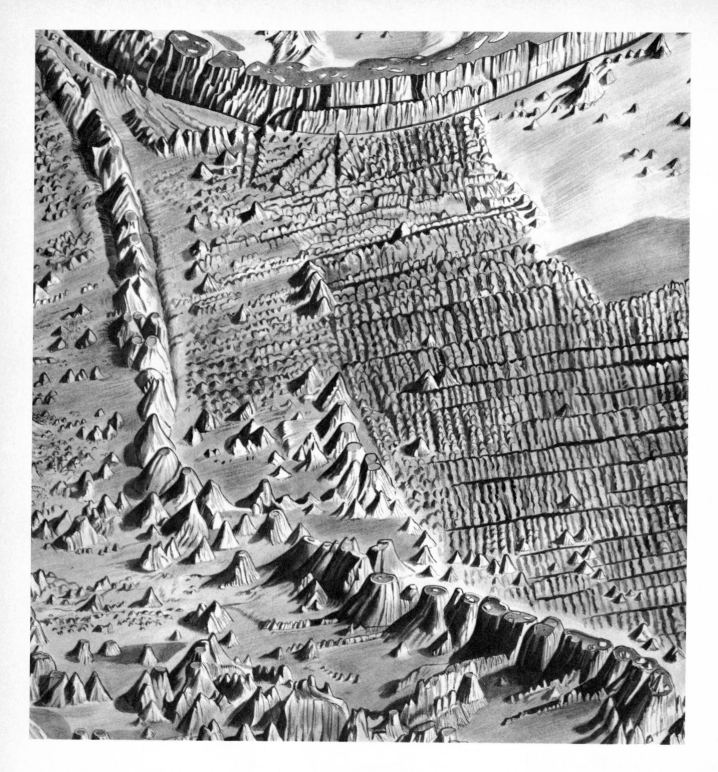

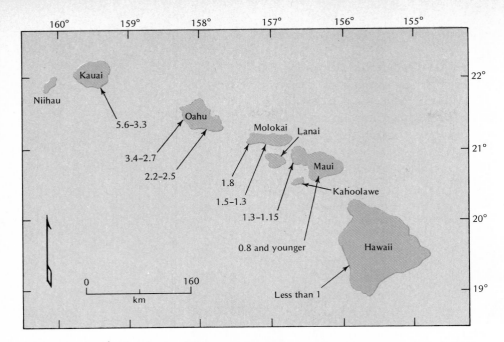

6-14. *Chain of volcanic structures running from Hawaii to Aleutian Trench. Radiometric age studies have indicated a progressive increase in the age of volcanic activity from Hawaii to the northwest (see diagram above, where the numbers indicate the age in millions of years of dated volcanic rocks).*

areas of intense magma generation or "hotspots," linear arrays of volcanic islands or submarine volcanos have been formed on its surface. Radiometric dating of the volcanic rocks from these islands has shown that their emplacement occurred at progressively younger ages from one end of the chain to the other. The Hawaiian volcanic chain represents a particularly good example of this phenomenon (Figure 6-14).

Seamounts and Atolls

As more details of undersea topography became available during the 1950s many examples of submarine volcanos with flat tops were discovered (Figure 6-15A). Their shapes seemed to suggest truncation and beveling of their tops by wave action, but this hardly seemed possible because many were as much as 2000 m below sea level. Marine geologists at first could not explain these enigmatic flat-topped seamounts. It is now clear that their tops were truncated at sea level by wave action, and the volcanic edifices have subsequently subsided to their present depth. This drowning of former volcanic islands can occur in two ways, either due to the subsidence that occurs as oceanic crust moves away from a spreading ridge or due to the depression of oceanic crust that results when the weight of a large volcanic structure or group of structures is superimposed upon it.

In the equatorial Pacific and other areas where conditions for the growth of coral reefs are favorable, upward growth of coral reefs that originally formed a fringe around volcanic islands has been able to keep pace with submergence of volcanic structures. In such areas the sites of drowned volcanos are now marked by the presence of

6-15.   A. *Profile of a typical flat-topped seamount. Details of the upper surface suggest that it was beveled by wave action. This interpretation is supported by the common presence of shallow water fossils on the upper surface of the structure. B. Diagrammatic cross section of typical atoll formed by growth of coral on flat-topped seamount. Note steep walls of coral reef and accumulations of sedimentary reef debris both within the central lagoon and on the outer edge of reef structure.*

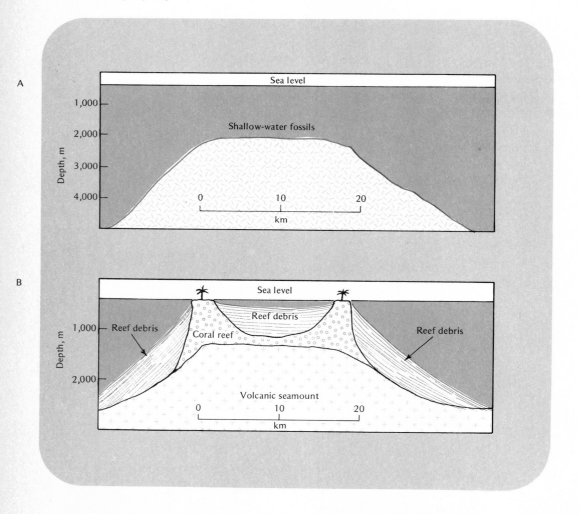

low lying, more or less circular, coral islands called **atolls** (Figure 6-15B). It is interesting to note that this explanation, involving compensatory upward growth of coral reefs over subsiding volcanic structures, was first proposed by Darwin during his celebrated voyage in the Beagle. Because of the northwestward movement of the Pacific Plate, reef coral remains have been found on Pacific seamounts that lie well to the north of the area in which present day coral reef growth can take place. This observation provides yet another piece of evidence in support of the plate motion hypothesis.

Throughout its lifetime the oceanic crust recedes from the spreading ridge where it was created. During this time interval it receives a gradually thickening blanket of sedimentary material and locally may have basaltic volcanos built upon it. The oldest oceanic crust present today beneath the oceans is estimated to be approximately 200 million years old, which, as we shall see, is but a small fraction of the average age of continental material. Thus, in terms of geologic time, the life span of oceanic crust is relatively short. Where then is the oceanic crust that must have existed beneath the oceans during earlier periods of earth history, and what was its fate?

THE DEATH OF OCEANIC CRUST

Trenches

The plate tectonic model dictates that the ultimate fate of oceanic crust is to undergo subduction back into the earth's mantle at convergent plate boundaries. As indicated in Chapter 5, the position of many active convergent plate boundaries is marked by deep oceanic trenches. These remarkable features are present in all the major oceans of the world, and in particular around the margins of the Pacific basin (Figure 6-16).

Typically trenches are arcuate in shape and have their convex sides towards the oceanic basins they border, but some trench systems do not display an arcuate form and extend in a straight line for thousands of kilometers (for example, the Tonga-Kermadec Trench system). The physical control of trench shapes is something not well understood at the moment but presumably relates in some way to the forces that cause downward movement of plates into the earth's mantle. The enormous depth of some trenches was not realized until

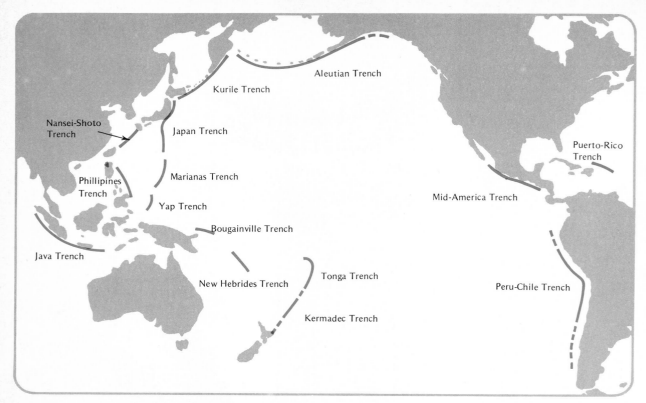

6-16. *Deep ocean trenches of the Pacific and neighboring areas.*

soundings made in the mid 1950s indicated a depth to their axial portions as great as 11,000 m, a distance well in excess of the height of Mt. Everest, the world's highest mountain. Since that time data on oceanic trenches have accumulated rapidly.

The deepest trenches have virtually no sediment in their axial portions, whereas others are partially or almost completely filled with sediment from adjacent land areas. Figure 6-17 illustrates the profiles of three different trench systems that exhibit variable degrees of sediment fill. The material in the bottom of partially filled trenches consists of flat-lying layers of sediments that exhibit no deformation. This fact initially puzzled marine geologists because, if trenches are sites of plate convergence, trench sediments should show abundant evidence of folding and distortion due to compression. It is now realized that the layers of sedimentary material that fill the bottom of some trenches are simply not cohesive enough to transmit stresses, and the compressional distortion of trench sediments occurs only on the inner trench wall. Detailed studies of trench topography indicate

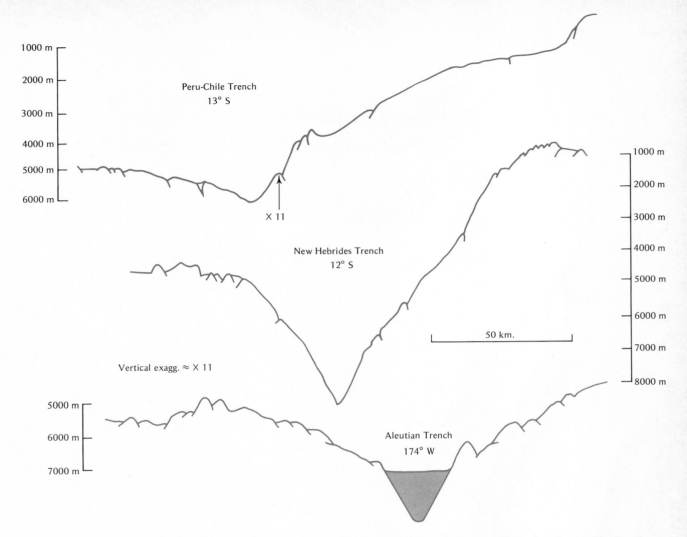

6-17. *Profiles of three typical trench systems. Note that the profiles involve marked vertical exaggeration and that, at true scale, trenches are rather broad features.* [*Modified from W. Menard:* The Marine Geology of the Pacific. McGraw-Hill Book Company, Inc., New York, 1964, (Figure 5.2, p. 100).]

that on the outer trench wall the downward bending of the ocean floor towards the trench causes tension resulting in steplike downfaulting of blocks toward the trench (Figure 6-18).

Two of the most remarkable phenomena associated with oceanic trenches are the strong negative gravity anomalies and the intense earthquake activity associated with them. In the vicinity of trenches, strong negative departures from the expected gravity field of the

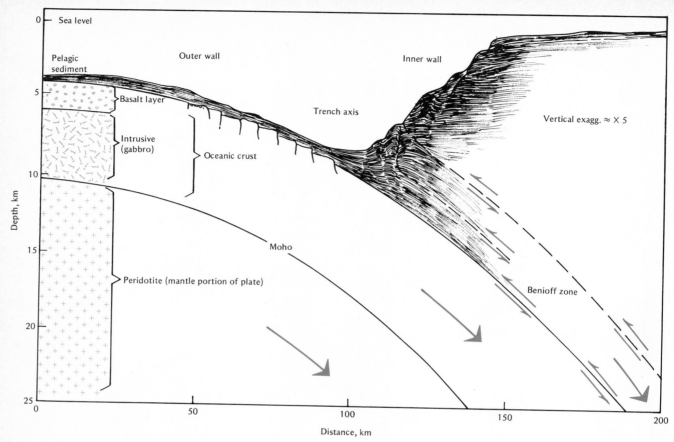

6-18. *Generalized diagram showing probable relationships of down-going oceanic lithosphere at a convergent plate boundary marked by an ocean trench. Note that some of the oceanic sediments become folded and plastered to the inner trench wall. Some probably descend at least part way down the Benioff zone. Outer wall of trench at flexure of plate is characterized by tension and normal faulting.*

earth are present. This indicates that some dynamic force must be acting to depress the trenches to their present depths. In the absence of such a force the trenches would simply cease to exist due to upward adjustment of their floors to normal oceanic depths. The observed negative gravity anomalies associated with trenches can be simulated very closely by geophysical models involving the downward movement into the mantle of oceanic lithosphere; thus, the gravity data from trench systems can be neatly integrated with plate tectonic theory.

Earthquakes and Subduction

The intense earthquake activity associated with trench systems provides yet further evidence for the plate tectonic concept and for the demise of oceanic crust. As oceanic lithosphere is underthrust, or slides below an adjacent plate, intense earthquake activity occurs along the upper inclined boundary of the down-going plate. Analyses of these zones of earthquake activity indicate that the inclination, or dip, of underthrusting can vary in individual cases from as little as 15° from horizontal to almost vertically downward. In fact, the occurrence of earthquakes generated at depths of greater than 100 km below the surface is virtually restricted to trench-associated inclined planes of earthquake activity. These inclined planes have been called **Benioff zones** in recognition of Hugo Benioff, a pioneer of the science of seismology (see Figure 6-18). Benioff zones extend to depths of approximately 700 km, the depth of the deepest known earthquakes. Below that depth either the down-going plate is completely resorbed into the mantle or the physical regime is such that the strength of mantle material is too low to allow the build up of strain energy, the release of which gives rise to earthquake activity.

All of these data on earthquakes fit well with, and in fact helped to shape, plate tectonic concepts as we now use them. More recently, detailed studies of the travel time of seismic waves generated by earthquakes at or near the Benioff zone have demonstrated the presence of high seismic velocity channels parallel to and just below the Benioff zones. This is precisely what one might expect if the cooler, more rigid material of a plate was reentering the mantle and provides still further support for the plate tectonic model of the demise of oceanic crust.

Subduction Rates

Rates of subduction at present convergent plate boundaries cannot be measured directly, but recently they have been calculated by analysis of worldwide plate motions and found to range from 4 to 20 cm/year. Average rates of subduction are distinctly higher than average rates of sea floor spreading because the total length of spreading ridge systems (~40,000 km) far exceeds the total length of plate boundaries at which convergence is taking place (~25,000 km).

As mentioned earlier, the oldest oceanic crust below the present-day oceans must have an age of approximately 200 million years.

From what has been deduced about plate motions in the past, this oldest oceanic crust must be present along the base of the continental slopes off the Atlantic margins of North and South America and Africa and in the extreme west equatorial part of the Pacific. In the latter case this old oceanic crust is being actively consumed along a subduction zone marked by the Marianas Trench. Along the borders of the Atlantic Ocean, modern convergent plate boundaries have yet to form, but it can be inferred from plate tectonic theory that eventually subduction zones will form at or near these continental margins and this old oceanic crust will end its life cycle with reentry into the mantle.

CONCLUDING STATEMENT

The conceptual model of the creation, life, and eventual destruction of oceanic lithosphere presented in this chapter is now accepted by a very large majority of earth scientists. Its strength lies in its ability to explain a vast array of both geological and geophysical data. Obviously refinements of the model can be anticipated as a result of current and future research, but it seems unlikely in the extreme that it will eventually be replaced by another, radically different, set of concepts.

QUESTIONS

1. Melting and crystallization temperatures of magmas can vary considerably. What are the main factors that influence the temperatures at which magmas form or crystallize?
2. Explain why magmas, once formed, tend to move upward toward the earth's surface.
★3. How do the Bowen Reaction Series and the concepts of fractional crystallization and partial melting explain the diversity of igneous rock compositions?
4. Oceanic crust is formed along spreading ridge systems largely by magmatic processes. By what mechanisms are the magmas

thought to be generated and why is the thickness of oceanic crust formed largely independent of spreading rate?

5. Discuss briefly the sources of sediments that form in oceanic areas. What influence does the depth of the ocean floor have on the sediments that can collect?

6. In the ocean basins the thickness of the layer of pelagic sediments increases with distance from oceanic ridge systems. What is the significance of this observation?

7. What is the generally accepted explanation for the formation of flat-topped seamounts and atolls?

8. Trenches are important features of most ocean basins. What is their relationship to the life cycle of oceanic crust?

seven

The Geology of Island Arcs

The demise of oceanic crust by subduction at a convergent plate boundary represents not only the closing of one cycle of geologic activity but the simultaneous opening of another. In this chapter we will investigate some of the important geologic consequences of the subduction process. As discussed in Chapter 6, the upper margin of the down-going plate at a convergent plate boundary is marked by an inclined plane along which earthquakes are generated. The upper surface of this inclined, earthquake-defined plane (Benioff zone) marks the zone along which differential movement between the two converging plates takes place.

What types of geologic events characterize the uppermost parts of the portions of plates that lie above Benioff zones? At the present time the areas above active Benioff zones (that is, on the inner side of subduction zones) are the sites of intense geologic activity. In large measure this geologic activity appears to be directly related to the subduction process that occurs at these convergent plate boundaries.

In general terms, plate convergence can occur in three distinctive plate environments:

1. Both converging plates consist of oceanic lithosphere.
2. The down-going plate consists of oceanic lithosphere and is descending beneath a plate consisting of continental lithosphere.
3. Both converging plates consist of continental lithosphere.

Examples of each of these plate environments can be identified at the earth's current stage of evolution, but in this chapter we will restrict ourselves to the environments where both converging plates consist of oceanic lithosphere (that is, where subduction is taking place within an ocean basin). The two latter plate environments will be discussed in Chapter 8, which deals with the geology of continents.

ISLAND ARC VOLCANOS

The majority of the world's active subaerial volcanos form belts of islands parallel to and along one margin of the great trenches of the western Pacific (Figure 7-1). Similar volcanic island chains or arcs, spatially associated with oceanic trenches, are found in the Antilles Arc, which flanks the Caribbean Sea, and in the East Indies. The volcanic activity in island arc systems tends to occur as vent-type eruptions that are quite distinctive from the fissure-type eruptions that characterize the basaltic magmatism at spreading center systems. More than 70% of the magma erupted above sea level along volcanic island arcs is of intermediate to felsic composition which, as was pointed out in Chapter 6, tends to erupt explosively. This explains why many of the worst natural disasters attributed to volcanic activity have occurred in island arc systems.

For example, the eruption of Mt. Pelee on Martinique Island, West Indies, in the year 1902 resulted in the loss of between 30,000 and 40,000 lives. On May 8th of that year a violent eruption shook the area, and shortly thereafter a fiery cloud of incandescent ash and volcanic gases was erupted (Figure 7-2). This dense, hot cloud swept down the mountainside and completely overwhelmed the city of St. Pierre.

Almost certainly one of the most impressive volcanic eruptions in recent history occurred in 1883 when the volcanic island of Krakatoa was literally demolished by a cataclysmic explosion. The volcanic center, which lies in the Indonesian island arc system, had been dormant for at least 200 years prior to this event. An impression of the enormity of this eruption can be gained from some of the observations recorded during the eruption. The sound of the main explosion was heard 5000 km away, and volcanic dust falls were recorded over an area of 4 million km^2. Following the main explosion,

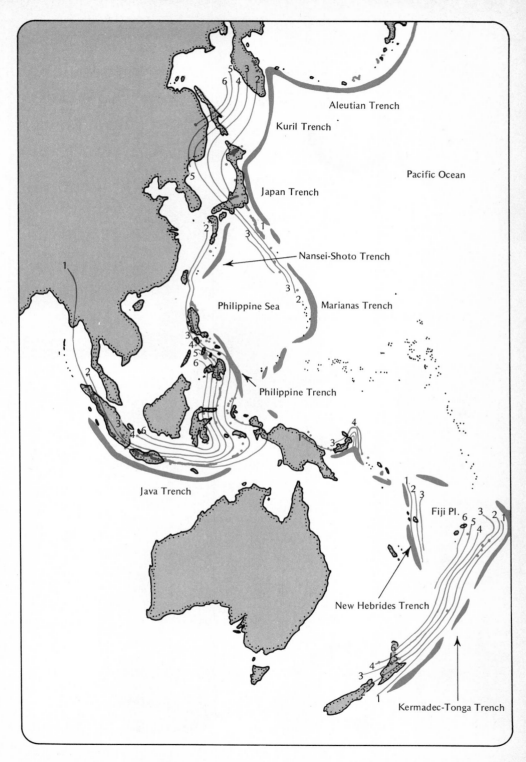

7-1. *Oceanic trenches, Benioff zones, and associated island arc systems of the western Pacific region. Lines behind arcs connect points on the surface that have the same elevation above the underlying Benioff zone. Numbers refer to the depth of the Benioff zone in kilometers. Note the close association of volcanos (solid dots) to trench systems.* [*From E. R. Oxburgh and D. Turcotte: Thermal structure of island arcs. Geol. Soc. of Amer. Bull, 81: pp. 1665–1688 (1970).*]

Aleutian Trench

Kuril Trench

Japan Trench

Pacific Ocean

Nansei-Shoto Trench

Philippine Sea

Marianas Trench

Philippine Trench

Java Trench

Fiji Pl.

New Hebrides Trench

Kermadec-Tonga Trench

A

B

7-2.   A. Fiery cloud of volcanic gas and ash erupting from Mt. Mayan, Philippines, April 27th, 1969. Note dense ash-laden material flowing down flanks of the volcano. [Courtesy of Smithsonian Institution.] B. From Orange Hill (N.E.) over dead St. Pierre, to steaming Mont Pelée—once beautiful and verdure clad—Martinique. [Courtesy of Library of Congress.]

a dust cloud rose an estimated 80 km into the upper atmosphere, and complete darkness reigned over an area of 40,000 km^2 for 24 hr. According to observers, residual dust from this eruption colored sunsets around the world for years. Although no persons were killed by lava or volcanic gases, the explosion generated a series of sea waves that drowned about 36,000 people on neighboring islands.

In 1815, 68 years prior to the Krakatoa explosion, an even larger volcanic eruption occurred at Tambora in the same island arc system. The earlier eruption is not as well documented as the Krakatoa event, but in fact blew three times as much volcanic ash (an estimated 150 million tons) into the stratosphere. The fine volcanic ash in the stratosphere spread worldwide within a few weeks, and after 8 months the average monthly temperatures in Europe had dropped by more than 2°C. The next 24 months in England were much colder than normal, and the coldest July on record in England was in 1816 (reasonably accurate temperature records in England go back to 1698).

What then was the relationship between this volcanic event and weather at the other side of the world? The excessive amounts of dust in the earth's upper atmosphere caused a sharp reduction in the amount of solar energy that normally reaches the earth's surface; hence, global weather patterns were affected. No good records are available from North America for this time, but old newspaper cuttings indicate extreme cold and poor harvests in Pennsylvania during 1816. The effects of all these changes on people were much more serious than just cold and unpleasant weather. Research using old newspapers indicates that the price of high grade flour in London more than doubled between December 1815 and June 1817. A year later flour had returned to more or less normal prices.

More recently (1912), a violent eruption of Mt. Katmai in the Alaskan Peninsula blew considerable volumes of fine volcanic ash into the earth's upper atmosphere. This ash persisted for several months and caused an estimated 20% reduction in the amount of direct solar radiation reaching the earth's surface during that time.

Spatial Relationships

Island arc systems and the trenches that parallel them are separated by a distance of 100 km or more in nearly all cases. In some instances the distance between the island arc and the trench may be as great as 300 km. The area between the island arc system and its associated trench has been named the **arc-trench gap** (Figure 7-3), and it is an important element in arc-trench geologic relationships.

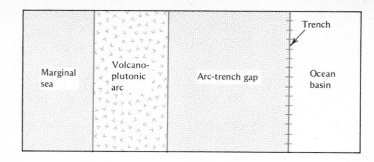

7-3. *Diagrammatic plan (A) and cross section (B) to illustrate the major components of island arc systems.*

Many of the present day volcanic island arc systems, such as the Kurile Arc, the Marianas Arc, the Hebrides Arc, and the Tonga-Kermadec Arc, only emerge above sea level sufficiently to form chains of small volcanic islands. Geologists have been able to gain only limited information about the composition and structure of these volcanic island arcs. Fortunately, some of the older island arc systems that are still active have been uplifted, and then sufficiently unroofed by erosion to allow geologists to gain a more complete picture of island arc geology. Examples of more mature active arc systems are the Japanese Islands (Figure 7-4), the Phillipine Islands, the Indonesian Archipelago, and the Alaskan Peninsula.

MAGMATISM IN ISLAND ARC SYSTEMS

By a careful integration of information from both young and more mature arc systems, a number of important general characteristics of island arc magmatism have been recognized. First, in the more

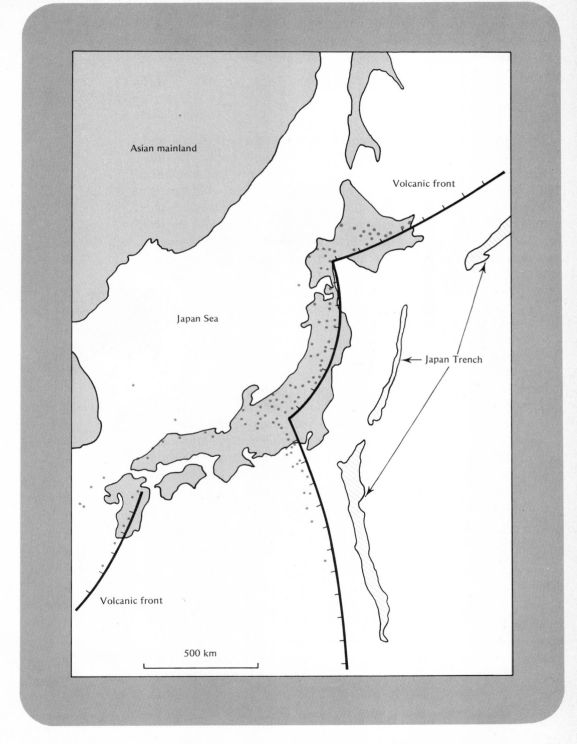

7-4. *Japan Island Arc system. Solid circles represent volcanos active during the last million years. Note well-defined volcanic 'fronts' and double front in southern area. [Modified from T. Matsuda and S. Uyeda: Tectonophysics, II: p. 8 (1971).]*

Asian mainland

Volcanic front

Japan Sea

Japan Trench

Volcanic front

500 km

deeply eroded portions of older island arcs, intrusive (plutonic) igneous rocks are widespread. Second, these intrusive rocks have essentially the same composition as the intermediate and felsic volcanic (extrusive) rocks so typical of arc systems.

Composition of the Magma

The bulk of these plutonic igneous rocks consists of **diorite** and **granodiorite** (see Figure 6-4) accompanied by only minor amounts of true granite. Despite this, the general term granitic has long been used to describe such suites of intermediate to felsic plutonic igneous rocks. In this book we will conform to this established usage, although the term granodioritic would be more appropriate. The important point is that, despite their limited exposure in modern arc systems, granitic intrusives (plutons) must be recognized as an important and integral facet of island arc geology. In fact, such systems would more properly be termed **volcano-plutonic** arcs, in recognition of the importance of plutonic (intrusive) igneous rocks in addition to the more obvious volcanic rocks.

Another, and particularly intriguing, characteristic of island arc magmatism is that, within any single arc system, the composition of the igneous rocks exhibits progressive changes with respect to both time and space. In 1960 H. Kuno observed that the composition of basalts erupted in recent times in Japan exhibited a progressive change across the islands. Knowing of the presence of an inclined Benioff zone beneath the Japan Arc, he attributed these minor differences in the basalt compositions to differences in the depth at which the original basalt magma was generated near the Benioff zone. As more and more chemical data on volcanic rocks along island arcs became available, other investigators were able to demonstrate similar progressive changes in the composition of arc andesites.

Andesites (Figure 7-5) are one of the most important volcanic rock types in island arc systems, and an empirical relationship between the potassium content of specific andesites and the depth to the Benioff zone directly below is demonstrable in nearly all volcanic island arc systems. This surprising relationship is illustrated diagrammatically in Figure 7-6. The interdependence of K (potassium content of andesites) and h (depth to Benioff zone) is of major importance to an understanding of island arc magmatism, for it indicates that there is a *causal,* as well as *spatial,* link between the subduction process and the generation of the magmas that build volcanic island arcs.

7.5. *Handspecimen of andesite. Note black phenocrysts of pyroxene in fine grained matrix. [Photo by M. Kauper.]*

⌐ 5 cm ¬

In addition to systematic variation in the composition of arc igneous rocks with respect to space, time-related variations in the composition of arc igneous rocks have also been identified. In modern arc systems only fragmentary evidence for this time-related change in the composition of magmas can be obtained. Added perspective on this feature of arcs has come, surprisingly enough, from some of the oldest rocks known.

In virtually all the ancient (ages 3.3–2.5 billion years) nuclei (shields) of the continental areas, highly distinctive discrete belts containing predominantly volcanic assemblages are found. These **greenstone belts,** as they are generally called, contain assemblages up to 10,000 m thick of volcanic rocks, which closely parallel the composition of many of those found in modern island arc systems. Due to a combination of erosion level and deformation, rather complete sequences of these ancient volcanics are exposed for study by geologists. As a result, it has been possible to reconstruct the sequential evolution of such volcanic assemblages (Figure 7-7).

Many workers now consider these ancient assemblages to be analogs of modern island arc systems. As mentioned earlier, such reconstructions are simply not possible in modern arc systems be-

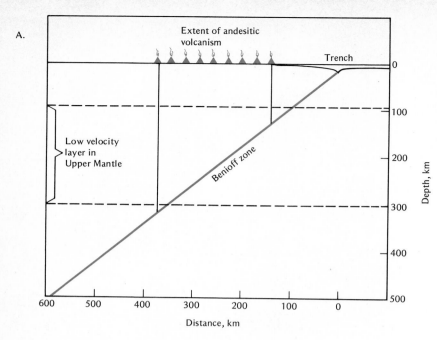

A.

Extent of andesitic volcanism

Trench

Low velocity layer in Upper Mantle

Benioff zone

Depth, km

Distance, km

7-6. A. *Diagrammatic vertical cross section to illustrate relationship of Benioff zone to andesitic volcanism in arch-trench systems.* B. *Variation of K_2O content of andesites containing 55% SiO_2 from various modern island arc systems in relationship to depth to center of underlying Benioff zone in each case. [Modified from T. Hatherton and W. Dickinson: The relationship between andesitic volcanism and seismicity in Indonesia, the Lesser Antilles, and other island arcs.* Jour. of Geophys. Res., 74: pp. 5301−5309 (1969).]

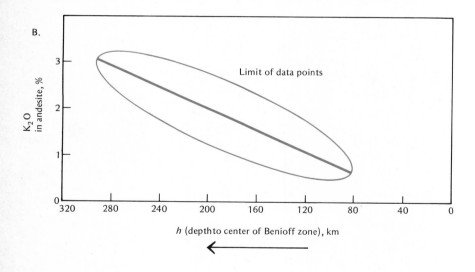

B.

K_2O in andesite, %

Limit of data points

h (depth to center of Benioff zone), km

cause so much of their volume lies below sea level. A more detailed discussion of greenstone belts and their related intrusive rocks can be found in Chapter 8. In general terms, a gradual time-related change in composition of magmas in arc systems, from basaltic to andesitic to dacitic and even rhyolitic, is now considered to be well documented.

7-7. *Generalized and highly schematic representation of the main stages in the evolution of an early Precambrian volcanic-sedimentary assemblage. Note that stages 2, 3, and 4 are essentially contemporaneous because, during construction of the felsic edifice, plutonic igneous rocks of similar composition to the volcanics are being emplaced and uplift and folding is occurring. As soon as the uppermost parts of the assemblage emerge above sea level, denudation and sedimentation will begin. [Modified from A. Goodwin: Evolution of the Canadian Shield. Geol. Assoc. Can. Proc., 19: [pp. 1–14 (1968).]*

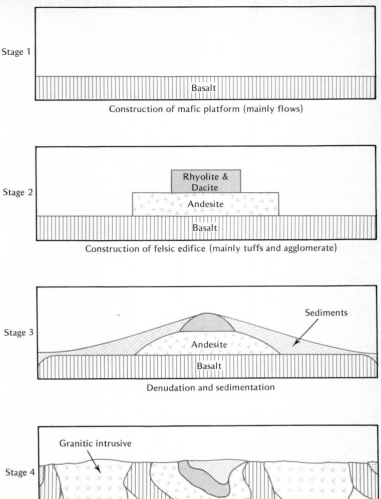

Stage 1

Basalt

Construction of mafic platform (mainly flows)

Stage 2

Rhyolite & Dacite

Andesite

Basalt

Construction of felsic edifice (mainly tuffs and agglomerate)

Stage 3

Sediments

Andesite

Basalt

Denudation and sedimentation

Stage 4

Granitic intrusive

Intrusion and folding

Generation of the Magma

Thus far we have attempted to demonstrate that volcano-plutonic island arcs are formed in oceanic areas by magmas whose generation is in some way connected with the underthrusting of a lithospheric plate along a Benioff zone. Next we must consider the processes by which arc-related magmas are generated.

First, what are the rates of magma production in arc environments? This is a difficult question to answer because it is virtually impossible to estimate with any degree of confidence the volumes of magma involved in building an island arc system. The volumes involved are unquestionably large. For example, studies of the Kurile Arc by Russian geologists have shown that, even in the small fraction of this arc system that lies above sea level, an area of 5345 km² has been covered by lavas during the last million years. The extent of submarine volcanism in the area during this same period can only be guessed at.

Detailed studies of the young volcanic rocks of the Kurile Arc have shown that andesite is the major rock type. Furthermore, although basalt, dacite, and rhyolite are all present in considerable quantity, the average composition of all the volcanics is andesitic. There is also a well defined increase in the potassium content of the volcanic rocks from east to west across the Kurile Arc, so that the K-*h* relationship discussed earlier holds true in this arc system. What then is the mechanism by which these very large volumes of andesitic and related magmas are produced?

From analysis of plate motions we know that the rate of underthrusting (subduction) of the Pacific Plate below the Kurile Arc is approximately 8 cm/year. From our discussion in the previous chapter we know that the upper 7 km of the Pacific Plate undergoing subduction must consist predominantly of material of basaltic composition. As this material moves deeper and deeper into the mantle as a result of the subduction process, it will be heated by friction along the Benioff zone and by transfer of heat from the adjacent hot mantle. Eventually, this basaltic material will be unable to adjust to the increased temperature regime by merely undergoing solid-state mineralogical transformations, and partial melting will be initiated.

As explained in Chapter 6, a magma will form from the constituents in the rock that melt at the lowest temperatures. The newly formed magma will be less dense than the surrounding solid material and will thus have a tendency to rise. Upward movement will only be initiated, however, when sufficiently large volumes of magma have been produced for their bouyancy to overcome the strength of the immediately overlying mantle. From chemical and experimental considerations, it can be demonstrated that a magma generated by a 40% partial melting of rocks of basaltic composition will have the composition of andesite. Smaller degrees of partial melting will generate more felsic magmas, with compositions equivalent to dacite or even rhyolite. Thus, the most likely source of the magmas of andesitic and more felsic composition that form island arc systems appears to be partial melting of oceanic crust in the vicinity of a Benioff zone (Figure 7−8). Rock types more felsic than andesite are probably

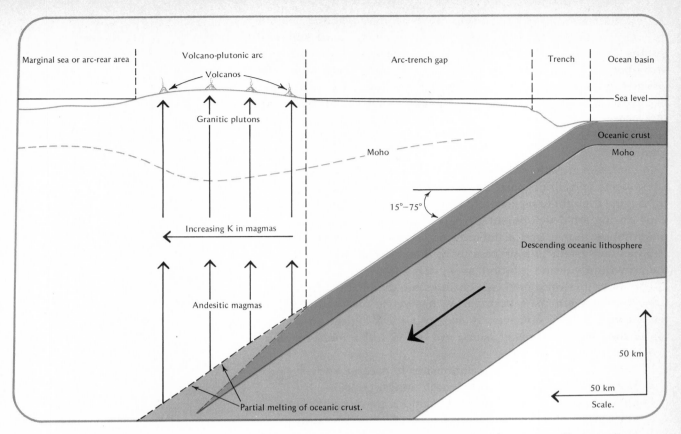

7-8. *Generalized cross section of an arc-trench system to illustrate the concept of generation of arc magmas by partial melting of oceanic crust subducted at a convergent plate boundary. Note that for clarity the basaltic magmas generated from the mantle overlying the Benioff zone have been omitted from the diagram. [Modified from J. Dewey: A model for the lower Paleozoic evolution of the southern margin of the early Caledonides of Scotland and Ireland. Scot. Jour. Geol., **7**: 219–240 (1970).*

created by the processes of fractional crystallization discussed in Chapter 6.

Hypothetical Model Linking Magmatism and Subduction

A rough calculation will serve to demonstrate that partial melting of subducted oceanic crust indeed provides a more than adequate mechanism to account for the volumes of andesitic magmas, both volcanic and plutonic, involved in building arc systems. Consider an arc system related to a convergent plate boundary where the rate of convergence is 5 cm/year. This means that over a period of a million years a strip of oceanic crust (7 km thick) 50 km wide will reach depths at which partial melting is taking place. If 20% of this material becomes partially molten and the resultant magma moves upward into the overlying arc system, it can be calculated that 70 km^3 of magma would be added to each linear kilometer of arc every million years. If the width of the arc system were 100 km, then the arc would grow in thickness by 0.7 km per million years over its entire width.

In fact, such growth rates are in excess of those observed in modern arc systems and do not take into account the large volumes of basaltic magma thought to be generated from the mantle above the Benioff zone during arc development. These considerations indicate that, in quantitative terms, partial melting of oceanic crust deep within a subduction system is a process more than adequate to account for the growth of volcano-plutonic arc systems.

This hypothetical model, which links arc magmatism to subduction at convergent plate boundaries, has yet to receive full experimental verification; however, it does provide a logical explanation for all the major features of island arc systems. The model explains

1. Their spatial and parallel relationship to ocean trenches.
2. Their spatial relationship to Benioff zones.
3. The volume and composition of island arc igneous rocks.
4. The empirical K-*h* relationship.
5. The arc-trench gap and the heatflow patterns in island arc systems.

Nevertheless, it would be erroneous to leave the impression that this model for island arc magmatism is either fully accepted by all geologists or without attendant problems. For example, it is not yet clear to what extent the sedimentary uppermost layer of oceanic crust becomes involved in the subduction process. The average composition of this sedimentary crust is markedly different from that of basaltic oceanic crust, and the presence or absence of this material at the site at which magma generation is taking place would have a measurable effect on the composition of partial melts formed close to the Benioff zone. Unfortunately, trace element and isotopic studies of igneous rocks in island arcs have as yet failed to clarify this point. Despite these uncertainties, the overall aspects of island arc development and the close relationship of such arcs to convergence of plates in oceanic areas are satisfactorily accounted for by the model presented.

7-9. *Sketch to illustrate concept of two-stage fractionation of mantle material to form a volcano-plutonic arc. In this manner original mantle material, which is low in silica, alumina and alkalies, is double distilled by processes involving partial melting to produce crust of continental thickness and composition.*

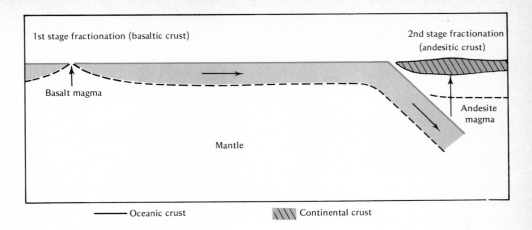

Two-Stage Fractionation.

In Chapter 6 we categorized the production of basaltic oceanic crust at spreading ridge systems as a first-stage fractionation of mantle material involving partial melting and uprise of magma. The production of a volcano-plutonic arc of andesitic composition thus represents a second stage, or refractionation, of a mantle "distillate" (oceanic crust) that also involves partial melting and magma uprise (Figure 7-9). The important point is that andesitic arcs have an average composition very similar to estimates of the average composition of continental crust. Thus we have evidence that a two-stage fractionation process involving plate tectonic mechanisms is capable of creating material of continental crustal composition from primary mantle material. Although the above represents a gross oversimplification, we consider the model of two-stage fractionation an important and fundamental concept in understanding the outermost portions of the earth.

SEDIMENTATION IN ISLAND ARC SYSTEMS

To this point we have restricted our discussion to the magmatic events that occur during island arc development. Sedimentary processes also play an important part in the formation of island arc systems. During the earliest stages of the growth of an island arc, the arc complex is gradually built up from the sea floor by submarine eruptions of basalts and andesites. Once growth has proceeded to the stage that volcanic islands appear, sedimentary processes start to play an important role.

Sources of Sediment

The fragmental volcanic material of these new-born oceanic islands undergoes rapid erosion, especially in the high energy coastal environment. Waves constantly batter these new impediments to their progress across the oceans. As a result, aprons of sedimentary material produced by shoreline erosion spread out below sea level around the volcanic islands. With continuing magmatic activity, the degree of emergence of the arc system increases, due both to construction of more volcanos and uplift caused by emplacement of granitic plutons within the arc system.

As the arc system grows, through continued magmatic activity and related uplift, plutonic igneous rocks in the core of the arc will be unroofed and also subjected to the ever-present processes of subaerial erosion. Therefore, during the later stages of island arc development, minerals derived from the weathering of plutonic igneous rocks will also become components of arc sediments. Thus, in contrast to the earlier sedimentary sequences, which are dominated by material of volcanic derivation, the younger sediments tend to contain a significant component of plutonic derivation.

Studies of modern island arc systems by marine geologists have indicated that the arc-trench gap is an important site of sediment accumulation (Figure 7-10). Because of the considerable distance

7-10. *Echo-sounding profile across the Marianas Arc system showing various areas of sediment accumulation. Note sediment fill in arc-trench gap and in behind-arc area. Sediment free area in center is due to splitting of arc and creation of new crust (see Figure 7-12). [From D. Karig: Structural history of the Mariana Island Arc system. Geol. Soc. of Amer. Bull,* **82:** *pp. 323–344 (1971).]*

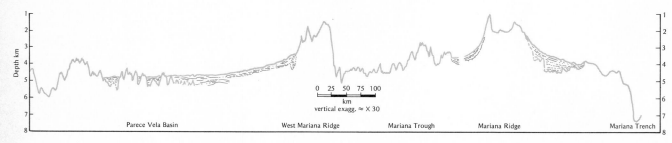

from the island arc to the trench, in most cases only minor amounts of arc-derived sediments actually reach the trench. Most of this material is trapped in subsidiary basins caused by local uplifts and downwarps between the island arc and the trench. These thick sedimentary sequences tend to contain in their rock and mineral fragments a sequential record of the progressive erosion of the magmatic arc complex.

Mechanism of Sedimentation

The large bulk of the sedimentary rocks consists of gray, poorly sorted sandstones that contain many small angular fragments of fine-grained volcanic rock obviously derived from the arc system. Sandstones of this type are called **graywackes.** One important feature of graywackes is that they commonly exhibit graded bedding (Figure 7-11); that is, each individual bed contains a gradation in particle size from coarsest at the base to finest at the top.

Graded bedding in a sediment results from the deposition of material of mixed grain size by a current of turbulent, particle laden water whose velocity is diminishing. As this takes place, the carrying capacity of the current will decrease and the coarsest material will settle out first. This process continues until the current even-

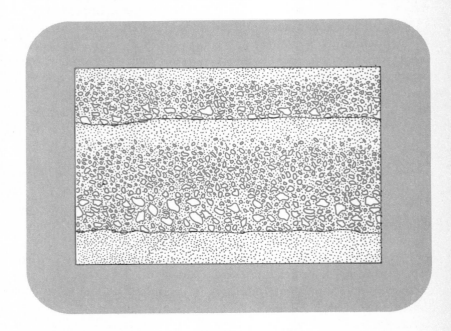

7-11. *Sketch to illustrate graded bedding in sedimentary units. Note well-defined break between top of one graded bed and base of overlying graded bed.*

tually ceases, and a sedimentary bed that grades from coarse material at the base to fine at the top has been laid down. The presence of graded bedding in graywacke sequences initially puzzled geologists because it indicated the action of strong currents during deposition, whereas most other features of graywackes indicate deposition in a deep water environment where strong currents were unknown.

This apparent contradiction was solved by the recognition of the key role played by **turbidity currents** in the deposition of graded sedimentary beds. Turbidity currents represent the subaqueous downslope movement of dense sediment laden (turbid) water. In the island arc setting, turbidity currents will be generated at sporadic intervals by the downslope submarine slumping of sedimentary material that has accumulated below the wave base on the flank of volcanic islands. The resulting dense slurry of sand and mud will move down even very small gradients until its progress is interrupted by some barrier or sediment trap present in the submarine topography. Obviously, the high incidence of earthquakes in island arc environments must aid in the initiation of the slumping that generates turbidity currents.

Marine geologists studying sediments in modern arc-trench gap areas have shown that the patterns of movement of turbidity currents are controlled by submarine topography, and paths of sediments transport from the island arc ridge outward are complex in detail. It is now known that very thick deposits of sediments (up to 15 km) can accumulate in the manner described above, although this can happen only if local subsidence is taking place in an area of sediment accumulation. The sedimentary rocks that are found in Precambrian greenstone belts consist mainly of graywacke sequences that exhibit close similarities to those from modern island arc systems. This provides further support for the concept that greenstone belts are ancient analogs of modern island arc systems.

Sedimentation Patterns

As stated earlier, sediment will accumulate and form aprons on both sides of the growing linear arc system once parts of it emerge above sea level. Sedimentation patterns on the inner side of the arc system are essentially similar to those pertaining in the arc-trench gap, but some complications exist. Many oceanic island arcs are separated from continental margins by marginal seas of relatively limited extent (see Figure 7-1). In these marginal seas, sediment derived from the arc system can reach the central parts of the mar-

Study of the patterns of sedimentation in young double arc systems such as the Marianas Arc and the Tonga-Kermadec Arc led initially to the formulation of the concept. Geophysical support comes mainly from seismic data and high heat flow observed in interarc basins or the marginal seas behind many island arc systems.

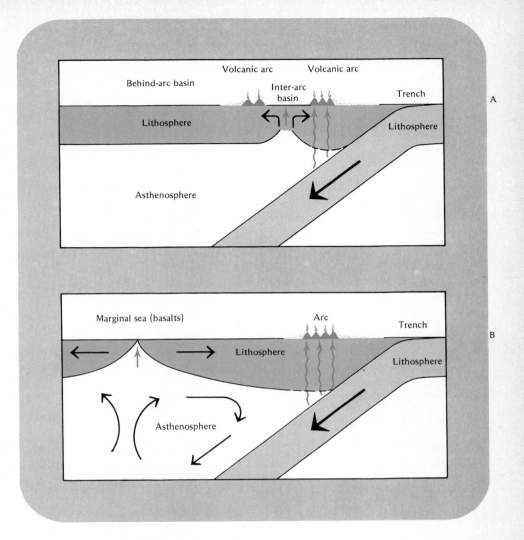

ginal sea basin and interfinger with sediments derived from the continent. As a result, complex sequences of sediments of contrasting type can accumulate.

Another complicating factor that can disrupt original sedimentary and magmatic patterns is a phenomenon known as behind-arc spreading (Figure 7-12). This phenomenon, the causes of which are as yet poorly understood, results from the development of local spreading center systems behind, or even actually within, young island arc systems. Despite these complications, the sediments found in young arc systems and the sedimentary rocks associated with older and ancient arc systems show abundant evidence in their texture and mineralogy of a limited sojourn in subaerial environments. They are predominantly immature, first cycle derivatives of the igneous materials from which the island arc system was initially built.

METAMORPHISM IN ISLAND ARC SYSTEMS

The model presented thus far deals with magmatic and sedimentary patterns of island arc geology. With the exception of the emplacement of plutonic igneous rocks, these are processes that occur in surficial and submarine environments. What kinds of processes are occurring at deeper levels within the arc system, several kilometers or more below the surface? Direct answers to this question obviously cannot be obtained from very young systems but must be sought in the more deeply eroded terranes of older arc systems, such as Japan. Plate convergence and subduction has apparently been

7-13. Cross-sectional sketch to illustrate the major environments of metamorphism in arc-trench systems. Distinction between low temperature (trench) and high temperature (arc) types is clear cut. Low pressure–high temperature and high pressure–high temperature environments within arc are transitional and merely dependent on depth.

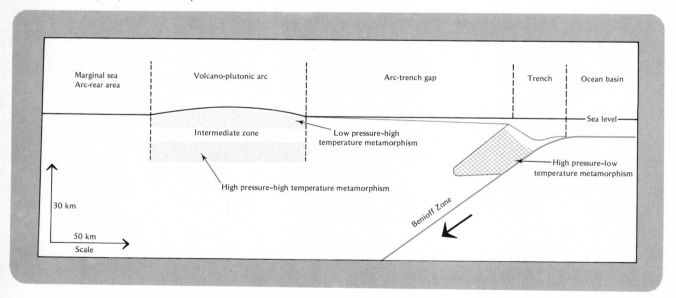

active along the Japanese Arc since Triassic time, and in many areas sufficient uplift and erosion have taken place to expose plutonic rock suites. These consist largely of plutonic intermediate and felsic igneous rocks and *metamorphic rocks.*

Generation of Metamorphic Rock

Metamorphism, the sum total of the processes by which metamorphic rocks are generated, involves the mineralogical and textural reordering of rocks subjected to burial in the earth's crust. As we saw in Chapter 2, all bodies of rock tend to change their mineralogic and textural character when subjected to new sets of physical and chemical conditions, whether within the earth or at its surface.

At convergent plate boundaries two quite distinct environments of metamorphism can be identified. One is characterized by high pressure and low temperature; the other is characterized by high temperature and varying pressures. Each environment of metamorphism occurs at an opposite side of the arc-trench gap.

High-Pressure Low-Temperature Metamorphism

The high-pressure low-temperature environment develops at the outer margin of the arc-trench gap (Figure 7-13). Below the inner wall of trench systems, oceanic sediment, pillow lavas, and generally some arc-derived graywackes tend to become complexly intermixed and be dragged downwards along the top of the Benioff zone. In this dynamic setting, cold oceanic crust and underlying plate material are moving downward due to subduction at rates of between 5 and 15 cm/year. These rates of movement are far in excess of the rate at which heat can be conducted through rocks. The net result is a considerable depression of isotherms (surfaces of equal temperatures) in this type of environment (Figure 7-14).

Blueschists. The products of metamorphism in such environments are thus characterized by high-pressure low-temperature mineral assemblages that are highly distinctive. They are typically dense, rather fine-grained schistose rocks that have a bluish color due to the presence of the mineral glaucophane. These blueschists, as they are commonly called, contain mineral assemblages indicative of high pressure and low temperature. Laboratory experiments on the miner-

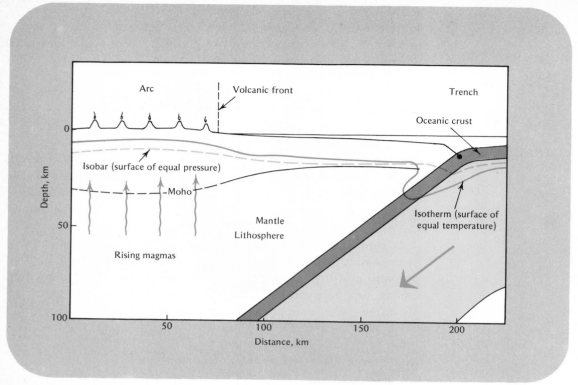

7-14. *Diagram to illustrate concept of depression and elevation of isotherms in an arc-trench system. Note marked departure of isotherms (surfaces of equal temperature) from an isobar (surface of equal pressure).*

als commonly present in these assemblages indicate that they have formed at pressures corresponding to depths of up to 25 km and temperatures that did not exceed ~300°C. This is a highly distinctive environment because normal geothermal gradients (rates at which temperature increases with depth) are known to be at least 20°C per kilometer; thus, at depths of greater than 15 km, temperatures in excess of 300°C would be expected.

Such high-pressure low-temperature metamorphic assemblages have been recognized and studied in many areas along the circum-Pacific belt and are also present in the Alps. No contemporaneous igneous rocks are associated with these blueschist terranes. However, in nearly all instances they are associated with numerous fragments, in some cases extremely large sheets (ophiolite complexes), of earlier-formed mafic and ultramafic igneous rocks chaotically mixed with oceanic sediments and arc-derived graywackes. There seems to be little doubt that these mafic igneous rocks represent oceanic crustal material and that blueschist metamorphic assemblages must record sites at which subduction has taken place.

One question has long puzzled students of these rocks. How did metamorphic rocks, formed at depths of up to 25 km in relatively young arc systems, undergo sufficiently rapid uplift to be now exposed at the surface? This problem has been clarified largely by the recognition that blueschist terranes are invariably bounded on one side by faults of large magnitude, although it is unclear by what mechanism uplift of these dense rocks can take place. Finally, it should be emphasized that blueschist metamorphic terranes represent a very small fraction of the metamorphic rocks exposed at the earth's surface and are virtually unknown in rocks of Precambrian age. In all probability, this is because such metamorphic assemblages are sensitive to temperature and, if ever heated much above 300°C, lose their distinctive mineralogy.

High Temperature Metamorphism

The high temperature environment develops on the inner margin of the arc-trench gap, within the arc itself. Here very different metamorphic assemblages are formed. Within the magmatic arc the temperature regime in the crust is such that relatively high temperatures occur at shallow depths (see Figure 7-14). This shallow high-temperature regime is the result of the introduction of heat by the magmas involved in the development of the volcano-plutonic arc. At deeper levels high temperatures will also be present, but the pressures will be high. Consequently, within arc systems proper, the volcanic and sedimentary rocks in this environment tend to be metamorphosed to mineral assemblages indicative of high temperature. Depending on the depth at which metamorphism occurs, the assemblages will vary from low pressure–high temperature to high pressure–high temperature types.

In detail, metamorphic terranes associated with the deeper parts of volcano-plutonic island arc systems are exceedingly complex, and much work remains to be done on them. In their attempts to study the types of metamorphism that occur within the deeper parts of island arc systems, geologists must, of necessity, limit themselves to those arc systems that have undergone sufficient uplift and subsequent erosion to lay bare their inner portions. Despite these complicating factors, the high temperature metamorphic belts in island arc systems can be demonstrated to be intimately associated with plutonic magmatism. In broad terms, the high temperature metamorphic areas can probably best be envisaged as elongate envelopes of metamorphic rock surrounding the numerous granitic intrusive bodies that make up the deeper parts of an island arc system.

Geologists who study metamorphic rocks have long realized that enormous amounts of thermal energy are required to convert sediments and volcanics of regional extent from their original condition into high-grade metamorphic rock assemblages. Clearly, the requisite heat must have been supplied from below, but by what physical means was it delivered to the site of metamorphism? The conduction of heat through rocks is so slow that thermal conduction of heat from deeper regions is simply inadequate to account for the geometry of many metamorphic belts.

The solution to this problem lies in the close association of high temperature metamorphic belts with plutonic intrusives. The emplacement and subsequent cooling of such intrusives apparently delivers enough thermal energy to the surrounding environment to cause the regional metamorphism of large volumes of the surrounding rocks. Whether a similar model can be proposed for the extensive metamorphic terranes in the Precambrian portions of the continents will be discussed in Chapter 8.

OROGENESIS IN ISLAND ARC SYSTEMS

Orogenesis is a general term used by geologists for processes by which mountain structures develop. Orogenic processes in most instances involve the folding, deformation, and uplift of large volumes of rock to produce generally elongate mountain belts. Island arc systems might strike the reader as distinctive from mountain belts such as the Andes and the Himalayas, which attain high elevations above sea level, but island arc systems do attain similar elevations above the floors of the ocean basin adjacent to them.

Folding

Where orogenesis has taken place, layered rocks such as sediments and volcanics are commonly found to be folded to varying degrees of intensity. (Figure 7-15A, B, and C). Individual folds can vary in

7-15. A. *Large fold structures developed in layered sedimentary rocks.* [*Courtesy of Geological Survey of Canada.*] B. *Fold structures in arc-related sedimentary rocks.* Note hammer for scale. [*Courtesy of P. Hudleston.*] C. *Small scale fold structures. Coin is about 2 cm in diameter.* [*Courtesy of P. Hudleston.*]

size from microscopic to structures involving large volumes of rock and extending in length for hundreds of kilometers. The presence of folds in rocks indicates that they have undergone permanent deformation without rupture (plastic flow) in response to applied forces. Under surface conditions, many rocks are brittle and will tend to rupture rather than undergo plastic flow when subjected to stress. Within the crust, however, under the influence of forces acting for long periods of time, rocks will tend to yield by folding. Many types of fold structures occur in deformed rocks; the most common of these are illustrated in Figure 7-16.

The presence of folds in a layered sequence of rocks indicates a

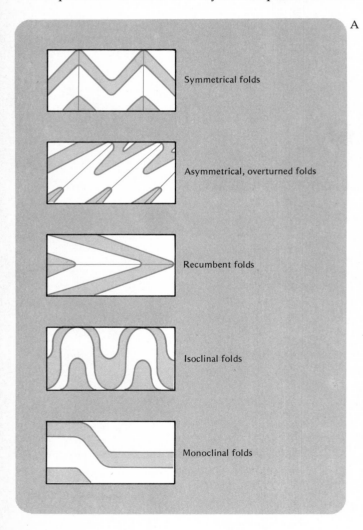

Symmetrical folds

Asymmetrical, overturned folds

Recumbent folds

Isoclinal folds

Monoclinal folds

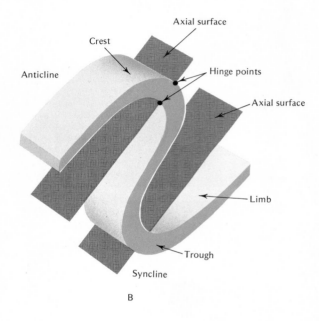

A

Crest

Axial surface

Anticline

Hinge points

Axial surface

Limb

Trough

Syncline

B

7-16. A. *Diagrammatic illustration of cross sections of various types of folds. Fold types are defined and named by structural geologists to aid them in describing deformation patterns in folded rocks. From the types of folds developed in layered rocks, geologists can learn of the stress conditions that caused the folding. B. Terminology used to denote the various features of fold structures.*

net shortening and thickening of that sequence, which must be due to compressional forces. Until recently, many geologists believed that episodes of major crustal compression and shortening were the prime cause of folding. Much folding is now realized to be related to vertical movements in the earth's crust, caused mainly by the emplacement of large bodies of igneous rock. In some instances it is the actual emplacement that gives rise to horizontally directed forces; in others the vertical uplift, combined with the force of gravity, provides the forces that produce folding. This concept will be enlarged upon in Chapter 8.

In island arc systems the gross trend of the fold structures tends to parallel the long axis of the arc, and it is becoming increasingly clear that much of the folding and deformation associated with volcano-plutonic arcs must be related to the emplacement of large bodies of granitic rock in linear belts. Consequently, the deformation that occurs in island arc systems is, like the sedimentation and metamorphism, intimately related to the development of elongate volcano-plutonic complexes.

Other Deformation Patterns

Within the trench segment of an arc-trench system deformation patterns of a different kind are developed. Within and below the inner wall of the trench, the Benioff zone is expressed at the surface generally as a complex zone of reverse faults. In this environment, oceanic sediments, fragments of basaltic oceanic crust, and arc-derived graywackes become strongly sheared and intermixed into rock assemblages called **melanges**. Melange terranes are extremely complex mixtures of the rock types present in trench systems and metamorphic equivalents of these rocks, which have become tectonically jumbled together in chaotic fashion. Prior to the advent of plate tectonic concepts, geologists were unable to explain satisfactorily the origin of melanges.

For the sake of clarity, we have treated each process (magmatism, sedimentation, metamorphism, and orogenesis) separately in this chapter. In fact, all four of these important processes are going on simultaneously at different levels within the arc system during its growth. Although the geology of island arcs can be exceedingly complex in detail, the broad aspects of their form, rock associations, and structure can be readily understood in terms of plate interactions at convergent plate boundaries.

RELATION OF ISLAND ARC SYSTEMS TO FORMATION OF CONTINENTAL CRUST

The reader might be puzzled at the emphasis that has been placed on the geology and genesis of island arcs, which at the present time represent strings of generally small and remote islands within the ocean basins. The significance of island arcs lies in the observation that, during their growth, new crust comparable in composition and thickness to continental crust is apparently formed in places where no continental crust existed before (Figure 7-17). This, as we shall see, is an important clue to the origin of the continents.

Continental crust, once formed, is apparently incapable of undergoing subduction due to its bouyancy. The newly created segments of continental crust are, according to the plate tectonic model, fated to ride the conveyor belts of plate movement until they eventually collide with other segments of continental crust. These may be either different active island arc systems or crustal segments of continental size. In general, these collisions will be accompanied by the imprinting of yet more structural complexity on the original arc system.

In this chapter we have seen that the theory of plate tectonics

7-17. *Diagrammatic illustration of thickening of crust under Japan. Note that crustal root under Japan is of continental thickness (~30 km).*

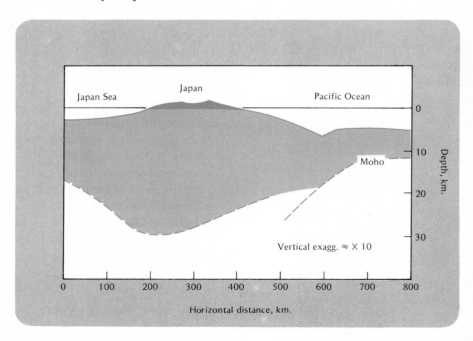

provides us with a frame of reference for understanding the dynamics of island arc formation. Nevertheless, enormous amounts of work lie ahead for geologists before meaningful reconstructions of the geological history of many ancient arc systems, now locked into continental interiors, can be made.

QUESTIONS

1. Why have most of the worst natural disasters associated with volcanic activity occurred in island arc systems?
2. Illustrate with diagrams the relationship between trenches, volcano-plutonic arcs, and associated geographic features.
3. Why is there always a distinct gap between the trench and arc elements of volcanic island arcs?
4. What is the K-h relationship? Why is it of significance with respect to island arc magmatism and plate tectonics?
5. Why have geologists been able to learn about island arc systems by studying ancient greenstone belts?
6. Explain why island arc magmatism can be thought of as a second stage of a two-stage fractionation of mantle material.
7. Greywackes are the most important sediment type associated with island arcs. Discuss their source, lithology and major mode of deposition.
8. Contrast the type of metamorphism that characterizes the area below the inner side of trenches, and that which characterizes the magmatic portion of an island arc system.
9. Why does the geology of island arcs provide a clue to the origin of continental crust?

The Geology and Evolution of Continents

In the last chapter we explored the hypothesis that subduction within an ocean basin results in a complex series of interrelated geological events, which together lead to the creation of an elongate segment of crust of continental type. As we have stressed, magmatic events related to subduction can be considered the second stage of fractionation in the two-stage process by which continental type crust is generated from mantle material.

According to the plate tectonic model, crust of continental type can also be formed when subduction occurs right at the edge of an ocean basin, that is, along a continental margin. Such areas are called active continental margins to distinguish them from inactive continental margins, where the transition from continental to oceanic crust is not marked by a plate boundary. Continental margins can be the sites of transform plate boundaries (for example most of the western coast of North America), but these are quite distinctive and are not to be confused with active continental margins as discussed in this chapter.

All continents have at least two fundamentally different geological environments: continental interiors and inactive continental margins. In addition to these, some continents may contain active continental margin environments (for example, western South America) or continental collision environments (for example, Himalayas, Asia) In this chapter we will examine the geological processes and events characteristic of each of these four environments.

The western edge of South America is the best-defined current example of an active continental margin. The presence of the Peru-Chile Trench, an active Benioff zone beneath the Andes, and numerous young volcanos in the Andes (Figure 8-1) all indicate that plate convergence and subduction are going on all along this coast. The geologic events that occur along the western margin of South America are essentially similar to those that characterize the development of island arc systems. The three major components of island arc systems, trench, arc-trench gap, and volcano-plutonic arc, are all present along this active continental margin and run essentially parallel to each other.

The late Tertiary volcanic rocks in the Andes contain progressively higher potassium in the east than the west, and this eastward increase in potassium content can be shown to correlate with increasing depth to the underlying Benioff zone. Thus the important K-h relationship, emphasized in the preceeding chapter, exists in the Andes. Furthermore, the intermediate and felsic igneous rocks from both volcanic and plutonic settings in the Andes exhibit strong similarities to their counterparts in island arc systems.

Magmatism at Active Continental Margins

When a plate consisting of continental lithosphere converges with a plate consisting of oceanic lithosphere, the resultant volcano-plutonic arc is emplaced upon and within continental crust. This fact accounts for most of the observable differences between the geology of active continental margins and the geology of island arcs, in which both converging plates consist of oceanic lithosphere.

One easily observed result is that the topographic expression of the two environments is markedly different. In the Andes, for example, the highest peaks attain elevations in excess of 6000 m above sea level, and extensive areas lie at elevations of greater than 4000 m. The mountain belts of Central America and western North America were also built along an active continental margin, although the situation at present is more complex.

Current continental margins along this sector that are now marked by convergence are largely limited to a section of Central America and the northern side of the Gulf of Alaska. Along nearly all the

Physiographic sketch of western South America and adjacent parts of the Pacific. The Andes form a well-defined elongate mountain chain close to the western margin of the continent. Note also close relationship of Peru-Chile Trench to the Andes chain. Strong vertical exaggeration.

remainder of this continental margin the boundary between the North American and Pacific Plates is marked by a series of transform faults. Analysis of former plate motions using sea floor magnetic anomaly patterns tells us that this entire belt must have been an active continental margin until relatively recent times. However, about 20 million years ago, collision of a spreading-center ridge system with a former trench system at the continental margin caused convergence to cease along much of this belt.

Another consequence of the high elevation of mountain belts formed at active continental margins is that the volcanic portion of the volcano-plutonic arc tends to be stripped off rapidly. Therefore, along the western flank of the Andes and in western portions of North America, large areas of plutonic igneous rocks are exposed at the surface (Figure 8-2). These plutonic igneous bodies, composed largely of granodiorites, diorites, and related rock types, represent huge volumes of igneous rock.

Origin of Batholiths

Detailed study of these large areas of plutonic igneous rock, called batholiths, has shown them to consist of many discrete intrusive bodies of predominantly intermediate composition that were emplaced at different times over intervals of time that spanned tens of millions of years. The elongate shape and the position of granitic batholiths with respect to subduction zones leaves little room for doubt that they represent the plutonic portions of volcano-plutonic magmatic arcs.

In places where erosion has reached just the right level, the field relationships between the plutonic granitic rocks and overlying volcanic rocks indicate that both were formed from essentially the same pulses of magma. Similar relationships between volcanic and plutonic igneous rocks can also be seen in island arcs where erosion has proceeded to the right level. In some instances it appears that plutonic igneous bodies of batholithic size have crystallized directly under a cap of volcanic rocks derived from the same magma.

The similarity in composition and tectonic setting between the plutonic igneous rocks of island arcs and those found in batholiths near continental margins seems to indicate that both types are the products of magmatism related to subduction. Apparently the magmas produced in conjunction with the subduction process can penetrate and rise through the overlying continental crust. One might expect that significant amounts of the continental crustal

8-2. *Map illustrating batholiths of western North and South America. Note elongate shapes of these bodies and their parallelism to continental margins. All are Mesozoic or Cenozoic in age and can be related in time to periods of subduction of oceanic lithosphere.*

rocks would be incorporated into the intruding magmas, altering their composition and eventual mineralogy. Isotopic studies, however, have indicated that in most instances the older continental crust has contributed very little to these magmas.

Prior to the development of theories of plate tectonics and subduction-related magmatism, the origin of granitic* rocks, and in particular granitic batholiths, was a highly controversial subject. Geologists were sharply divided on this matter. One group strongly maintained that all large granitic bodies were of igneous origin and had been emplaced as magmas (magmatic theory). Another group

---

*The term **granitic** is used in deference to common usage. It should be clearly understood, however, that the composition of most granitic rocks and batholiths more closely approximates granodiorite than granite.

maintained, equally strongly, that all large granitic bodies were the result of in-place recrystallization of preexisting rocks, aided by the migration of chemically active solutions (granitization theory). Both groups were able to amass a considerable body of field data to support their particular theory. The data supportive of a magmatic origin for granitic rocks (for example, sharp contacts of intrusive against country rock) came mainly from granitic bodies of Mesozoic age (see Figure 8-2). The data supportive of a nonmagmatic origin for granitic rocks (for example, diffuse contacts between granitic rock and surrounding high grade metamorphic rocks) came mainly from deeply eroded Paleozoic and Precambrian terranes. The controversy between the two schools of thought produced the most bitter and heated debates ever witnessed in geology.

It is now generally accepted that most granitic rocks are of magmatic parentage, but most geologists accept the idea that, marginal to intrusive granites and in high grade metamorphic terranes, a certain amount of granitization can occur, especially at deeper crustal levels.

Another important controversy regarding granitic rocks has centered on whether they are of crustal or subcrustal origin. The large bulk of granitic igneous rocks occur in well-defined belts characterized by thick accumulations of volcanic and sedimentary material. These are also the sites of extensive deformation and metamorphism. Conscious of the problem of deriving large volumes of granitic magma by differentiation of subcrustal ultramafic material, many geologists favored the concept that granitic magma was produced by the melting of deeply buried sedimentary rocks.

The whole process of granite emplacement and orogeny in mobile belts was envisioned as a series of events initiated by the development and filling of a **geosyncline**. The concept of the geosyncline evolved to account for the thick elongate sedimentary sequences of regional dimensions that accumulate in certain areas. The later discovery of oceanic trenches strengthened the geosynclinal concept, for the trenches were visualized as typical unfilled geosynclines.

The advent of the theory of plate tectonics has led to a reevaluation of the concept of geosynclines, and a considerable restriction in the use of the term by many geologists. It has also largely clarified the relationships between magmatic arcs, thick accumulations of sediments, metamorphism, and orogenesis (see Chapter 7). Furthermore, studies of the strontium isotopic chemistry of many granitic rocks have indicated that they could not have formed by melting of preexisting sediments in crustal environments. Some granitic rocks, however, have clearly been formed as the consequence of melting of deeply buried sediments.

Granitic igneous rocks are widely distributed in Precambrian shield areas, especially along the margins of greenstone belts. In much the same way that the volcanic rocks in greenstone belts resemble volcanic rocks in modern arc systems (see Chapter 7), so many of the granitic intrusives associated with greenstone belts exhibit compositional ranges, textures, and isotopic chemistry similar to granitic intrusives of island arcs. Furthermore, the form and structural configuration of greenstone belts appear to have been determined mainly by the diapiric (see Chapter 6, page 000) emplacement of large granitic bodies along their margins (Figure 8-3).

Careful radiometric dating of the granitic intrusives and the volcanic rocks in greenstone belts has indicated that both are closely associated in time as well as space. Thus, both the plutonic and volcanic components of these probable ancient arc systems are apparently preserved in Precambrian shield areas. Much work remains to be done before the formation of these complex, ancient shield areas is well understood. Nevertheless, the origin of large bodies of granitic intrusive rock is now better understood than before, although further, broadranging studies are needed.

Sedimentation at Active Continental Margins

The fact that the mountain belts formed at active continental margins tend to stand high with respect to sea level strongly influences erosion and sedimentation patterns associated with such belts. Much of the arc-trench gap might lie above sea level and be subject to the processes of erosion, rather than acting as a sediment trap. As a result the supply of sedimentary material from the mountain belt can be sufficient to fill the trench completely.

The Andean Mountain–Peru-Chile Trench system provides an excellent example. Here the trench is a well-defined submarine topographic feature off the arid coast of Peru and northern Chile, one of the driest areas on earth. Erosion is slow and the rivers that drain the western flank of the Andes in this area do not carry major volumes of sediment into the sea. In southern Chile, however, rainfall is high, erosion more rapid and potent, and, as a result, the trench as a topographic feature of this section has been practically obliterated by sediment fill (Figure 8-4).

Erosion of the mountain belts formed at active continental margins

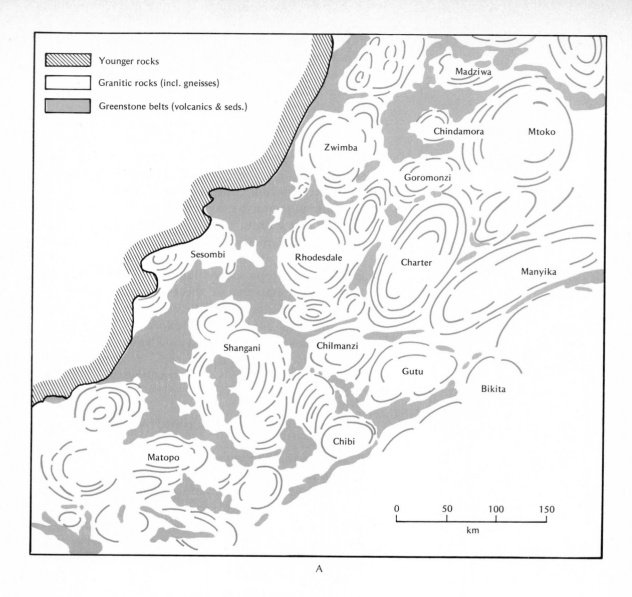

Younger rocks

Granitic rocks (incl. gneisses)

Greenstone belts (volcanics & seds.)

Madziwa

Chindamora

Mtoko

Zwimba

Goromonzi

Sesombi

Rhodesdale

Charter

Manyika

Shangani

Chilmanzi

Gutu

Bikita

Chibi

Matopo

0   50   100   150

km

A

leads to the formation and transport of large volumes of sediment. Some of this material will be trapped in intermontane basins, but the bulk will be shed off in two directions: outwards toward the oceanic side of the belt or inward toward the continental side. Probably the best studied example of sedimentation and tectonic patterns on the ocean side of a mountain belt formed at an active continental margin is provided by the geology of north-central California (Figure 8-5). This area has three distinctive geologic units exposed in parallel bands, each of which formed in a distinct en-

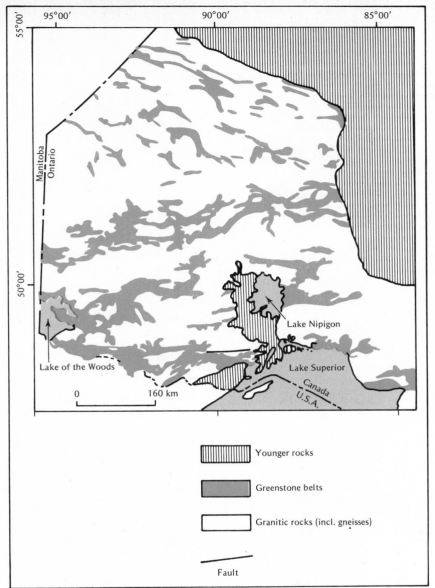

8-3. *Two maps showing relationship of greenstone belts to surrounding areas composed of granitic intrusive and gneisses. The greenstone belts consist of infolded screens between domelike intrusions of granitic rock. A. A portion of the Precambrian Shield exposed in Rhodesia. [From A. MacGregor: Some milestones in the Precambrian of Southern Africa. Trans. Proc. Geol. Soc. S. Afr., 54: pp. 27-71 (1951).] B. A portion of the Precambrian Shield exposed in Ontario, Canada.*

Lake Nipigon

Lake of the Woods

Lake Superior

Canada

U.S.A.

0   160 km

Younger rocks

Greenstone belts

Granitic rocks (incl. gneisses)

Fault

B

vironment. From east to west the three units are the Sierra Nevada Batholith, the Great Valley sequence, and the Coast Ranges.

The Mesozoic Sierra Nevada Batholith is surrounded by an envelope of metamorphosed volcanic and sedimentary rocks, which together with the batholith form a typical magmatic arc assemblage.

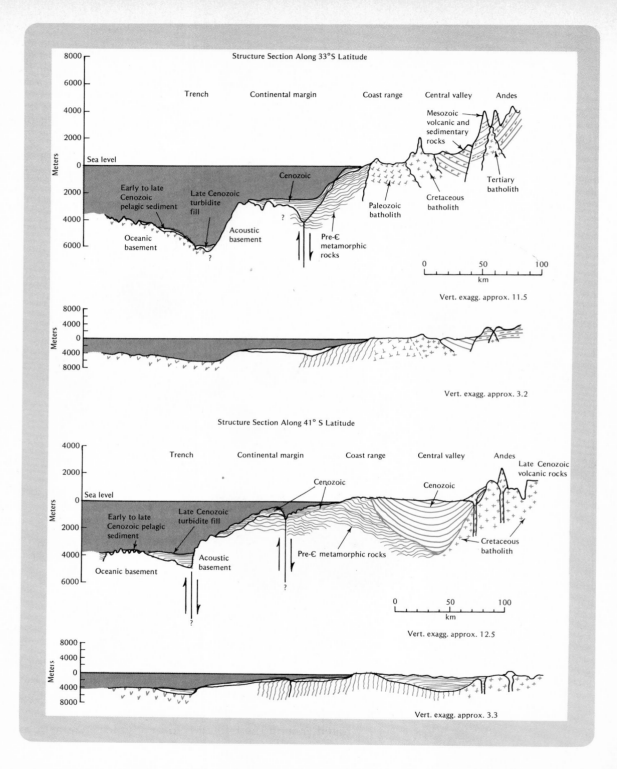

Structure Section Along 33°S Latitude

Trench   Continental margin   Coast range   Central valley   Andes

Mesozoic volcanic and sedimentary rocks

Sea level

Cenozoic

Early to late Cenozoic pelagic sediment

Late Cenozoic turbidite fill

Oceanic basement

Acoustic basement

Pre-Є metamorphic rocks

Paleozoic batholith

Cretaceous batholith

Tertiary batholith

0   50   100
km

Vert. exagg. approx. 11.5

Vert. exagg. approx. 3.2

Structure Section Along 41° S Latitude

Trench   Continental margin   Coast range   Central valley   Andes

Late Cenozoic volcanic rocks

Sea level

Cenozoic

Cenozoic

Early to late Cenozoic pelagic sediment

Late Cenozoic turbidite fill

Oceanic basement

Acoustic basement

Pre-Є metamorphic rocks

Cretaceous batholith

0   50   100
km

Vert. exagg. approx. 12.5

Vert. exagg. approx. 3.3

8-4. (OPPOSITE) *Idealized structural cross sections from Pacific sea floor to crest of the Andes, central Chile. Structure of offshore area based on interpretation of acoustic (sound) reflection profiles. Note strong vertical exaggeration (upper sections) and rather gentle slopes when this exaggeration is reduced (lower sections).* [From D. Scholl et al.: Peru-Chile Trench sediments and sea floor spreading. Bull. Geol. Soc. Amer., 81: pp. 1339-1360 (1970).]

8-5. *Simplified map and cross section of the geologic relationships between inferred arc, arc-trench gap, and trench assemblages in north-central California. Note parallelism of three assemblages on map.*

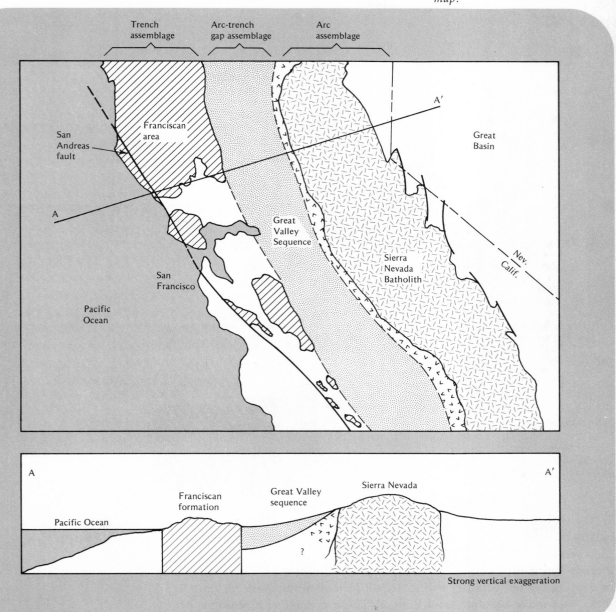

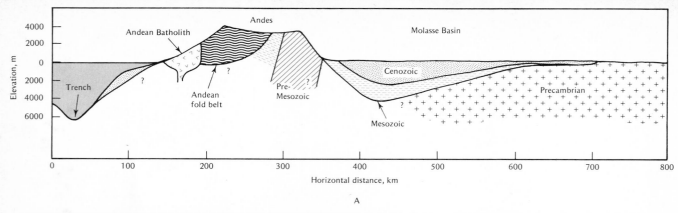

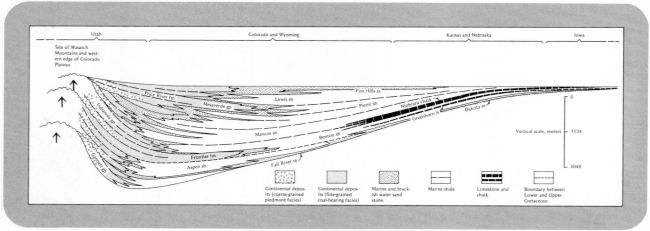

8-6. A. *Diagrammatic cross section across the central Peruvian Andes to illustrate the formation of a molasse-filled basin between the organic belt and the stable area of Precambrian continental crust (Brazilian Shield) to the east. Note vertical exaggeration of × 25. B. Cross section of molasse sediments that filled a basin between an active continental margin and the stable continental interior in the Rocky Mountain area during Cretaceous time.* [*From P. B. King: The Evolution of North America. Copyright © 1959 by Princeton University Press, Princeton, N.J. (Fig. 62, p. 109). Used with permission.*]

To the west lies the Great Valley sedimentary sequence that consists mostly of graywackes and attains thicknesses of 12–15 km in places. This sequence represents a typical arc-trench gap assemblage and there is abundant evidence that most of its sediments were deposited in deep water by turbidity currents. Still further to the west are the rocks that make up the Coast Ranges. These consist of a strongly deformed **melange** sequence in which shreds of oceanic crust and high pressure–low temperature blueschist metamorphic rocks are widespread. The entire three-unit assemblage is thus representative of the magmatic arc, arc-trench gap, and trench elements we have described from island arc systems. Along the western flank of the Andes similar patterns are presumably present, but only the upper part of the magmatic arc lies above sea level.

The inner (continent facing) flank of orogenic belts formed along active continental margins is also an important site of sediment accumulation (Figure 8-6). Areas such as this are underlain by continental crust and, as a consequence, only relatively shallow basins can form. Accordingly, the sedimentary materials derived from erosion of the adjacent mountain belt are deposited in shallow marine, nonmarine, or even subaerial environments. Despite this, a certain amount of subsidence must accompany the sedimentation along the inner margins of orogenic belts because some of these basins contain shallow water sedimentary rock sequences up to 5 km thick.

The rock units typically consist of an assortment of poorly sorted conglomerates, sandstones, and shales that, in many instances, exhibit textures and structures (Figure 8-7 A, B and C) indicative of deposition by streams and rivers on land or in shallow marine environments. The term **molasse** is used for such sedimentary assemblages, in which **arkose**, a feldspar-rich, **immature** sandstone, is a typical rock type. Molasse sediments are generally reddish in color due to the iron oxides they contain, which in turn, reflect the oxygenated environments in which these sediments are laid down. It is worth noting that the molasse basins on the inner side of the mountain belts of North and South America contain important oil and coal reserves in several places.

Orogenesis at Active Continental Margins

The mountain belts that form along active continental margins are sites of major deformation. The study of deformed rocks and analysis of the forces that cause their deformation are the primary interest of **structural geologists**. Studies of such deformed (**orogenic**) belts have shown them to contain two main types of structures; folds and faults. Folds can vary considerably in size, shape, and attitude (see Figure 7-16). Their presence indicates the shortening and thickening of an originally flat lying stratified rock assemblage due to compressive forces. Faults are classified mainly on the basis of the angle of the fault plane and the direction of movement of the blocks on either side of the fault (Figure 8-8).

A

B

C

8-7. *Photographs of (A) "ripple marks", (B) "cross bedding", and (C) mud cracks. Such structures are typical of sediments laid down by running water in shallow basin environments. Sedimentary structures such as these in (A) and (B) are of considerable aid to the geologist because they provide information on current directions at the time of deposition. [Photographs courtesy of U.S. Department of the Interior Geological Survey.]*

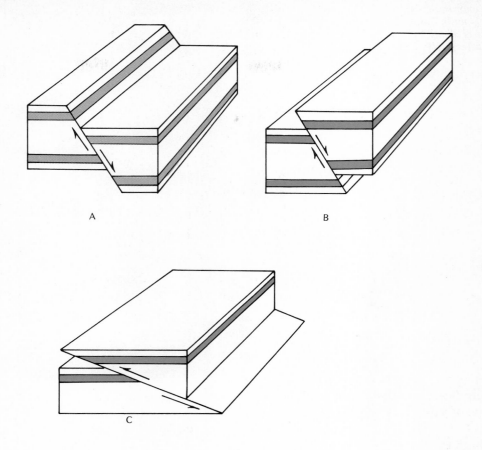

A

B

C

8-8. *Classification of faults commonly found in orogenic belts. Transform faults are not included.* A. *Normal fault.* B. *High angle reverse fault.* C. *Thrust fault (low angle reverse fault). Note that in normal faults (with the exception of vertical normal faults) the two blocks move away from one another to some extent. In reverse faults the two blocks actually move towards each other to some extent.*

Permanent deformation of rock units occurs when they are subjected to stresses that exceed their limits of elasticity. At very shallow levels within the crust, these stresses will tend to cause failure of rocks by rupture or faulting, especially if the stress buildup is rapid. Under conditions of slower stress buildup or at greater depths, where temperatures and confining pressures are higher, the brittleness of rock units tends to be reduced. Accordingly, rocks subjected to these conditions tend to respond to the forces applied to them by

8-9. *Photograph of high grade complexly deformed gneiss, typical of metamorphic rocks formed at deeper levels within orogenic belts. Coin for scale is about 2 cm in diameter. [Courtesy of P. Hudleston.]*

plastic flow; therefore, in layered sequences of rocks such as sediments or volcanics, folds of various types will develop. At depths in excess of ~5 km, especially along magmatic arcs, the temperatures are high enough that the rocks soften and behave like highly viscous liquids when stressed. Consequently, the metamorphic rocks formed in the deeper parts of orogenic belts tend to exhibit complex flow deformation structures. (Figure 8-9)

As mentioned in Chapter 7, the common occurrence of folded rocks in orogenic belts (both island arc systems and active continental margins) was long considered to indicate some process of crustal shortening. The origin of the forces that led to crustal shortening and concentrated this shortening along well-defined linear belts (mountain belts) remained mysterious, however. The search for a plausible mechanism led some workers to postulate that, due to cooling, shrinkage of the earth was taking place. This of course would cause the crust to wrinkle in much the same way that the skin of a drying apple does. The advent of the theory of plate tectonics and the recongition of the role of gravity forces in orogenic belts have led to a much clearer understanding of the mechanisms that control the complex patterns of deformation found in orogenic belts.

Shallow Level Deformation. It is now realized that the occurrence of folding and thrust faulting in rocks does not necessarily indicate the former presence of *regional* compressive forces. Only *local* compressive forces, induced by downslope movement of large rock units and diapiric emplacement of granitic rocks, are required. A careful analysis of the extent, thickness, and inherent strength of large **thrust sheets** (Figure 8-10) indicated that whatever force was responsible for the horizontal displacement must have been acting on the entire sheet and not merely one edge of it. In other words, these thrust sheets simply did not have the strength to transmit forces from one edge to the other without deforming internally.

The only conceivable force that could act equally on an entire thrust sheet is the force of gravity. Initially the mechanism that could overcome the friction between the moving thrust sheet and the underlying rocks was unknown. Movements of this type are now known to be facilitated by the low inherent shear strength of certain shales and limestones and are also promoted by high fluid pressures that markedly reduce the strength of certain beds. Furthermore, it is now realized that gravity induced thrusting can occur down slopes with gradients of only a few degrees. The thrust planes always develop along the weakest unit in a sequence and only cut across beds to intersect the surface at their leading edges (see Figure 8-10). In some instances deformation within the thrust sheet may take place by folding.

If much of the shallow level thrusting and folding of rocks is induced by downslope movement on the flanks of uplifts, what is the prime cause of these uplifts? Studies of mountain belts along active continental margins indicate that major periods of thrusting and folding are virtually always associated in time and space with major periods of plutonic magmatism. In other words, uplift and deformation are closely tied to the emplacement of batholiths and the **isostatic** uplift resulting from the thickening of the continental crust that accompanies the formation of a volcano-plutonic arc (see Chapter 7).

The large volumes of granitic magma that are generated at active continental margins via the subduction process (see Chapter 7) move upward into the crust mainly as diapirs (Figure 8-11). Emplacement of these large igneous bodies in this manner causes large vol-

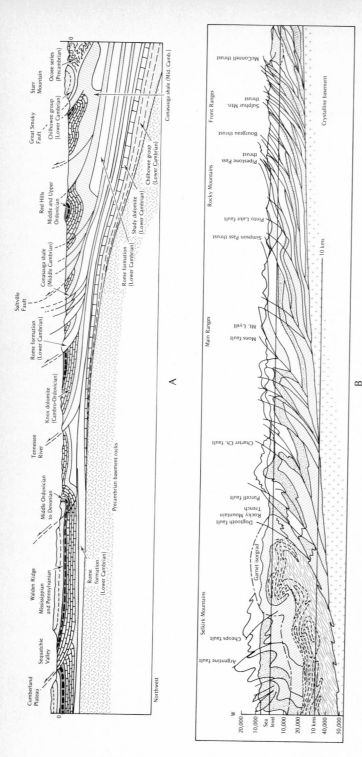

8-10. *Two sections illustrating thrust faulting away from an orogenic belt toward the continental interior. Although thrust planes now slope uphill, it should be emphasized that at the time thrusting took place the thrust planes were predominantly downhill. A. Sections across Valley and Ridge Province in eastern Tennessee, U.S.A. Note that thrusting and folding is restricted to cover rocks and does not involve underlying Precambrian basement rocks. The sedimentary units involved in the deformation are largely of Cambrian and Ordovician age. Minor vertical exaggeration. Length of section is approximately 80 km. [From P. B. King: The Tectonics of Middle North America. Copyright © 1951 Princeton University Press, Fig. 32, p. 128. Used with permission of Princeton University Press.] B. Section across the southern Canadian Rockies in western Canada. Note that elevations are in feet, not meters. [From R. Price and E. Mountjoy: Geologic structure of the Canadian Rocky Mountains between Bow and Athabaska Rivers (a progress report). Geol. Assoc. Canada, Sp. paper No. 6: pp. 7–25 (1970).]*

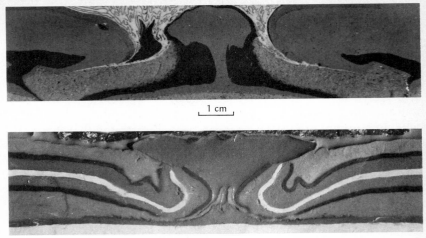

8-11. *Two photographs of cross sections illustrating deformation patterns developed in laboratory modelling experiments to simulate diapirism and deformation in orogenic belts* [*Courtesy H. Ramberg.*]

1 cm

1 cm

8-12. *Diagrammatic cross section of Andes showing deep crustal root indicated by geophysical studies in southern Peru. Note vertical exaggeration of × 2.* [*From D. James: Plate tectonic model of the evolution of the Andes. Bull. Geol. Soc. Amer., 82: pp. 3325–3346 (1971).*]

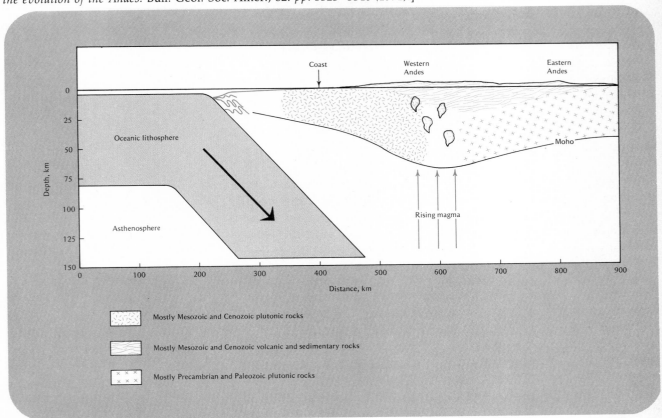

umes of preexisting rocks to be forced aside and upward. Essentially, it is this action that sets in motion the processes of mountain building along active continental margins.

In addition to uplift, the accompanying deformation may be divided into two broad categories, shallow seated and deep seated. The more or less horizontal movement of rock sequences by thrust faulting and folding are confined to the shallow portions of orogenic belts. In a number of instances it can be demonstrated that the rocks underlying folded and thrust sequences do not reflect the overlying structural features. This has given rise to the concept of "thin-skinned" tectonics (see Figure 8-10).

Deep Level Deformation. In the deeper portions of orogenic belts deformation is caused by the upward passage of large bodies of magma and the emplacement of deep seated plutons. The high temperatures and pressures, and consequent mineralogical re-crystallization (metamorphism) that characterize this environment, have the effect of softening the rocks to the extent that they deform by flowage when stressed. The complex structures of metamorphic rocks formed under such conditions (see Figure 8-9) provide striking evidence of such flowage. Deep within orogenic belts even homogeneous bodies of igneous rock, which at higher crustal levels tend to resist deformation, develop planar or linear structures due to reordering of their constituent minerals.

In geologically young mountain belts such as the Andes, which stand high above sea level, the observable structures are mainly those that form at shallow depths, that is, folds and thrusts. Geophysical studies have shown, however, that the thickness of crust under some parts of the Andes is in excess of 60 km (Figure 8-12), which well exceeds the average thickness of continental crust. As the forces of erosion continue to wear down the Andes, this deep root of low-density continental crust will almost certainly allow isostatic uplift to continue and erosion to progress until the crust eventually reaches a more normal thickness. At this stage, the deeper portions of the Andean orogenic belt will be exposed at the surface.

Paleozoic and Precambrian orogenic belts that have been through such cycles of uplift and deep erosion, and consequently exhibit little, if any, marked topographic expression are underlain by crust of normal thickness. Furthermore, they contain large quantities of high grade, complexly deformed metamorphic and igneous rocks and exhibit no particular relationship to present day plate boundaries or continental margins.

Continental Growth

In much the same way as subduction-related intraoceanic arc systems lead to the creation of new continental crust, the geologic events at active continental margins lead to a net addition to the volume of the continents. This occurs even if the new magmatic arc develops largely within preexisting continental crust. Subduction and consequent magmatism at an active continental margin can be considered to represent another example of the second stage of the two-stage fractionation process by which continental crust is generated from primary mantle material.

CONTINENTAL COLLISIONS

As explained in Chapter 6, the fate of virtually all oceanic lithosphere is to return to the mantle by the subduction process at a convergent plate boundary. Rafts of continental lithosphere are apparently simply too thick and continental crust too bouyant to undergo subduction. Thus continental lithosphere, once created, will remain at the outer surface of the earth and endlessly ride the conveyor belts of plate motions. Under certain circumstances continental lithosphere will be rafted to a subduction zone along an active continental margin (Figure 8-13). When this happens the inability of continental lithosphere to undergo subduction will cause it to collide with the island arc or continent that is present on the inner side of the subduction zone.

Himalayas

An excellent example is provided at the present time by the collision of India with Asia. Northward movement of the Indian Plate over the last 100 million years has resulted in the loss of the oceanic lithosphere formerly separating India and Asia and the eventual impact of these two continental masses. Analysis of present day plate motions indicates that the Indian Plate is still moving northward relative to the Asian Plate (see Figure 5-2) at approximately 4 cm per year. As a consequence, along the Himalayan belt India

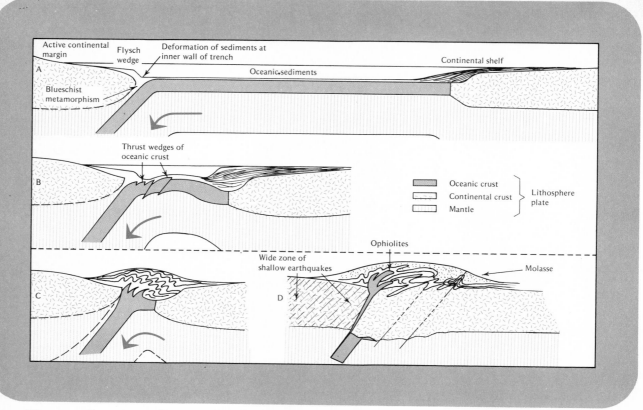

8-13. *Diagrammatic sequence of cross sections to illustrate the approach and eventual collision of two continents. Note thickening of continental crust that results from continental collision. [From J. Dewey and J. Bird: Mountain belts and the new global tectonics. J. Geophys. Res., 75: pp. 2625–2647 (1970).]*

is being ground against Asia. The topographic results of this collision are indeed spectacular—the highest mountains on earth and the extensive highlands of the Tibetan Plateau.

In geologic terms, forcing together these two continental blocks has created an abnormally thick section of continental crust (>70 km) overlain by a belt in which the rocks are complexly thrust and strongly folded. All that remains of the oceanic crust that originally lay between India and Asia is a zone of mafic and ultramafic rocks. These oceanic rocks apparently were caught and squeezed upward just prior to collision and mark the **geosuture** along which the two continental blocks are being welded together (Figure 8-14).

Although earthquakes are frequent in the Himalayan belt, they are virtually all shallow seated (<100-km depth) and occur in a broad diffuse zone. The presence of earthquake activity off the southern

8-14. *Generalized map A and cross section B of Himalayas and adjacent regions. Note Indus suture zone composed of rocks of oceanic affinities, and that structural trends in northernmost peninsular India are mostly at a high angle to those in the Himalayan belt. The line of the cross section S–N is shown on the map. [Modified from A. Gansser: Geology of the Himalayas, New York: Interscience Publishers, 1964.]*

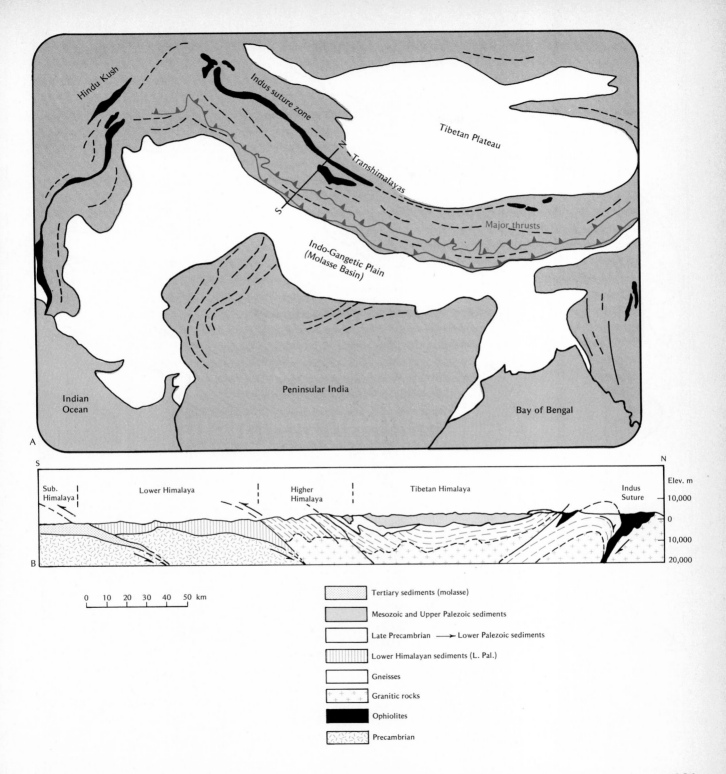

Hindu Kush

Indus suture zone

Tibetan Plateau

Transhimalayas

Major thrusts

Indo-Gangetic Plain
(Molasse Basin)

Indian
Ocean

Peninsular India

Bay of Bengal

A

B

S   N

Sub.
Himalaya   Lower Himalaya   Higher   Tibetan Himalaya   Indus
                                            Himalaya   Suture

Elev. m

10,000

0

10,000

20,000

0   10   20   30   40   50 km

Tertiary sediments (molasse)

Mesozoic and Upper Palezoic sediments

Late Precambrian ——▶ Lower Palezoic sediments

Lower Himalayan sediments (L. Pal.)

Gneisses

Granitic rocks

Ophiolites

Precambrian

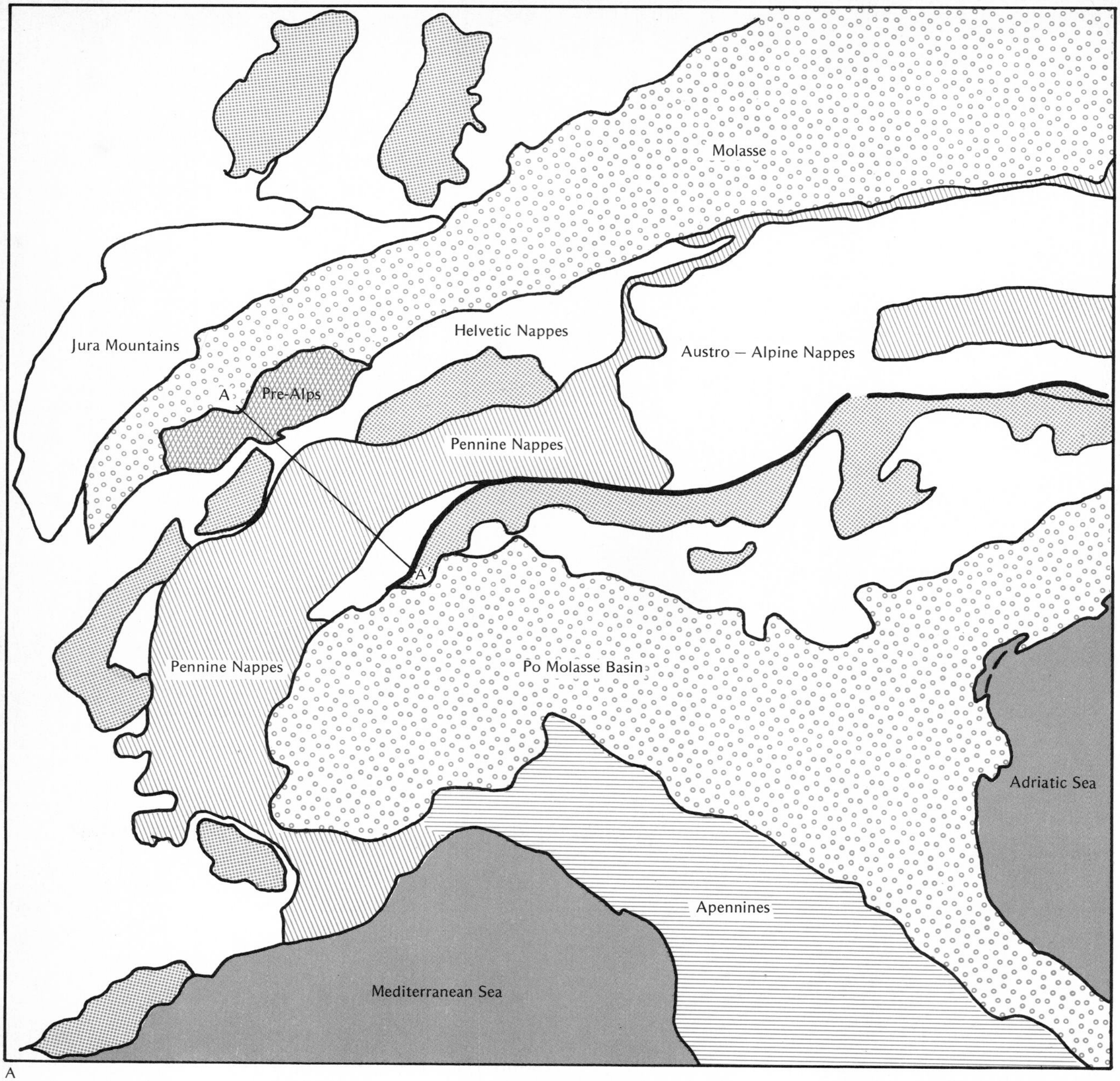
Molasse
Jura Mountains
Helvetic Nappes
Austro — Alpine Nappes
A
Pre-Alps
Pennine Nappes
A'
Pennine Nappes
Po Molasse Basin
Adriatic Sea
Apennines
Mediterranean Sea

A

8-15. *Generalized map* A *and cross section* B *of the Alps. The line of the cross section A-A' is shown on the map. Note complex structure shown by cross section and northwestward movement of large thrust structures* (nappes).

tip of India has led some workers to suggest that a new convergent plate boundary might be in the process of forming in this area. One result of this collision between India and Asia was rapid erosion and the consequent formation of a thick sequence of molasse sediments in northern India during late Tertiary time. At present the bulk of the material that is eroding from the Himalayas is being dumped into the Indian Ocean by the Ganges and Brahmaputra river systems.

Alps

Another important example of continental collision tectonics is the Alpine mountain belt. The origin of this highly complex and carefully studied area has long puzzled structural geologists, but since the advent of plate tectonic theory the Alpine belt has been interpreted to be the result of convergence between the African Plate and the Eurasian Plate. The details, however, remain obscure. The central Alpine belt is characterized by huge thrust sheets which slid mainly northwards after being squeezed upward from the suture zone, which is now marked by the shreds of oceanic crust (Figure 8-15). Along the northern and southern margins of the Alpine belt are a series of basins containing molasse sediments, which were derived mainly from the orogenic uplift in the center. The Alpine belt, which is now more deeply eroded than the Himalayas, contains a considerable percentage of high-grade metamorphic rocks in its core zone but a very small volume of plutonic igneous rocks.

Older Examples of Continental Collisions

Examples of continental collision that are geologically much older than the Alps and Himalayas have also been postulated. The Urals, for example, probably represent an eroded mountain belt formed by

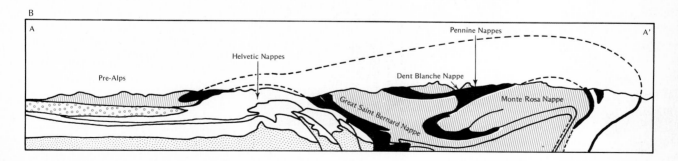

collision during late Paleozoic time of the Siberian continental mass to the east and the European continental mass to the west.

Another important example of continental collision is represented by the Paleozoic, Appalachian-Caledonide orogenic belt (Figure 8-16), which has an aggregate length of several thousand kilometers. Along this extensive belt are many ophiolite complexes, a series of major thrust faults, and local areas of blueschist metamorphism, all indicative of an earlier subduction event. The patterns of igneous, metamorphic, and sedimentary rock units and the structural relationships all can be interpreted in terms of continental collision. These plate events must have been of a complex nature, for orogenic events occurred at different places within the Appalachian-Caledonide belt at various times during the Paleozoic era, and culminated in the closing of a "Proto-Atlantic Ocean" at least 100 million years prior to the initial opening of the present Atlantic Ocean (Figure 8-17).

In Precambrian terranes much of the evidence for subduction and continental collision, such as ophiolite complexes, has been removed by erosion or obscured by metamorphism. Nevertheless, it has been suggested that the boundaries between some of the Precambrian

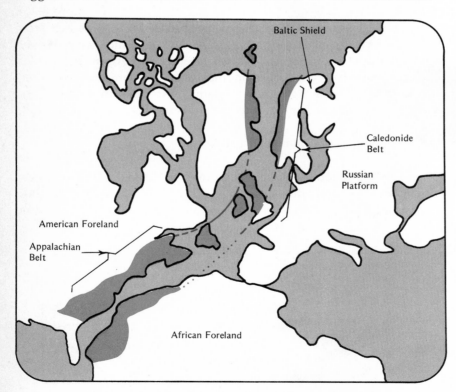

8-16. *Sketch map showing the extent of the Caledonide-Appalachian orogenic belt formed in Paleozoic time by the closing of the proto-Atlantic Ocean and subsequent continental collision.*

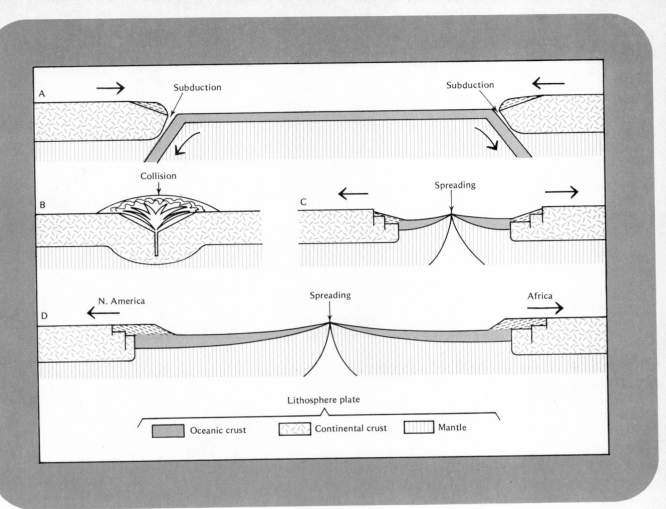

8-17. *Inferred plate tectonic history of area surrounding North Atlantic Ocean (highly diagrammatic). A. Closing of proto-Atlantic Ocean. Note that subduction is indicated at both margins of the ocean basin. Interpretation of the geologic relationships in the Caledonides suggests subduction on western margin of proto-Atlantic, whereas subduction on the eastern margin of the proto-Atlantic is suggested by the geology of the southern Appalachians. B. Continental collision due to complete closure of proto-Atlantic Ocean. This event occurred in Paleozoic time, and resulted in the formation of the Appalachian-Caledonide orogenic belt (Figure 8-16). C. Breakup of large continental mass to initiate formation of present Atlantic Ocean in early Mesozoic time. D. Present day situation. Spreading is still active (although slow) along the mid-Atlantic Ridge, and North America and Africa are still receding from one another.*

structural provinces of Canada represent ancient suture zones that mark the sites of former continental collisions (Figure 8-18). These very old apparent suture zones are characterized by the presence of major faults and mafic and ultramafic igneous rocks or their metamorphosed equivalents. In fact, the puzzle of how different

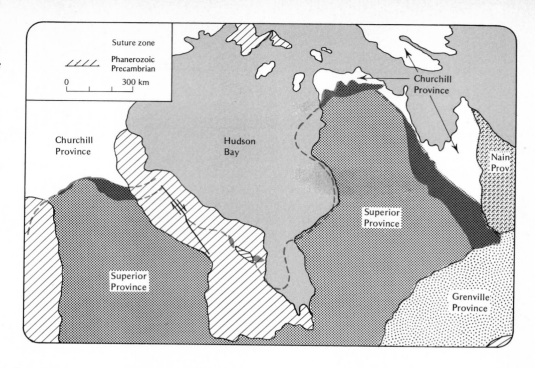

8-18. *Sketch map to illustrate position of proposed suture zones around the northern margin of the Superior Province of the Canadian Shield. [From: Earth and Planetary Science Letters, vol. 10: p. 418, Fig. 1. Gibb and Walcott: A Precambrian Suture in the Canadian Shield.]*

large blocks of continental crust, each with distinctive structural and age patterns, come to be juxtaposed in the Precambrian terranes of continents largely disappears if one accepts the possibility of Precambrian plate motions and continental collisions.

The tectonic style of mountain belts associated with continental collision events is somewhat distinctive from those formed at active continental margins. Thin-skinned tectonics account for some of the structures present, but regional compression is obviously an important factor. The diapiric uprise of granitic magma is of lesser importance with respect to deformation. In fact both the Himalayas and the Alps are characterized by limited magmatism in comparison with the Andes.

In summary, the main geologic events associated with continental collision orogenic belts can be tabulated as follows (see Figure 8-13):

1. Subduction of intervening oceanic lithosphere at one continental margin leading to the approach of another continental block.
2. Overthrusting of slices of oceanic crust and underlying upper mantle just prior to collision and strong deformation of the various types of sediments between the blocks.
3. Thickening of the crust in the collision zone, uplift, and sub-

sequent gravity thrusting of large sheets outwards from the central uplifted mountain belt.

4. Deposition of thick molasse sedimentary sequences in basins marginal to the zone of maximum deformation, uplift, and thrusting.

5. Eventual cessation of plate convergence along the zone of collision due to a reshuffling of plate boundaries and margins.

INACTIVE CONTINENTAL MARGINS

Inactive continental margins represent areas where there is no relative motion at the margin between a continent and its adjacent oceanic crust. In other words the continent-ocean interface is within a plate rather than at a plate boundary. Figure 8-19 indicates that, at the present time, about half the continental margins are of this type.

8-19.   *World map showing present day location of majority of inactive continental margins.*

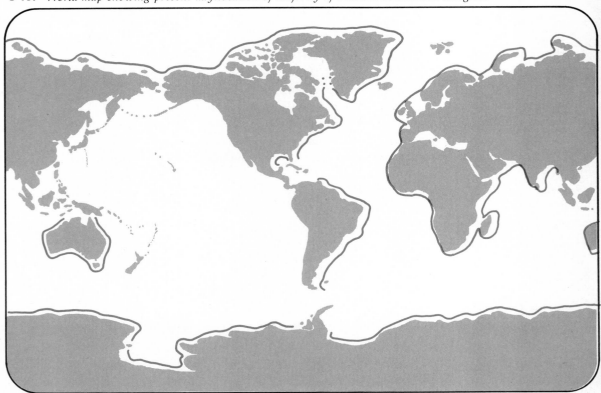

Before embarking on a discussion of the geological features of inactive continental margins, let us examine how most such margins come into existence. In most instances, the initial stage is the rifting apart of a large continental mass and creation of intervening oceanic lithosphere. However, inactive continental margins can also come into existence when subduction at an active continental margin ceases for some reason. The precise cause of continental breakup is not clear, but upwellings deep within the earth's mantle have been suggested to trigger the process.

The Atlantic Ocean is an example of an ocean basin created by the rifting apart of a supercontinent. The breakup was initiated in late Triassic times. The Atlantic margin of North America has been subjected to considerable study by marine geologists and geophysicists and illustrates the main physiographic characteristics of inactive continental margins (Figure 8-20).

Sedimentation is the major geologic activity that takes place at such margins. On the continental shelf adjacent to the coastline, clastic sediments are deposited. In shallow water, where the sea bottom is affected by wave motions, the sediments are mainly clean sands, but further away from the coast the continental shelf is the site of deposition of finer grained sediment such as mud. In areas that do not for some reason receive much clastic sediment, deposits of carbonate sediments (limestones) will tend to form. These develop by accumulation of the carbonate skeletons of marine organisms or by direct precipitation of calcium carbonate from sea water. This latter process is important on the Bahama Banks. Calcium carbonate, unlike most substances, is less soluble in warm water than cold water; thus, as cold ocean water passes over the shallow Bahama Banks and is heated by solar radiation, calcium carbonate precipitates.

Considerable thicknesses of shallow-water sediments can be deposited on continental shelves, indicating that subsidence must accompany sedimentation along many inactive continental margins. For example, drilling off the coast of the eastern United States has indicated that a sequence of sediments nearly 4000 m thick has been deposited since Jurassic times (Figure 8-21.) Not all the sedimentary material delivered to the ocean along inactive continental margins is deposited on the continental shelf. A significant amount of this material manages to bypass the shelf area to be eventually deposited on the continental rise and even the deep ocean floor. By what method is this submarine transport of sedimentary material achieved? Continental shelves, and particularly continental slopes, are furrowed in many places by major topographic features called sub-

8-20. (OPPOSITE) *Physiographic sketch to illustrate the transition from the eastern coast of North America to the deep floor of the Atlantic Ocean. Note that the sketch involves considerable vertical exaggeration.*

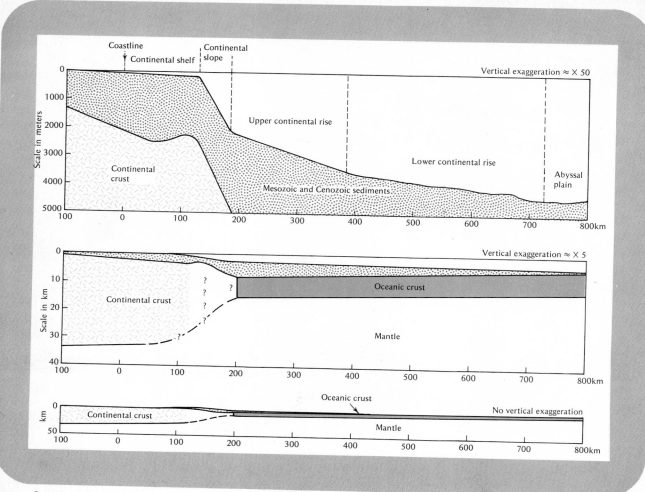

8-21. *Stratigraphic section of the sedimentary rocks underlying the Atlantic coastal plain, shelf, and continental slope in the vicinity of Cape Hatteras, N. C., U.S.A. Note thickening of units oceanward and extreme vertical exaggeration (1100:1). Lower two sections show continent—ocean crustal interface at a vertical exaggeration of ~×5 and at no vertical exaggeration.*

marine canyons (Figure 8-22). These canyons, some of which are closely related in space to river systems, although some are not, are now known to be cut by turbidity currents.

Striking confirmation of the movement of turbidity currents down continental slopes was furnished by the series of events that followed the 1929 Grand Banks earthquake (a rare event on inactive continental margins). The earthquake was followed by the rupture of a number of trans-Atlantic submarine telegraph cables in the area.

8-22. *Drawing of southern tip of Baja California showing the numerous submarine canyons that have been cut into the continental shelf and slope. Strong vertical exaggeration. [Courtesy W. Normark.]*

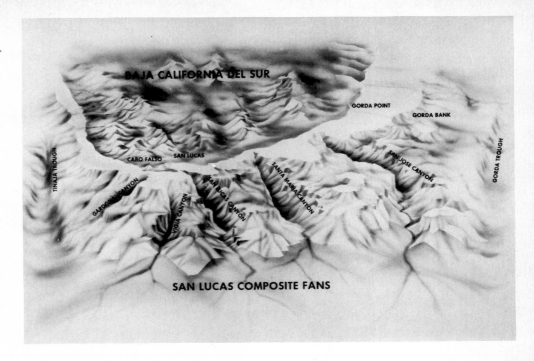

Many years later the precise time at which interruption of trans-Atlantic communication took place along various cables was analyzed. The analysis revealed that the cables to the south of the earthquake epicenter were broken in a regular progression (Figure 8-23). It was then realized that the breaks well to the south of the epicenter must have been caused by turbidity current action. Apparently this turbidity current was initiated by downslope slumping of unconsolidated sediments dislodged by the earthquake from the upper edge of the continental slope. The time-space data on cable breaks indicated that the current must have attained speeds of 72 km/hr as it rushed down the 4°–6° gradient of the continental slope.

Sedimentary material carried down the continental slope by turbidity currents is delivered to the deep ocean basins. At the base of major submarine canyons huge **deep-sea fans** (Figure 8-24) spread outward on the continental rise and the ocean floor. The coarser material will tend to be deposited at the head of the fan, where initial deceleration of the feeding turbidity currents takes place; the finer material can be spread outwards for hundreds of kilometers along the floor of an ocean basin. The largest deep-sea fan known in the ocean basins is the Bengal Fan in the Bay of Bengal. It receives the major portion of the sedimentary material produced by erosion of the Himalayas and has attained a total length of

2500 km and a maximum thickness of greater than 15 km. This enormous feature is still growing and presently contains at least 20 million km³ of sediments.

What then is the eventual fate of the major sediment accumulations that are formed along inactive continental margins? Their life expectancy prior to major disruption is no longer than that of the oce-

8-23. *Sketch to illustrate the epicenter of the 1929 Grand Banks earthquake and the timing of cable breaks to the south (1855 6:55 p.m.; 0033 12:33 a.m.). [From W. Menard:* Geology of the Pacific *(Fig. 9.7). McGraw-Hill Book Co., New York, 1964.]*

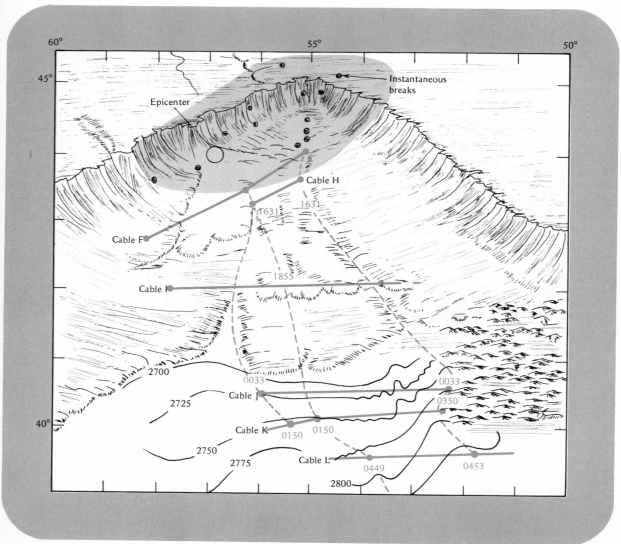

anic crust on which they lie. As was demonstrated in Chapter 6, oceanic crust has a relatively short life cycle and eventually reenters the mantle by the process of subduction. Thus, inactive continental margins are fated either to become active continental margins or to become involved in collision events. In fact the thick sedimentary prisms that form along inactive continental margins might be instrumental in initiating new subduction zones. By heavily loading the oceanic crust at the continent-ocean interface they may pro-

8-24. A. *Generalized sketch to illustrate typical features of a deep-sea fan. B. Actual echo sounding profile of a deep-sea fan built out over oceanic crust.* [*Courtesy W. Normark.*]

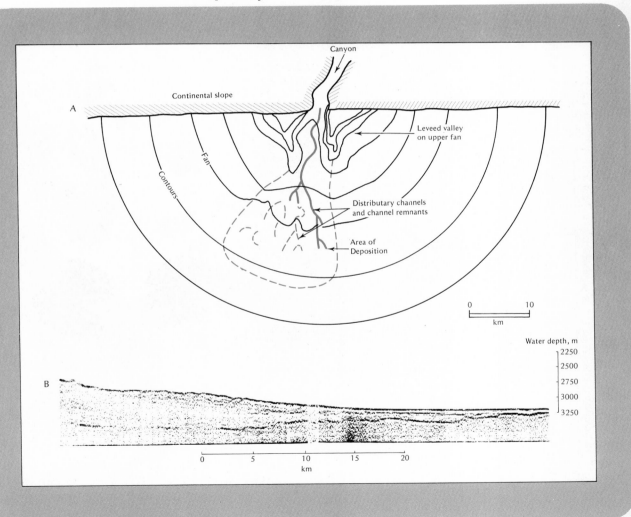

vide the impetus for the initial downward descent of a plate that marks the creation of a new convergent plate boundary.

CONTINENTAL INTERIORS

The plate tectonic model for crustal evolution dictates that continental-type crust and lithosphere form only at convergent plate boundaries. If this is correct, then the igneous and metamorphic

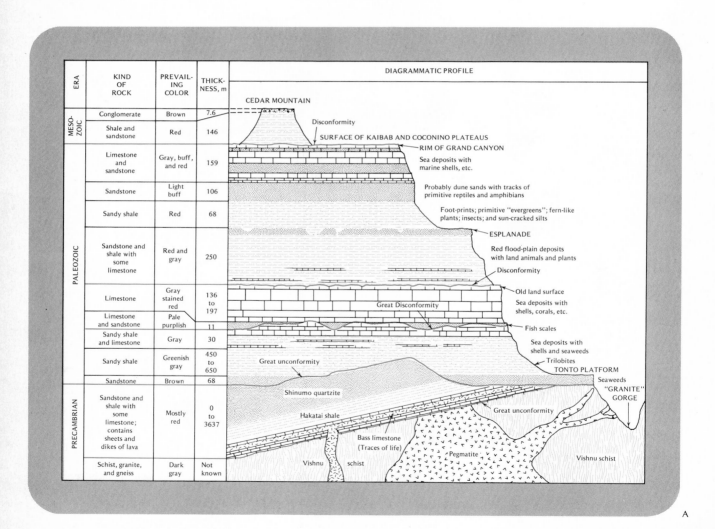

ERA	KIND OF ROCK	PREVAILING COLOR	THICKNESS, m	DIAGRAMMATIC PROFILE
MESOZOIC	Conglomerate	Brown	7.6	
	Shale and sandstone	Red	146	
PALEOZOIC	Limestone and sandstone	Gray, buff, and red	159	
	Sandstone	Light buff	106	
	Sandy shale	Red	68	
	Sandstone and shale with some limestone	Red and gray	250	
	Limestone	Gray stained red	136 to 197	
	Limestone and sandstone	Pale purplish	11	
	Sandy shale and limestone	Gray	30	
	Sandy shale	Greenish gray	450 to 650	
	Sandstone	Brown	68	
PRECAMBRIAN	Sandstone and shale with some limestone; contains sheets and dikes of lava	Mostly red	0 to 3637	
	Schist, granite, and gneiss	Dark gray	Not known	

A

rocks that make up the bulk of the interior regions of continents must have been created mainly at former convergent plate boundaries. Continental crust is thick and light; therefore, the average elevation of the continents is such that they stand above sea level. The forces of weathering and erosion are constantly working to wear down this emergent material, and it is remarkable that the continents, at least away from active plate margins, stand as high above sea level as they do now.

To some extent continental height is related to the amount of the world's water that is locked up as ice in the polar regions. If all this ice were to melt, sea level would rise approximately 70 m worldwide. The rise would be sufficient to drown significant areas of the continents and, incidentally, most of the world's major cities. In terms of

8-25. A. (Opposite) *Section of rock units exposed in Grand Canyon, Ariz. U.S.S. Note flat-lying sedimentary cover rocks overlying older basement rocks, and examples of two unconformities and three disconformities. Note also extensive time span represented by the section—from early Precambrian to early Mesozoic. [From A. Eardley:* Structural Geology of North America, *2nd ed., (Fig. 26.4, p. 410). Harper and Row, New York, 1962. Modified from Noble.] B. Old pen drawing of Grand Canyon, Ariz., showing sequence of rock layers depicted diagrammatically in A. [Courtesy U.S. Department of the Interior Geological Survey. Drawing by William H. Holmes.]*

B

total earth history, the presence of large ice caps in the polar regions is almost certainly an anomalous situation, and during much of the geologic past large areas of the continents must have been covered by shallow seas. A modern day example of this situation is provided by Hudson Bay, a large shallow marine basin underlain entirely by continental crust.

Abundant evidence for the former presence of extensive shallow seas over the continents can be obtained by studying the younger (<600 million years old) sedimentary rocks that form a relatively thin veneer over the igneous and metamorphic "basement" rocks of many continental areas. The older basement rocks are invariably strongly deformed, whereas the sedimentary cover rocks in stable continental interiors are largely undisturbed. This break between underlying deformed rocks and virtually flat-lying cover rocks provides an example of an **unconformity** (Figure 8-25). Long periods of geologic time can pass between initiation of erosion of the basement and eventual deposition of the lowermost beds of the cover rocks, and such breaks in the geologic record are called major unconformities. The term unconformity, however, has broad usage and is used for any surface that represents a period of erosion or nondeposition between the formation of an underlying rock unit and the rocks deposited above it (see Figure 8-25).

Sedimentation in Continental Interiors

The sedimentary units that form along the margins and at the bottom of shallow continental seas are typically extensive, thin, and well-bedded. Detrital material delivered to the shallow marine environment by rivers draining adjacent areas of the continental interiors tends to undergo considerable reworking by wave action along shorelines prior to eventual deposition. As a result, sand-size material is effectively separated from finer detritus, and the non-resistant sand-size materials are broken down. The end products of these processes are sedimentary deposits consisting of clean quartz sands, siltstones, and shales. Small changes in sea level or minor uplifts and downwarps have the effect of causing specific depositional environments to migrate long distances laterally, and in this manner distinct beds of sandstones, siltstones, and shales are formed (Figure 8-26A and B).

In areas starved of detritus, beds of limestone will tend to be deposited on the shallow sea floor. These limestones might consist of chemically precipitated calcium carbonate or be of biologic origin

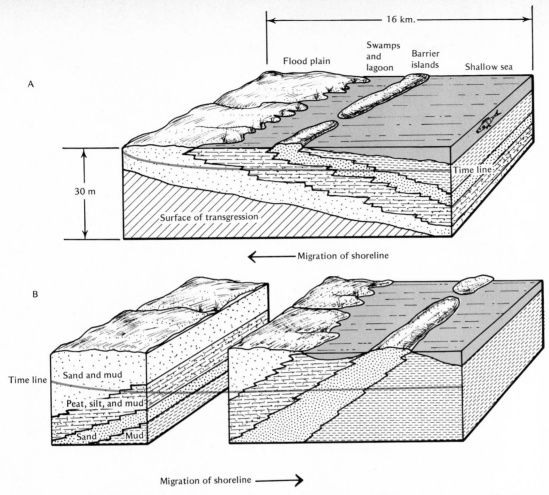

8-26. *Two diagrams to illustrate how nearshore environments of deposition of sediments will migrate with time in a shallow continental sea in response to changes in relative sea level or outward building of shorelines by sedimentary processes. Note how time lines cross sedimentary units, but are not truly horizontal. Vertical scale greatly exaggerated (500 ×). A. Migration of shoreline towards land area. B. Migration of shoreline depositional basin. [From D. L. Eicher:* Geology Time. *Prentice-Hall, Inc., Englewood Cliffs, New Jersey, 1968, p. 44. Used with permission.]*

(that is, composed of fragments of the skeletons of marine organisms). Reef complexes and the clastic debris from the erosion of reefs by waves may also contribute to the formation of limestone beds in a warm intracontinental shallow sea environment. Under favorable conditions important coal and oil deposits will also form during sedimentation in continental interiors.

Geologists who study sequences of sedimentary rocks, such as

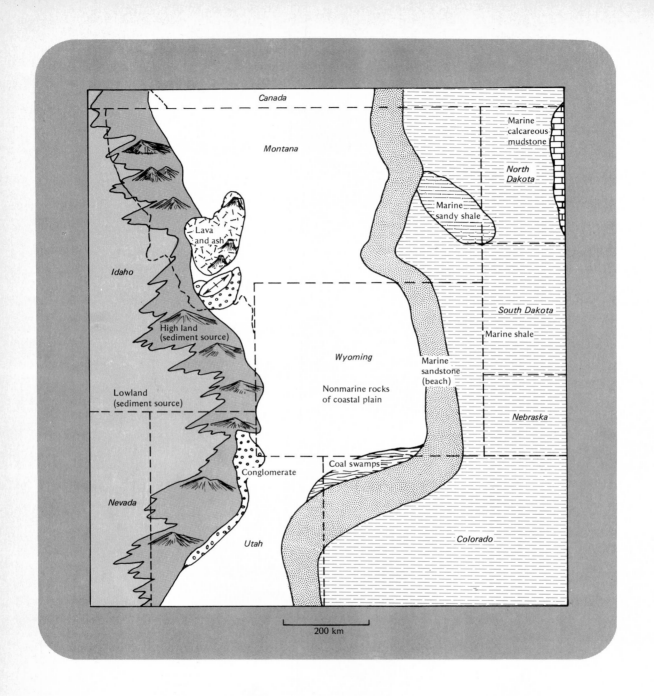

8-27. *Paleogeographic map of Rocky Mountain region, western United States, in latest Cretaceous time. Note progression from positive land areas in west, to coastal plain, beach, and continental sea environment in east. Note also wide variety of rock types being deposited synchronously within the map area. Paleogeographic maps are a considerable aid in building up portions of the more detailed geologic history of an area, and great amounts of field data are required for their construction. [From: Geologic Atlas of the Rocky Mountain Region.* Rocky Mountain Assoc. of Geologists, Fig. 45 (1972). Used with permission.]

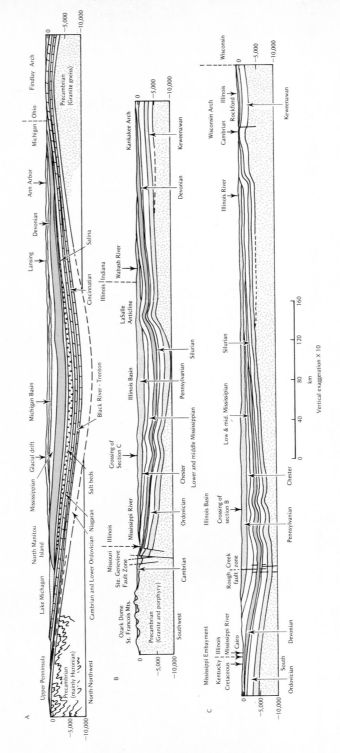

8-28. *Cross section showing stratigraphic relations of thin sedimentary units that cover the stable interior of the United States. Note gentle warping of basement surface and very gradual change in thickness of sedimentary units. Vertical exaggeration × 10. The lines A–A', B–B', C–C', along which the sections have been constructed are shown in Figure 8-29. [From Phillip B. King: The Tectonics of Middle North America. Princeton University Press, Princeton, N.J., (p. 40, 1951). Used with permission.]*

those formed in continental seas, are called **stratigraphers**. Stratigraphers find the sedimentary cover rocks of continental interiors particularly attractive for detailed studies because they generally contain abundant fossil remains, occur over extensive areas, and in most cases are little deformed by structural events. Consequently, the age of individual units can be closely determined from fossil assemblages, and correlations of individual units may be made over

8-29.   *Central stable region of United States. The contour lines on map represent the depth to the surface of the underlying basement rocks in feet. Note series of broad domes, arches and basins indicated by these contour lines. Lines A–A', B–B' and C–C' indicate geographic position of sections shown in Figure 8-28.*

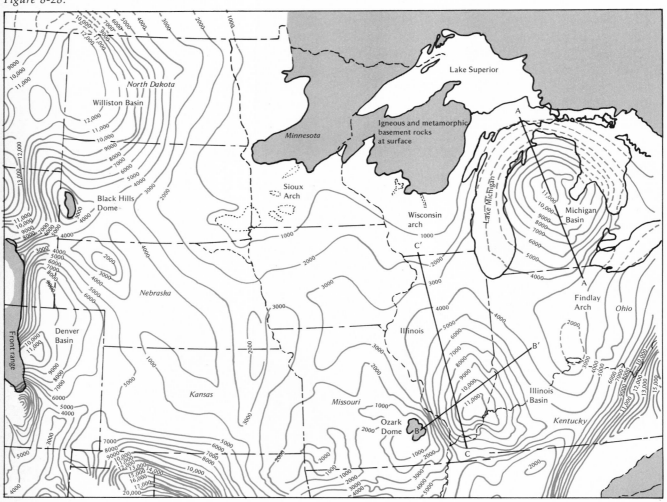

long distances. All this is in marked contrast to arc-related sedimentary sequences, which tend to contain few identifiable fossils, exhibit considerable variability, and be strongly deformed. Careful work on the sedimentary rocks of continental interiors permits the reconstruction of ancient depositional environments with some exactitude, and the assembly of **paleogeographic maps** (Figures 8-27).

The geology of the interior lowlands of the United States provides a particularly good example of intracontinental marine sedimentary rocks overlying much older basement rocks. In this area series of sedimentary units only a few hundred meters thick may represent almost continuous, but very slow, deposition over time periods that extend from late Precambrian to Early Cretaceous times (~500 million years) Figure 8-28. During the remainder of Cretaceous time (~30 million years) a sequence of sedimentary rocks 2500–3000 m thick was deposited in parts of the Rocky Mountain area. The dramatic change in sedimentation rate was a direct consequence of mountain building events associated with an active continental margin along a previous Pacific coast during the Cretaceous. This observation also serves to illustrate that much information relating to the location and timing of plate tectonic events can be gained from the investigation of sedimentary rocks.

Studies of the intracontinental sedimentary cover rocks over sufficiently broad areas reveal that earth movements do occur, although such areas are essentially stable in comparison with orogenic belts. The structural activity is manifested as faulting and gentle broad warping movements that tend to throw the underlying basement into a series of basins separated by arches and domes (Figure 8-29). Such movements are clearly reflected by changes in thickness and type of contemporaneous sediments. The cause of these movements is not well understood, but presumably they reflect processes at work in the underlying upper mantle.

Magmatism in Continental Interiors

Although of relatively minor importance, magmatism in continental interiors is widespread in time and space. Intracontinental magmatic rocks must be distinguished from the older volcanic and plutonic igneous rocks, now locked within continental interiors, that presumably formed at ancient convergent plate margins. Two rather distinctive compositional types are known. In volumetric terms, the most significant igneous rocks of the continental interiors

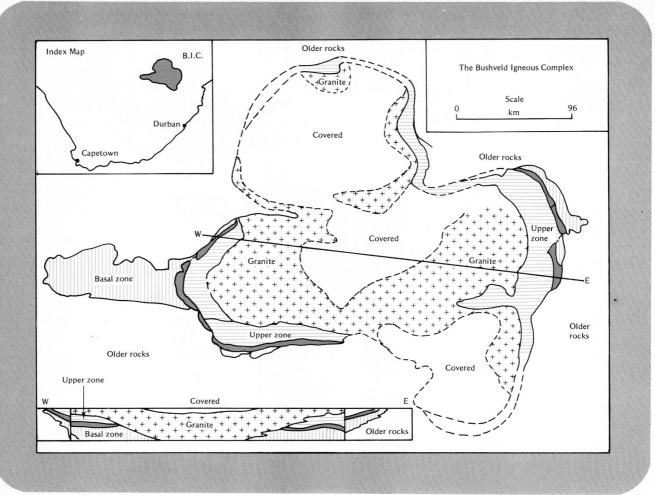

8-30. *Simplified map and cross section of the Bushveld Igneous Complex. Note huge size of this intrusive body. The extensive granitic rocks probably represent in large part melted crustal material and are distinct from the mafic rocks, which have an average composition of basalt. The Critical Zone separating the Basal and Upper Zones of the Complex contains important chrome and platinum deposits.*

are extrusive flood basalts and the layered intrusive sheets of overall basaltic composition. These latter basaltic rocks are probably the intrusive equivalents of flood basalts.

In a number of continental areas (for example, Southern Brazil, northwest India, southeast Siberia, northwestern United States) huge outpourings of basaltic lava have occurred in the past. The original magmas must have been derived by partial melting of sub-continental mantle, perhaps as a result of regional tension, but as yet little is known about the factors that cause these major magmatic events. In some instances, the basaltic magma apparently failed to reach the earth's surface, but instead formed large intrusive sheets within the upper crust. A number of these intrusive sheets are known around the world (for example, the Duluth Complex, Minn.), but by far the largest is the Bushveld Igneous Complex in South Africa (Figure 8-30). This enormous body of igneous rock is exposed over an area of almost 70,000 km^2 and estimated to be at least 8000 m thick in its central parts.

Although studies have shown that the Bushveld Igneous Complex was probably emplaced by several pulses of magma, during much of its formation the magma chambers were so large that active convection currents were set up during the time crystallization was taking place. Dramatic evidence for the existence of these currents are the discrete beds of crystals of the same mineral that were laid down on the floor of the magma chamber and actual sedimentary structures, such as graded bedding and crossbedding, that are present in some exposures of the igneous rocks. In this manner numerous monomineralic layers of pyroxene, plagioclase, and even chromite ($FeOCr_2O_3$), which have the gross appearance of water-laid sedimentary beds, were formed (Figure 8-31). As a result of these processes of magmatic differentiation and sorting, the Bushveld Igneous Complex contains major economic deposits of chrome, iron, titanium, and platinum.

8-31. *Photograph showing "sedimentary" layers of chromite-rich and feldspar-rich units in the Bushveld Igneous Complex* [Courtesy E. Cameron.]

Quite distinct from these examples of basaltic magmatism are the volcanic and subvolcanic alkaline igneous rocks that occur in discrete, but widely scattered, complexes in intracontinental areas. These igneous rocks are mainly mafic and ultramafic in composition but are characterized by high alkali (sodium and potassium) contents and are therefore highly distinctive from the more common low-alkali mafic and ultramafic rocks (see Figure 6-4). In some instances even carbonate rocks of magmatic origin, called **carbonatites**, are present in these alkaline complexes.

Of great interest, both in economic and scientific terms are **kimberlite pipes**, which are generally spatially associated with alkaline complexes (Figure 8-32). Kimberlite is soft bluish material consisting of mica minerals and numerous fragments of ultramafic rock that have been dragged up from relatively deep within the earth's mantle. One of the materials found in kimberlite is the valuable crystalline form of carbon, diamond.

8-32. *Photograph of the Kimberley Diamond Mine, South Africa. All the diamond-bearing kimberlite that filled the upper parts of this pipe has been removed by mining, leaving this gaping, steepsided hole. [Courtesy of the DeBeers Company.]*

8-33. [*Right*] *Sketch map to illustrate extent of African Rift System.* [*Below*] *Young fault scarp at edge of the Lake Manyara Rift, northern Tanzania.* [*Photograph F. J. Sawkins.*]

Mediterranean Sea

Cairo

Nile River

Khartoum

Aden

Afar depression

Addis Ababa

Nairobi

Lake Victoria

Mombasa

Dar es Salaam

Indian Ocean

0   500   1000

km

Breakup of Continents

The most radical change that can occur in a continental interior is rifting and subsequent continental breakup. Earlier in this chapter the formation of the Atlantic Ocean by the breakup of an extensive continental mass incorporating North and South America, Europe, and Africa was briefly discussed.

The Red Sea is a modern example of continental breakup. Although

8-34. *Sketch map to show extent of the Keweenawan basaltic province and its extension in the subsurface as a feature known as the Midcontinent Gravity High. The Midcontinent Gravity High is due to the denser basaltic rocks that occupy this buried elongate zone, and probably resulted from attempted continental break-up approximately 1 billion years ago.*

the Red Sea has a very slow rate of spreading (less than 1 cm/year), it is underlain by thin oceanic-type crust and is characterized by high heat flow, both typical features of a young ocean basin. An incipient stage in the process of continental breakup may be represented by the African Rift System (Figure 8-33 A and B). Development of this extensive rift system has proceeded intermittently from Triassic to Recent times and is still going on. Whether this rift system will eventually lead to the creation of a new ocean basin is uncertain.

It does appear, however, that rifting, accompanied by extensive basaltic magmatism, can occur without subsequent continental breakup. The 1 billion year old Keweenawan basaltic province and associated Midcontinent Gravity High (the buried southward extension of this province) of the United States (Figure 8-34) probably represent just such a situation.

TRANSFORM FAULTS

Transform faults can occur within continental areas as well as oceanic areas. Large amounts of relative horizontal movement (hundreds of kilometers) can occur along transform faults, so they cause major disruptions in the original geologic patterns and juxtapose rock formations that were initially formed at considerable distances from one another. Movement along transform faults will also strongly disrupt landform patterns (Figure 8-35), and this disruption of surface features, such as ridges and stream valleys, allows geologists to locate the position of such faults in areas where they are not exposed at the surface. Several large transform faults occur along the margins of the Pacific Ocean (Figure 8-36), but the best known and most closely studied is the San Andreas Fault System in California. This fault system marks part of the boundary between the Pacific and American Plates.

In some areas, especially where it passes through hard rocks, the fault trace is marked by a narrow (~10 m) zone of finely crushed rock. In other areas the fault "feathers out" into a large number of smaller, more or less, parallel structures that may occupy zones as wide as ~10 km. Careful geologic work on the rock assemblages on either side of the fault have indicated that at least 260 km of movement have taken place along this structure since Oligocene time. Magnetic anomalies at the mouth of the Gulf of California indicate

8-35. *Aerial photograph of the trace of the San Andreas Fault in southern California. Note strong effect of fault on the topography of the area.* [*Courtesy of C. R. Allen.*]

that this much motion has, in fact, taken place within the last 4 million years—equivalent to an average rate of movement of 6 cm/year.

The San Andreas fault is still very active, at least along some of its length. In the area between San Francisco and Los Angeles movement along the fault, accompanied by frequent small earthquakes, is currently taking place at approximately 6 cm/year. In the San Francisco and Los Angeles areas no movement along the San Andreas fault is apparent at present; therefore, the 6 cm/year relative motion almost certainly is being absorbed in the form of elastic strain in the rock units on either side of the fault.

Under San Francisco (Figure 8-37) the fault appears to have been essentially "locked" since the earthquake of 1906 brought large scale destruction to that city (see Chapter 12). Obviously the accumulation of strain in the crust below San Francisco will eventually exceed the strength of the rocks in which the strain is stored; then, if its release occurs suddenly, a major earthquake will undoubtedly result. Much the same is true for the Los Angeles area in Southern California where the San Andreas fault is also currently "locked." In terms of plate tectonics and the earthquake hazards associated with plate margins, these two large cities could hardly be located in less favorable areas.

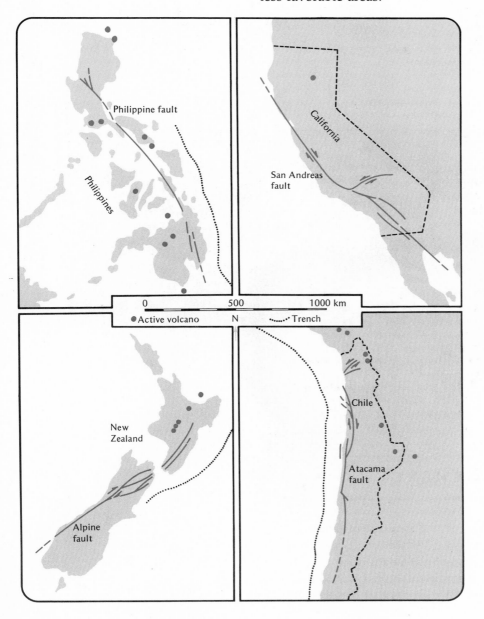

8-36. *Series of sketch maps to illustrate four areas along the margins of the Pacific that contain large transform fault systems. [From C. Allen: Circum-Pacific faulting in the Phillipines-Taiwan region. J. Geophys. Res., 67: 4795–4812 (1962).]*

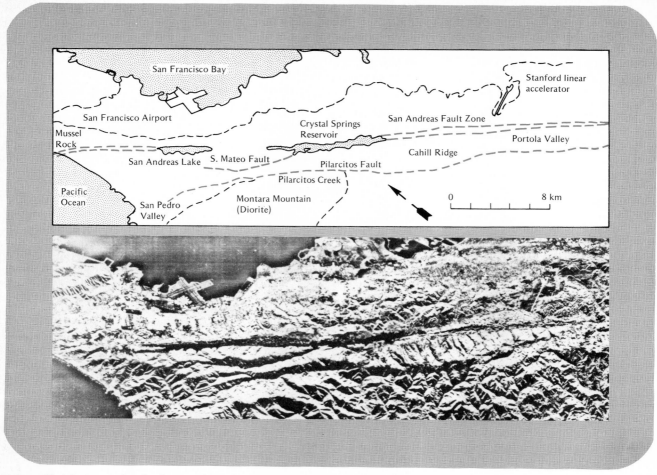

8-37. *Sketch map and radar image of San Francisco Peninsula showing trace of San Andreas Fault in close proximity to the city of San Francisco. [NASA image courtesy of U.S. Department of the Interior Geological Survey.]*

CONCLUDING STATEMENT

As a conclusion to this chapter on the geology of continents, it is necessary to add a word of caution. Plate tectonic models provide a satisfactory and internally consistent explanation for the generation, deformation, and breakup of continental crust during the latter portions of earth history. However, the events that occurred during the earlier times are shrouded with uncertainty because the record is much more difficult to read and because the conditions prevailing at and near the surface of the earth may have been somewhat different.

For example, at the time the oldest known rocks formed, about 3.8 billion years ago, the amount of heat being produced within the earth by radioactive decay of potassium, uranium, and thorium (the main heat producing radioactive elements in the mantle and crust) must have been approximately double the present value. Therefore, even if lithospheric plates formed and tectonic interactions of plates did occur at this early time, the nature of these ancient plates and their interactions might not have been strictly comparable to those we observe today.

QUESTIONS

1. List the geological features found along the western margin of South America that indicate this to be a convergent plate boundary.

2. Active continental margins tend to be elevated much higher above sea level than island arcs. Why is this and what geologic consequences does it have?

★ 3. Discuss the controversy that existed amongst geologists regarding the origin of batholiths and develop a model for the formation and emplacement of batholiths along the western margins of North and South America.

4. Sedimentary basins containing molasse tend to form on the inner side of active continental margins. What can be learned by a study of the sediments and sedimentary structures found in these basins?

5. The mountain belts that form along active continental margins are sites of major deformation. Contrast the types of structures formed at shallow depth in such belts versus those formed at depths in excess of ~5 km.

6. Explain the concept of "thin-skinned" tectonics. What is the fundamental force that leads to the deformation of rocks by thin-skinned tectonics?

7. Give two examples of continental collision. Explain with the help of diagrams what type of plate tectonic situation leads to continental collision.

8. Outline briefly the history of the north Atlantic Ocean and its borderlands since the end of Precambrian time.

9. Illustrate with a diagram the main features of inactive continental margins. By what means are sedimentary materials derived from erosion of the adjacent continent transported beyond the continental shelf?

10. What is meant by the term unconformity? What does the presence of an unconformity in a rock sequence indicate to a geologist?

★11. Explain why the sedimentary units that were deposited on continental interiors tend to be thin but extensive.

★12. Extensive basaltic magmatism in an intracontinental environment may indicate attempted continental breakup. Explain the basis for this statement, citing any examples you can.

13. A major earthquake hazard exists in both San Francisco and Los Angeles. Explain as fully as you can why.

★14. The average age of continental material is much greater (about ten times) than oceanic material. Explain as clearly as you can why this is a predictable consequence of plate tectonic theory.

Ocean and Atmosphere

nine

If we were to look down on the earth from outer space, most of what we would see would be water—water in the oceans and water suspended in the air of the atmosphere as clouds. We usually think of geology as concerned with rocks, but those two fluids, water and air, are also very important to surficial geology. Water and air serve to transmit and distribute solar energy over the surface of the earth and are the means by which solar energy can drive the processes of weathering, erosion, and sediment transport. Much more immediately important to us, water and air are essential to maintaining life as we know it. Air shields us from the deadly solar radiation of outer space and enables us to breath, and water is an irreplaceable fluid in the workings of our cells. Without those two fluids, the surface of the earth would be as lifeless as that of the moon.

HYDROSPHERE AND ATMOSPHERE

The earth's **atmosphere** extends upwards, ever thinner, to heights of about 100 km above the surface. The atmosphere is made up of all the free gases that are prevented from escaping into space by the pull of the earth's gravity. The total supply of free, chemically un-

Location	Hydrosphere, %
World ocean	97.2
Ice (ice caps, glaciers, sea ice)	2.15
Subsurface water (soil moisture, ground water)	0.625
Surface fresh water (lakes, rivers)	0.009
Surface saline water (salt lakes, inland seas)	0.008
Atmosphere (water vapor, clouds)	0.001

Table 9−1 Estimated Distribution of The World's Surficial Water Supply*

*Note that less than 1% of the approximately 1.4×10^9 km³ of water is liquid fresh water directly useful for supplying human water needs. Total mass of the hydrosphere is about 1.4×10^{24} g. Data from U. S. Geological Survey.

combined water near the earth's surface is called the **hydrosphere.**

A quick glance at Table 9−1 shows that almost all the water of the hydrosphere can be found in the oceans. The next most important location of water is in the earth's ice—the great continental ice caps, in the glaciers, and in sea ice. Geologically we live in a rather unusual age, and through much of geologic time the amount of water tied up in ice has been much less. Only a small part of the hydrosphere, less than 1%, is in the form of fresh (nonsaline) liquid water. Of that, only a portion can be tapped to supply our personal and industrial needs. Although atmospheric water in clouds is the most spectacular feature of the earth as seen from space, the amount of water in the atmosphere is tiny compared to that in the oceans and elsewhere on the surface.

ORIGIN OF THE OCEAN AND ATMOSPHERE

The geologic history of the hydrosphere and atmosphere is by no means easy to reconstruct. We have only the present state of the ocean and atmosphere and the record left behind in rocks to go by. The present state of our fluid environment could conceivably be reached by many paths, and the rock record is patchy, incomplete, and often ambiguous. Yet the question of how the volatile elements, which make up water and air, have behaved with time is very important in models of the earth's early and subsequent history. The history of water and air is not separate from the history of the rest of the earth.

It seems reasonably well accepted by earth scientists today that water and the other volatiles of the hydrosphere and atmosphere were not accreted as free gases in the early stages of earth formation. Their composition is not the same as the average composition of solar system gases, although the relative scarcity of hydrogen and helium on the earth cannot really be used as evidence. The light atoms of hydrogen and helium can escape because they reach sufficient velocity in thermal motions to break free from the earth's gravitational pull. More significant is that the heavier noble gases, neon, krypton, and xenon, are also depleted in the earth's atmosphere relative to solar system abundances, and these gases are not likely to have leaked away. Thus, at present the earth's atmosphere is quite different from the solar system composition.

In addition, the proportion of gas to rocky material on the earth is far lower than the solar system average. This is also true of the other three inner planets. For these reasons most workers postulate that the earth's present atmosphere and hydrosphere were accreted either as gases absorbed on the surfaces of solid particles or in chemical combinations such as water of hydration. At this point the agreement ends.

Degassing Model for Ocean and Atmosphere

One school of thought holds that the volatile elements are gradually being leaked out of the interior of the earth as volcanic gases (Figure 9-1). This process is called **degassing** of the earth. The gases that accompany volcanic eruptions are very similar in composition to that part of the hydrosphere and atmosphere that cannot have come from weathering of igneous rocks. Isotopic studies show that a large part of the water in volcanic gas was absorbed into the magma from the surface, but some of the volcanic volatiles must be of mantle origin.

The degassing model carries with it two consequences. One is that a cold origin for the earth must be assumed because, if a significant part of the mantle had ever molten, the volatiles included in it during accretion would surely have escaped. The other consequence is a gradual but steady increase in the volume of the oceans and the mass of the atmosphere. The early thermal history of the earth is very much open for speculation, but the second consequence raises a tangible problem.

Continental Freeboard Problem. The problem has to do with the "freeboard" of the continents—the amount by which the continents stand above sea level. Geologic evidence tells us that the

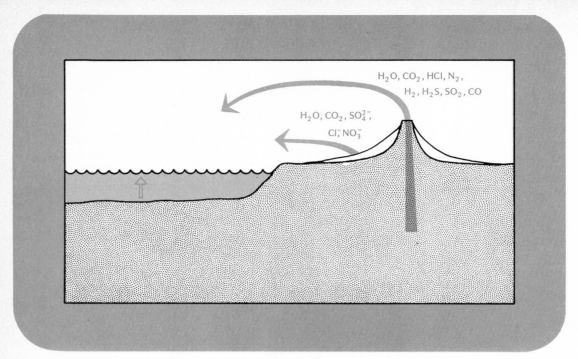

9-1. *One model of the history of the atmosphere and hydrosphere—gradual growth by degassing of the earth's interior during volcanic activity. Volcanic gases contain mostly water, with lesser amounts of nitrogen, carbon dioxide, hydrogen, sulfur dioxide, carbon monoxide, and other gases. At least some of these gases may come from mantle sources. Weathering of igneous rocks can provide minor amounts of water, carbon, chlorine, sulfur, and nitrogen. The open arrow depicts the effect on sea level.*

surfaces of most of the Precambrian continental platforms, or cratons, have remained near sea level for at least the last 600 million years. The extensive cover of Paleozoic and older marine sediments of very shallow water origin on the stable cratons is sufficient evidence that the relative elevation of the continents has not changed much for a long time. If the ocean has been growing in volume, the continents should be increasingly submerged; yet they are not. If one believes that lateral continental accretion takes place, encroaching on the ocean basins, the problem is that much worse.

Recycling Model for Ocean

There are various ways to avoid the continental freeboard problem. The simplest and most direct is to assume that the present ocean and atmosphere were formed very early in earth history and have remained more or less stable in volume. Belief in hot origin of the earth or in early core formation with melting caused by release of gravitational energy goes naturally with this viewpoint. However, the model of a very early ocean and atmosphere also has its problems. A xenon isotope that is a decay product of a short-lived, now extinct (Chapter 4) isotope of iodine has been detected in deep gas wells. The xenon most likely was incorporated in the earth very early in its history and is just now escaping. This indicates that at least some kind of degassing is now going on.

Both of the models above fail to consider the possibility of leakage of water and other volatiles back into the mantle. However, plate

9-2. *Recycling of water through the mantle during plate tectonic processes. The colored arrows show the path of water through the system. Water can be removed from the ocean during sea floor spreading (A), carried down in subduction zones (B), and returned to the surface by volcanism (C). If the rate of A is greater than C, the ocean will decrease in volume and the mantle will become more hydrated. If C is faster than A, the oceans will grow and the mantle become dryer. Neither rate is well known at present.*

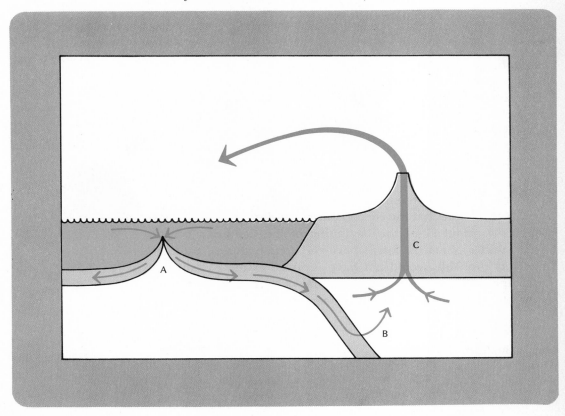

tectonics provides just such a possibility. Consider the example of water. In plate tectonic theory, there are two ways water can communicate between mantle and surface (Figure 9-2). Degassing of mantle water might accompany island arc volcanism, as in the growing ocean model. The new water brought up would eventually enter the ocean. However, sea water can return to the mantle after incorporation into the oceanic crust during metamorphism and hydration of the dry basalts formed by sea floor spreading. Crust hydrated in this way and perhaps some wet oceanic sediments would then be carried back into the mantle in a subduction zone. There the water would be released by the higher temperatures at depth.

Given the freedom of water to travel in both directions between mantle and surface, anything can happen. The volume of the oceans would increase if the first process is faster, would remain constant if the two balance, or would actually decrease if the latter process dominates. It is still not clear which of these three possible histories has actually been true of the earth's ocean. Furthermore, it is best to be very cautious in relating the history of water to that of other volatiles. Substances with different physical and chemical behavior are quite likely to have different histories.

Composition of Sea Salt

The composition of sea salt seems less likely to have suffered drastic changes than the volume of the oceans. It is understandable that ancient evaporite deposits should be scarce because they can be so readily dissolved by ground water. However, those that have survived are not much different from modern examples. It is fair to conclude from this that the mix of elements dissolved in the sea has not changed much, although the total amount may be different. One can also argue that the concentration of salt in the sea has been relatively constant since the beginning of the Cambrian because certain species of marine fossils have very long histories, and these organisms probably would not have survived extreme changes in the composition of sea water. This kind of long term stability in composition does not seem to have been true for the atmosphere.

Oxygen in the Atmosphere

As with so many other things about the early earth, the exact nature of the early atmosphere is not known. One thing is clear—it could not have contained even a fraction of its present concentration of oxygen gas. One popular model for the composition of the early atmosphere is similar to the atmosphere of Jupiter; there carbon, nitrogen, and oxygen are combined with hydrogen and exist as methane, ammonia, and water, and possibly there is some free hydrogen. Other workers believe that the atmosphere must have been composed of carbon dioxide, nitrogen, and water by 3.3 billion years ago, the age of the oldest known sedimentary rocks.

In either case, a very small amount of free hydrogen and oxygen gases could be formed from water molecules broken up by ultraviolet light in the upper atmosphere. Some of the resultant hydrogen gas would escape from the earth's gravity, leaving residual oxygen. Free oxygen would have had a very short lifetime because it would combine rapidly with the methane-ammonia atmosphere, which would act as a reducing agent.

No matter what the original state of the atmosphere, the development of life began to change its composition. One advantage of the methane-ammonia-water-hydrogen model for early air is that organic molecules are relatively easy to form in such a mix. Experiments have shown that a spark discharge or ultraviolet light can synthesize amino acids, the basic building blocks of life, from that mixture of reducing gases. Thus, life on earth may have started with strokes of lightning. Free oxygen would have been a poison to early organisms. After photosynthesizing plants evolved and the oxygen content of air began to rise, mechanisms protective against oxygen also had to be evolved.

Only the outline of the time scale for evolution of life and of the atmosphere is known. The oldest known earth rocks at the time of this writing are some metamorphic rocks in Greenland aged about 3.8 billion years. They tell us little about any early atmosphere. The oldest known sedimentary rocks are located in Africa and have an age of about 3.3 billion years. The hydrosphere must have existed by that time, and these old rocks bear signs of organic remains. By 2.7 billion years ago, photosynthetic plants are known to have existed.

In Precambrian rocks older than 1.8 to 2 billion years, banded iron formations are common. The iron in these formations must have been transported in the reduced state as ferrous ions. The reduced ferrous (Fe^{2+}) ions of iron are soluble in water, while the oxidized ferric ions (Fe^{3+}) are not. In younger rocks, banded iron formations are practically nonexistent. The end of deposition of the iron formations signalled some event in the chemistry of the ocean and atmosphere, probably a significant increase in the amount of oxygen. The appearance of abundant hard-shelled fossils at the beginning of the Cambrian, 600 million years ago, may also mark an atmospheric event. However, objects that look very much like the remains of hard skeletal parts have recently been discovered in 1.2 billion year old rocks in Arizona.

THE PRESENT STATE OF THE ATMOSPHERE

Composition of the Atmosphere

At least one set of facts about the atmosphere is not plagued by the uncertainties that dominated the previous discussion. The average composition of the present atmosphere is very well known. As can be seen from Table 9-2, nitrogen gas makes up almost 80% of the atmosphere by volume. Since the volume of any gas depends only on the number of molecules and not on the molecular weight, the volume percentages are also equivalent to per cent of total molecules. Nitrogen and oxygen together account for 98–99% of the atmosphere. The amount of water vapor is variable, ranging up to about 1%. The most abundant of the noble gases, argon, is interesting for its origin. Most atmospheric argon has been derived from decay of radioactive potassium-40, with subsequent outgassing of the resulting argon-40. Carbon dioxide, although present in rather small amounts, is as important to life on earth as oxygen because it is needed for photosynthesis by plants.

Of all the assorted gases of the atmosphere, perhaps only the unreactive noble gases can be described as stable, inert components. The other gases all are involved in a series of complex cycles by which

Gas	Volume, %	
Nitrogen (N_2)		78.1
Oxygen (O_2)		20.9
Argon (A)		0.934
Water (H_2O)	variable, up to	1.0
Carbon dioxide (CO_2)		0.031
Neon (Ne)		1.8×10^{-3}
Helium (He)		5.2×10^{-4}

Table 9–2. **Present Composition of the Earth's Atmosphere** *

*Other gases that exist in trace amounts are ozone (O_3), hydrogen (H_2), xenon (Xe), krypton (Kr), methane (CH_4), carbon monoxide (CO), and nitrous oxide (N_2O). The total mass of the atmosphere is 5.3×10^{21} g.

they are removed from and returned to the air in a series of steps. For example, water enters the atmosphere as water vapor by evaporation from the oceans and leaves again as precipitation—rain, snow, hail, and dew. This cycle is powered by solar heat energy. The importance of temperature to the cycle and to the atmosphere is obvious if we note that a surface temperature of greater than 100° C would boil the oceans. Since the mass of the oceans is 250 times the mass of the atmosphere (Tables 9-1 and 9-2), a very dense, practically pure steam atmosphere would be the rather unpleasant result.

Carbon dioxide and oxygen participate in both organic and inorganic cycles. The two gases are involved in the organic processes of respiration, photosynthesis, and decay. The concentration of carbon dioxide in air is controlled mostly by its solubility in the vast oceanic reservoir as bicarbonate (HCO_3^-) ions. Carbon dioxide concentration is also influenced by weathering of carbonate minerals in rocks. Figure 9-3 demonstrates how complicated the carbon cycle can be.

Oxygen is used up in weathering of rocks and released by break up of water in the upper atmosphere. Any excess of photosynthesis over respiration will also produce extra oxygen (Figure 9-4). Even the relatively inert nitrogen gas passes through a cycle of incorporation into organic compounds and release by decay.

Three gases play an especially important part in controlling energy transfer through the atmosphere. Ozone (O_3), formed by recombination of oxygen molecules broken up by ultraviolet light, serves as a shield against ultraviolet by absorbing it strongly. Without a layer of ozone in the upper atmosphere, sunlight would be deadly to all creatures and plants that ventured beyond the protection of a body of water.

The Greenhouse Effect. Water vapor and carbon dioxide are very important in another process, called the **greenhouse effect**. Most of the energy in solar radiation lies in the visible wavelengths of light, which pass readily through the atmosphere. When the earth's surface absorbs solar energy, it heats up and reradiates infrared light. Water vapor and carbon dioxide absorb infrared strongly and reradiate the infrared light they absorb in both directions, up to space and down to earth. Thus, a good part of the reradiated solar energy is not allowed to escape and remains trapped in the ground and atmosphere. The greenhouse effect helps to maintain the surface temperature of the earth in a particularly favorable range, which allows liquid water to exist at the surface.

On the planet Venus, the surface temperature is much higher,

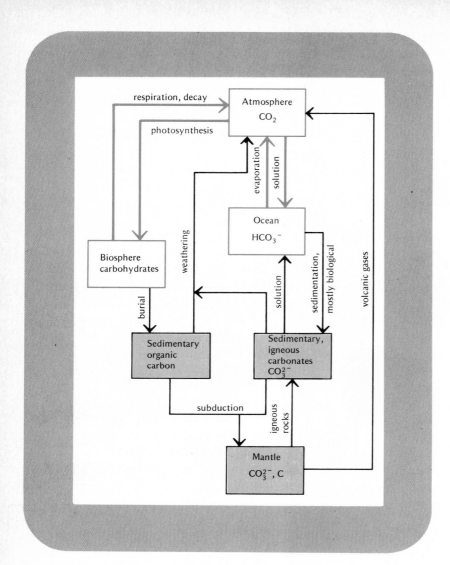

9-3. *The complex cycle of carbon (C) and carbon dioxide (CO_2) between atmosphere, ocean, biosphere, lithosphere, and mantle. The boxes denote carbon reservoirs, and the arrows indicate processes that transfer carbon between reservoirs. The more important processes are marked with heavy colored arrows.*

around 700° C. The high temperature is probably partly caused by a stronger greenhouse effect, because Venus's atmosphere contains much more carbon dioxide than the atmosphere of the earth. The fact that Venus is closer to the sun and receives twice as much solar radiant energy as the earth also contributes to a higher atmospheric and surface temperature.

Recently, some chemists and meteorologists have become worried that human activities might upset the temperature balance of the earth. Fossil fuels are being burned at a tremendous rate, and carbon dioxide released into the atmosphere much more quickly than before

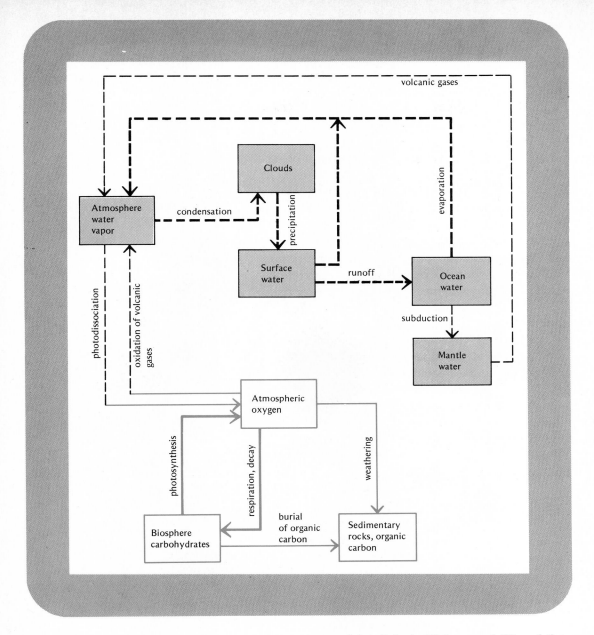

9-4. *The interlinked cycles of water and oxygen are as complex as that of carbon. Colored arrows represent the path of oxygen, black arrows transport of water. The more important processes are marked with heavy arrows.*

human use of fossil fuels. If the capability of the oceans to keep the atmospheric level of carbon dioxide constant were exceeded, the surface temperature of the earth would rise due to a stronger greenhouse effect. A change of only a few degrees in average worldwide temperatures could cause the polar icecaps to melt and sea level to rise. This would have disastrous results on low-lying human population centers, as well as on the habitats of more innocent land-dwelling animals and plants. Fortunately, at present the ocean seems able to keep up with the increased production of carbon dioxide.

Motions of the Atmosphere

Wind and weather are the means by which the atmosphere moves trapped solar energy around the world. There are three main physical processes operating to drive winds and control their direction. Solar radiant energy converted to heat by absorption causes masses of air to expand and become less dense. Then gravitational forces cause the expanded air mass to rise and denser air to flow in underneath and take its place. The resulting circulation is then twisted into involved patterns by the rotation of the earth. These primary processes are well understood, but the overall effect is so complicated and uneven that even today weather prediction is a very risky business.

The amount of solar radiation falling on a given patch of ground varies with latitude (Figure 9-5), time of day, season, cloud cover, and type of ground. Masses of air in different places thus get very different heat inputs. Air over land is heated more in the daytime and cooled more at night than air over water. With its greater heat capacity, water absorbs more solar energy in the day and releases more at night. Influenced by all these effects, the pattern of warmth and therefore of density of the air can be very complex.

One additional factor that is a bit surprising at first glance is that wet air is less dense than dry. The nitrogen molecule (N_2) has a molecular weight of $2 \times 14 = 28$, whereas the oxygen (O_2) molecule has molecular weight 32. Thus, the water molecule (H_2O) with its molecular weight of 18 is only a little more than half as heavy as the average for air. Because the water molecules will take up just as much space as the heavier nitrogen and oxygen, wetter air will be less dense. Not only will warmed air tend to rise but, if other things are equal, so will air containing appreciable water vapor.

The rise of warm air and inflow of colder air can lead to precipitation as well as wind. As a parcel of air rises, it moves into the lower pressures existing at greater elevations above sea level and, as a result, expands even farther. The expansion leads to a drop in temperature. This in turn can cause the air to become oversaturated with water vapor, because cold air can hold less water. Depending on the temperature and other conditions, either rain or snow can be formed.

If a wind runs into a topographic obstacle and is deflected upwards, the effect can be the same—expansion and cooling of air with the release of pressure and resulting condensation of moisture. The processes just mentioned are all ways in which heat energy derived from sunlight is transformed into gravitational, kinetic, and internal energy of air masses.

The Coriolis Effect. The already turbulent flow of air caused by unequal solar heating and the resulting buoyancy forces is then

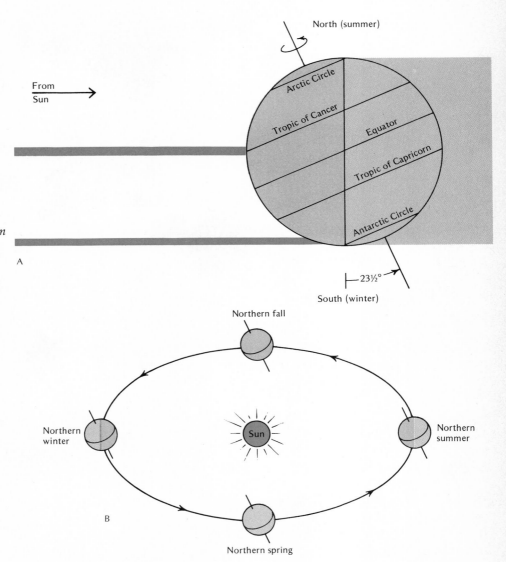

9-5. A. *The causes of variation of solar radiation input with latitude are partly due to the shape of the earth and partly due to the fact that the axis of rotation is inclined at 23.5° to the plane of earth's orbit. Drawn for the middle of the northern summer. A given amount of sunlight is spread out over a greater area near the poles than directly under the sun. In addition, in a given day the summer hemisphere receives more radiation than the winter hemisphere because more of it is exposed. The first effect causes the polar regions to be colder than the tropics, and the second causes the seasons. B. How the tilt of the axis causes first the northern then the southern hemisphere to be exposed to the sun in the course of a year's orbit about the sun.*

9-6. *The Coriolis effect. The length of the colored arrows shows how fast the eastward motion due to the earth's rotation is at each latitude. As a wind blows away from the equator toward one of the poles (thick black arrows), it passes over ground that is not moving eastward as fast as that from which it came. Seen from the ground, the wind will appear to be deflected to the east. Wind blowing toward the equator suffers the opposite effect; it is left behind by the faster moving surface over which it passes and appears to be deflected to the west.*

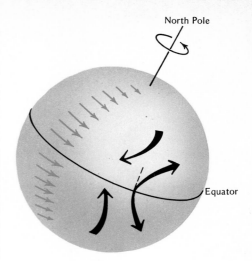

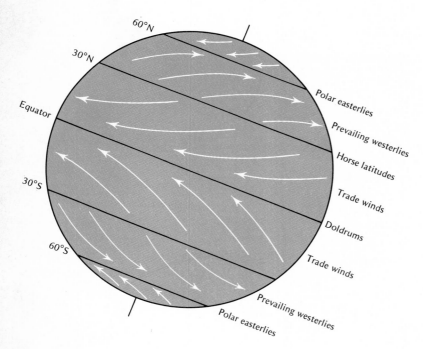

9-7. *The general, average circulation of the near-surface winds. For any particular time, an actual map of the wind directions would be much more complicated. Winds are named for the direction* from *which they blow. The Coriolis effect forces the winds into their predominately east-west pattern. Thus, the prevailing westerlies of the midlatitudes blow towards the east because they move, on the average, away from the equator.*

complicated by the rotation of the earth. The rotation leads to the Coriolis effect, which is explained in Figure 9-6. This effect causes any flow of air in the Northern Hemisphere to be deflected to the right, whereas flow in the Southern Hemisphere is bent to the left. The deflection itself makes the winds circulate *around* the high and low pressure areas created by unequal heating rather than flowing directly away from the high pressure or directly toward the low pressure areas. As air seeks to flow away from a high pressure area, the winds are bent into a clockwise spiral in the Northern Hemisphere. Winds will spiral counterclockwise into an area of low pressure. In the Southern Hemisphere, the directions are reversed. Superimposed on the flow around high and low pressure systems is the general pattern of surface winds shown in Figure 9-7.

The effects of weather on surficial geology are many and varied. One effect might be guessed from the term for rock decomposition—weathering. Climate and supply of atmospheric moisture are very important in weathering processes. The winds themselves cause erosion and transport sediments, forming dunes, aeolian sandstones, and the fine silt called **loess**. The most important factor in erosion and sedimentation on the continents, however, is runoff of water from precipitation. We will discuss these processes in more detail in Chapters 10 and 11.

THE PRESENT STATE OF THE OCEAN

The ocean is as important as the atmosphere in controlling the climate of the earth. We have already mentioned that water acts as a thermal moderator because its great heat capacity smooths out variations in temperature. In addition, water has a built in temperature regulator—evaporation. The hotter ocean water comes, the greater tendency there is for evaporation to occur if the overlying air is not already saturated with water vapor. The process of evaporation of water involves the absorption and conversion of heat energy into internal energy of the water vapor produced. Thus, water vapor carries heat away as stored energy, acting to cool the remaining liquid water. The energy is not released until condensation of the vapor occurs. For the same reason, perspiration cools our bodies.

Motions of the Ocean

Deep sea sailors have long had to contend with a complicated pattern of ocean currents (Figure 9-8) and winds. Most realized that winds and currents are related, but probably few of them realized that both are ways of redistributing solar energy on a rotating globe.

Currents. To a certain extent, the surface currents show in Figure 9-8 are driven by wind. One can see that the great **gyres**, or closed patterns of circulation in the North and South Atlantic and Pacific Oceans and in the Indian Ocean, conform to the average wind directions of Figure 9-7. In addition, variations of density due to temperature salinity differences and the Coriolis effect are important in driving the oceanic circulation.

Motions of the deeper waters are almost entirely due to density differences and the Coriolis effect. Even the deep waters tend to acquire their characteristic densities at the surface. Solar heating causes the water to expand, becoming less dense. Mixing in of fresh water dilutes the salt of sea water and also decreases the den-

9-8. *The ocean currents. The generally circular pattern in the oceans above and below the equator are called gyres. The Gulf Stream is the warm current originating in the Caribbean, passing up the coast of North America, then crossing the North Atlantic.*

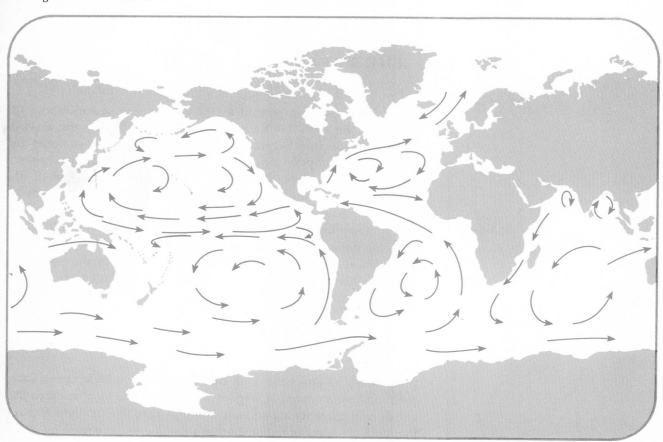

sity. On the other hand, evaporation concentrates the salts in water, making it denser. Exposure to cold and ice, as in the arctic regions, has the same effect. The various masses of water, having acquired a typical density in these ways, will try to find their own levels in the ocean, with denser water below and less dense water above. Then, as denser masses of water seek to flow beneath less dense masses, their motions are twisted by the Coriolis effect.

The surficial currents serve to modify the climate of coastal areas. Northwestern Europe is spared a sub-arctic climate by the warm waters of the Gulf Stream, which run all the way across the Atlantic from the Caribbean. The western coast of South America is rendered much cooler than usual by the Humbolt current coming north from the Antarctic waters.

Upwellings of cold subsurface water, where winds blow away from the coasts, not only affect the local climate but lead to greatly enriched biological productivity. The supply of unconsumed nutrients that accumulate in deep water are brought up in upwelling areas and stimulate biological activity. The upwelling itself is caused by the wind moving the warmer surface water away, out to sea. The fisheries off Peru and Chile owe their especial richness to such coastal upwelling.

Tides. The ocean currents move vast quantities of water. There are two other very important kinds of motion in the sea, tides and waves; these involve water moving back and forth, more or less in place, with very little net transport. There is more energy, however, in the motions of the tides and waves than in the currents. We will first consider the tides.

Tides are quite different from the oceanic and atmospheric processes discussed so far because they do not rely upon solar energy. They are purely a gravitational effect. The moon is the most important source of the ocean tides. As shown in Figure 9-9, there is one tidal bulge toward the moon and another on the other side of the earth. The sun also raises tides, causing a tidal force about half as strong as the lunar force. As the earth turns on its axis, the peak of the lunar tidal force will sweep past a point on the surface twice every 24 hr and 50 min. The extra 50 min comes from the fact that in 24 hr the moon will have moved on in its orbit and the earth must turn a little further to catch up. It is easy to see that the highest tidal forces will occur when the sun and moon are lined up with the earth, at either full or new moon.

The tide producing force varies quite regularly, depending on the position of the moon and sun. The tides themselves do not.

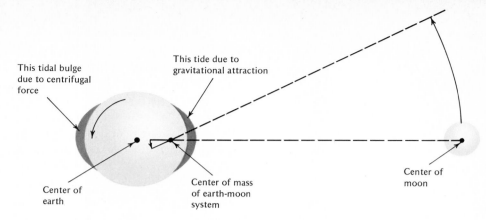

9-9. *The origin of the tides. The gravitational attraction of the moon provides a force that tends to pull the oceans into a bulge facing the moon. There is a second tidal bulge on the opposite side of the earth that is due to centrifugal force. The earth and moon rotate once a lunar month around their common center of mass, the balance point of the system. This balance point lies more than 4500 km from the center of the earth toward the moon. The eccentric motion of the earth as the moon revolves about it causes the second bulge to be thrown outward by centrifugal force. The two tidal bulges are of nearly equal height. The earth rotates beneath the bulges, causing the daily rise and fall of tides. Distances in the diagram are not to scale.*

The actual rise and fall of sea level at any point on the sea shore can be very complicated. Because water is a fluid and its flow is modified by the shape of its container, the shape of the ocean basin strongly affects the height of the tides. In the deep ocean basins, the tidal change in sea level is usually less than a meter. When the tide approaches the shallower water near a coast line, its height can be greatly amplified. In addition to the ocean tide, there is a small tide in the not-quite-rigid solid earth that causes the surface to rise and fall by as much as 3 cm in a tidal cycle.

The energy of tidal motion is capable of doing surficial geologic work. Erosion in both the near-shore and deep sea environments is caused by tidal currents where they run at sufficient speed. One of the most interesting tidal effects, however, acts on a much larger scale. Friction acting on the motions of the tides causes the dissipation of the rotational energy of the earth into heat. Thus, the earth's rotation rate is gradually slowed. Daily growth rings in Devonian corals show that 370 million years ago there were 400 days in the year. Since that time, the rotation rate has slowed to the present 365 days per year. Simultaneously with the slowing of rotation, energy of angular motion is transferred to the moon, which is speeded up and moved outward to a higher orbit.

Waves. The most relentless geologic work of ocean water is done by waves. Wind waves are created when a wind blows over the surface of the water and piles up ridges, which move in the direction of the wind. Over a large stretch of water and with high winds, the waves can pick up considerable energy and transport it to be released upon the shores. As a wave travels, the water particles go through a circular motion (Figure 9-10) that extends down only about 10 m. When the wave arrives in shallow water, the rising bottom causes it to heighten and break. It may break several times before washing up on the beach as surf. Seismic sea waves, triggered by submarine earthquakes, landslides, or volcanic explosions, travel at great speed and can break with enormous fury and destructive potential upon an open coast.

The erosion of shorelines by wave action, especially by storm waves, can be quite dramatic (Figure 9-11). Great sea cliffs are formed where hills have been worn half away. Usually, however, the shore is somewhat protected by a rocky or sandy beach, which absorbs part

9-10. *Waves in the ocean. As a wave passes by, the water moves in circular patterns that become smaller and smaller down to a depth called the* **wave base.** *When the wave enters water shallower than the wave base, the motion of the water begins to be affected by the bottom, causing the wave to heighten and eventually break. Waves travelling in deep water do not involve any net transport of water.*

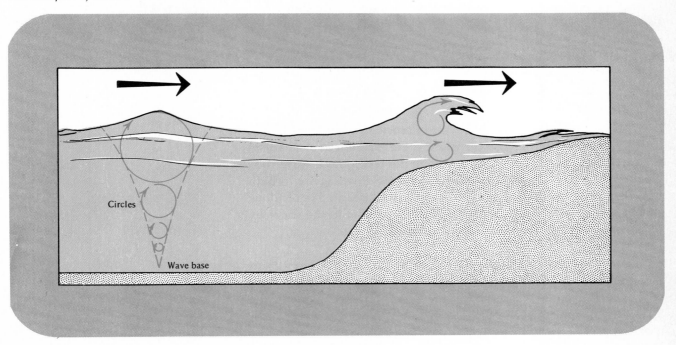

Circles

Wave base

9-11. Sea cliffs *and other shore line erosional features on the Oregon coast. The rocks standing up out of the water are isolated remnants of the sea cliffs and are called* stacks. *In the background, the flat surface of the land is probably an uplifted* marine terrace. *In the foreground, a new marine terrace is forming at sea level* [Photo by B. Dahlin.]

of the wave energy and slows the rate of erosion. When the crests of the waves strike obliquely on the shore, there is a built in mechanism for moving the products of erosion along the beach. Each wave runs diagonally up the beach, then diagonally back down again; in both cases the wave moves the same way along the beach. Tremendous amounts of sand grains can be transported parallel to the shore by the constant angling of the waves.

Sea Level. Although the waves and tides can carve wide terraces on the shore and can level volcanic islands, the most important effect of the level of the seas and oceans on the continents lies above sea level. Rivers cut down the landscape, but only to sea level. Thus, changes in sea level can greatly change the aspect of the land. During the height of the Pleistocene glaciation, enough water was locked into the continental icecaps to lower sea level significantly. As a result, rivers cut down their valleys into the newly exposed land. When the sea returned, the new valleys were drowned and became submarine canyons.

There are factors other than glaciations that can change the level of the oceans globally. Changes in the average rate of sea floor spreading could affect the volume taken up by the midoceanic ridges within the ocean basins and cause sea level to fluctuate. Superimposed on these comparatively rapid worldwide changes would be any long term growth or shrinkage of the amount of sea water. Local sea level changes can also result from vertical movements of the crust over restricted areas. The isostatic rebound of Scandanavia, mentioned under ''Isostasy'', in Chapter 4, has lifted new land from beneath the waves. In any case, the rivers will continue cutting the land down to the level of the sea and depositing the sediments beneath the water.

Composition of the Ocean

Although sea water is, on the average, 97.5% water by weight, the dissolved salts that make up the other 3.5% are of considerable interest. The **salinity**, or saltiness, is usually quoted in **parts per thousand (ppt)**, which is the same as grams of dissolved salt per kilogram of sea water. The relative amounts of the major dissolved ions in sea water are remarkably constant for ocean water in various areas (Table 9-3), although the total salinity varies somewhat from place to place. The usual range of salinity in the open sea is from 33 to 38 parts per thousand.

Two of the processes that affect the salinity, evaporation and dilution by fresh water, have already been mentioned in connection

Table 9-3   Concentration of the Most Important Dissolved Ions in Sea Water*

*The concentrations shown are for a total salinity of 35 ppt equivalent to 35 g of dissolved salts per kilogram of sea water.

Ion	Concentration, g/kg	Per Cent of Total Ions Present
Chloride (Cl^-)	19.3	48.7
Sodium (NA^+)	10.8	42.0
Sulfate (SO_4^{2-})	2.7	2.5
Magnesium (Mg^{2+})	1.3	4.8
Calcium (Ca^{2+})	0.41	0.9
Potassium (K^+)	0.39	0.9
Bicarbonate (HCO_3^-)	0.14	0.2

with deep currents. In addition, ice formed from sea water is relatively salt free, so the unfrozen water becomes more saline. If the ice is detached from its place of origin and blown elsewhere, it will reduce the local salinity when it melts. Rain falling on the ocean will do the same.

Every year about 3.5×10^{15} g of dissolved materials are brought into the ocean by the rivers of the world. We know that the salinity of the sea is not changing rapidly with time; therefore, an equivalent amount of material must be removed from ocean water each year. Thus, the dissolved ions in the sea must be involved in cycles analogous to the cycles the water itself travels through (Figure 9-4). Each type of ion is being added to and removed from the ocean in a way that keeps constant the total amount of that ion present at any time.

The most abundant ion, chloride (Table 9-3), is recycled rather directly. Most of the chloride dissolved in river water comes from salts deposited in sedimentary rocks or from sea salt that has been blown inland by winds and removed from the atmosphere by rain. Sodium and potassium are freed from igneous rocks, in particular from feldspar minerals, by weathering and then carried to the sea by streams. These two ions are probably removed from the sea mainly by clay minerals. Sulfate is formed by oxidation of the common accessory mineral pyrite and washed into the sea from which it is removed by sulfur-reducing bacteria. These bacteria thrive in stagnant, oxygen-poor sea water and form hydrogen sulfide gas and sedimentary pyrites.

Magnesium and calcium are mainly derived from weathering of limestone and dolomite. The calcium is removed from sea water by deposition in the shells of calcite-secreting marine organisms. It is somewhat of a mystery where the magnesium goes. Carbonate and bicarbonate follow mainly an organic path, with contributions from the atmosphere. Carbonate produced by soil bacteria or atmosphereic carbon dioxide make ground water acid and help dissolve calcite and dolomite, yielding bicarbonate ions. At the same time that marine organisms secrete calcite, they also release carbon dioxide to the atmosphere. Silica, which is not one of the major dissolved constituents of sea water, is present in small amounts and is also biologically precipitated.

When ocean water becomes isolated and evaporation outruns addition of fresh water, salty brines and then evaporite deposits are formed as the salt becomes more concentrated. Although sodium and chloride are the most abundant ions of sea water, calcium sulfate is much less soluble so that it is the first salt formed in major amounts when the salinity becomes high enough. Because of the different solubilities of the various salt compounds, evaporite deposits tend to consist of a definite layering of salts, deposited sequentially as the brines became more and more concentrated. Discovery by the Deep Sea Drilling Program of such an evaporite sequence in the sediments of the Mediterranean Sea is one of the pieces of evidence that has recently led to the startling conclusion that the Mediterranean was completely dried up in Pliocene time.

The Marine Biomass

Just as the oceans cover most of the earth, marine organisms constitute most of the mass of life on earth, both by weight and by total numbers of plants and animals. Much of marine sediment is provided by the shells of silica or calcite-secreting organisms. The petroleum that is so important a fuel to modern civilization comes from the decay of the soft parts of marine organisms. Even more immediate, most of the oxygen in the air we breathe is provided by the photosynthesis of marine plants. Humanity is finding out that the ocean, despite its size, cannot be considered an endless and harmless dumping place for wastes and pollution. The delicate chain of life in the sea may be vulnerable to our chemical offcasts, and we are vulnerable with it.

QUESTIONS

1. Why is so little of the water present at the surface of the earth directly useful for supplying human water needs?
2. What are the ways that water can travel back and forth between the mantle and the surface, and the consequences of each?
3. How has the evolution of life affected the composition of the atmosphere?
4. Explain how ozone, water vapor, and carbon dioxide make the earth hospitable to life.
5. What is the basic energy source that powers weather? In what ways does the rotation of the earth affect the weather?

★6. Why are there two high tides a day at many coastal locations?

7. Explain how waves can transport energy without much net transport of water.

8. Weathering of igneous rocks releases many dissolved ions which are washed into the sea. Why doesn't the composition and amount of salt in sea water change much with time?

ten

Rocks in the Surface Environment

The daily weather does not affect most town dwellers very much. A warm rain in spring may seem to do no more than cause tulips to bloom or ruin someone's picnic plans. Most of the people on earth live in places where nature has provided a comparatively mild environment. Thankfully, it is unusual when the weather is so severe as to destroy people's lives and works. Of course we all realize that the edible products of ranches and farms are tied closely to weather, but few of us are aware that the geological nature of the land's surface is just as closely tied to the same weather.

In this and the next chapter we will discuss the relation between weather and the earth's crust. When you finish these chapters, you should be able if you travel any distance and look at rocks, rivers, hills, and valleys, to see them in a completely new light. The simple presence of a car-sized boulder lying on a flat field in Iowa can be ignored, or it can indicate to an aware observer a history as fascinating as if it were an extinct volcano. You may not have the eye of an experienced field geologist, but you can develop a much deeper appreciation and understanding of the history and changing nature of the surface on which we live.

Although anyone can, after some education, understand the processes operating on rocks at the earth's surface, an idea of geologic time must be coupled with this understanding. A human life is so infinitesimally short in terms of geologic time that the daily changes in the crustal surface are almost too small to be measured and go unnoticed. What you can appreciate is the impermanence of rock, and what you can see is the effect the processes of rock decay have had on the land during the many centuries before you had the opportunity to see.

CHEMICAL WEATHERING AND WEARING DOWN THE LAND

As soon as rock is first exposed at the earth's surface as an emergent island arc, as a mountain range being born, or as lowlands left by a retreating continental sea, its spontaneous destruction begins. In the context of this chapter, destruction can be viewed as complete when the emergent land is reduced to sea level. Two major sources provide the energy to bring about this destruction. The sun provides the heat energy necessary for effective chemical activity, for the evaporation process by which moisture enters the atmosphere, and for winds to distribute that moisture over the earth. Gravitational energy acts to bring this moisture down in the form of rain and to cause stream flow and other downhill phenomena. With these two sources of energy as their main driving forces, three sets of processes are at work.

1. **Weathering** involves the most fundamental processes of rock disintegration and decomposition. The term refers to the breakdown of rock into chemical components, fragments of the original rock, and sometimes particles of a new crystalline nature. The other two activities involve processes not always separable from weathering, and are usually dependent upon it in terms of their effectiveness.
2. **Erosion** is the physical or chemical removal and transportation of rock debris. It is effected mostly by liquid water, and to a much lesser degree by ice and wind.
3. **Mass wasting** refers to the gravitational downward movement of rock fragments without the involvement of a transport medium such as water. If you fall off a cliff, you are mass wasted, geologically speaking.

These surficial processes not only disintegrate and decompose crustal material and move it about, but they give the land its shape or **topography**. Mountain ranges and continents are primary landforms and are the result of the earth's internal energy, but what happens to them at the surface is the result of external surficial processes. The secondary landforms, valleys for instance, imposed on primary ones are visible stages marking the present progress of all those processes that lead to a lowering of the land's surface. Since land is basically crustal material that is above sea level, the surface of the sea is a practical level below which nearly all surficial processes become ineffective.

In the remainder of this chapter, we will be concerned principally with weathering. Yet, if it is unrealistic to divorce erosion and mass wasting from weathering, the reverse holds true. We may think of them as separate processes in concept, but this is seldom true in practice. If erosion or mass wasting does not act on weathered rock, little change would be apparent. Just as one may think of men and women as separate sexes, but if they do not act together there are no visible results.

The rates at which surficial processes operate on rock are governed by a variety of factors. The type of rock exposed, the elevation, the vegetation, the climate, even the actions of man influence the rate of decomposition and disintegration. The surficial processes that are dominant in the Amazon basin may be inconsequential in the Sahara. The shifting of continents into different climatic realms, worldwide climatic changes, and nonuniform rates of weathering, erosion, and mass wasting make it impractical to estimate the life span of most landforms.

10-1. *A system of nearly horizontal and nearly vertical joints in a plutonic igneous rock of the same approximate composition throughout.*

If the future is difficult to predict, there are also great problems in conceiving what occurred in the past. The idea of uniformity must be used with the knowledge that the surficial conditions on earth have varied widely. What were the active weathering and erosional processes on a very early earth with a different atmosphere? What were the rates of erosion on a landmass prior to the evolution of rooted plants? What is important is to realize that in our comparatively gentle environment, there is a continuum of process and change at the crustal surface. As a man wrote of one of the great early geologists, James Hutton, who fully appreciated this continuum: "So preturnaturally acute was his sense of time that he could foretell in a running stream the final doom of a continent."

PHYSICAL WEATHERING: DISINTEGRATION

The term weather, as we commonly use it, refers to the state of the atmosphere in respect to wind, temperature, moisture, and pressure. Geologists have separated weathering processes into two categories; however, the two do not act completely independently—they interact. Temperature and moisture are the most basic factors involved with the **chemical weathering** of rock and pressure is the most basic factor in the **physical weathering** of rock.

Jointing

Anyone who wants to quarry an igneous rock, such as granite for use as building stone, is faced with one large problem in addition to finding the granite. It is difficult to find an igneous rock body of significant size that is not laced with cracks. Nearly all exposed rock bodies are cut by a system of fractures (Figure 10-1). A close examination of a rock exposure will reveal that no motion has taken place along these cracks; obviously, then, the cracks are not faults. Because we would have to join the pieces along the fractures to put the rock back together, the term **joint** is used for the crack itself.

Joints are commonly centimeters to meters apart and may or may not form some regular pattern. However, reasonably close observation usually will reveal a tendency for some preferred orientation of joints in a rock, at least locally. The cooling of a plutonic body, the removal of water from some sediments as they lithify, and tectonic forces of compression and tension all produce stresses in a rock. As the rock reacts to these stresses, joints are one common result. A physical weathering process is also a cause of stress and, hence, of jointing.

When a deeply buried rock body is uplifted and erosion removes material from above it, the confining pressure on the rock is greatly reduced. The relief from the great burden of overlying material results in the expansion of the rock mass, and joints are visual evidence of this relief. The expansion of rock near the surface can be detected. If a body of plutonic rock in a quarry is carefully measured, removed, and remeasured after a time, it will have increased in volume a small amount.

A more spectacular, and even dangerous, result of pressure relief occurs in some mines. When rocks are first exposed at the working face, expansion can be so rapid that pieces break off and fly through the air. One would expect that a rock that had been near the surface for a long time would have adjusted to the pressure conditions. But in some mines pieces of rock may fly off when exposed to atmospheric pressures at a depth of less than 100 m. Obviously, our knowledge of a rock's reaction to stress is somewhat imperfect. One type of joint system seems to be a very evident result of pressure reduction; the joints are subparallel to the land surface. This joint pattern is most evident in large coarse grained igneous bodies, and the rock may split into curved tabular sheets several meters thick (Figure 10-2).

The net result of jointing is to provide access routes for surficial water down into otherwise impermeable rock. Thus, one type of weathering makes another more effective.

Thermal Expansion and Contraction

Some areas of deserts are covered with angular fragments of the underlying rock, and it was once thought likely that the fragments were produced by daily temperature variations. The intense heat of day and the cold nights were thought to produce enough stress in a rock to cause its breakage. The rock's inability to absorb the expansion and contraction produced by the rapid radiation of heat after sundown was considered analogous to the inability of ordinary glass to go from a hot oven into cold water without cracking. Tales are told of foreign legionnaires being called to their Saharan battle

10-2. *Sheet type joints developed in an igneous pluton by relief of pressure. Note the parallelism with topography.* [*Courtesy U.S. Air Force.*]

stations when the bursting of rocks in the night was mistaken for the gunfire of their camel-riding involuntary hosts.

Hot-cold cracking probably does occur, but apparently it takes centuries of daily expansion and contraction to cause fracturing. In the laboratory, rocks have been subjected to temperature changes over 100° C every few minutes, to simulate an equivalent of over two centuries of nature's activities. No observable changes were noted. Although no confirmation of the expansion-contraction theory could be obtained, such experiments do not invalidate it either. Thus, we still are unsure of the nature of the fracture mechanism.

An additional problem is that a desert area today may not have been a desert a geologically short time ago. The remains of nondesert dwelling plants and animals plus archeological data indicate that parts of African and American deserts, for example, were more humid only centuries ago. Thus, the appearance of a rock may be due to weathering processes which acted upon it under a different climate.

Crystal Growth and Frost Action

The growth of crystals within a relatively confined space can exert enough pressure to cause rock disintegration. Although one incident of this process may completely disintegrate a bottle of milk or beer left outdoors in subzero weather, the process is most effective on rock when freezing and thawing is repetitive. At atmospheric pressure, water increases in volume about 9% when it freezes. If a fracture in rock is filled with water, the surface water freezes first as the temperature drops and, to some extent, confines the water deeper in the crack. Under the increased confining pressure, the freezing point is lowered. In addition, the forces of surface tension on water in very fine fractures inhibit freezing so that, even at temperatures below 0°C, some water can migrate and feed growing crystals within the fracture.

The growth of these ice crystals causes the rock to break apart, and in some mountainous areas the ground may be covered with loose angular rock fragments. This type of physical weathering is termed simply **frost action**. The process is most effective at high elevation where freezing and thawing may recur daily. It is less important at lower elevations in mid and high latitudes where it occurs primarily in spring and fall. The result of frost action on rock ranges from the widening of joints to such complete breakup that angular rock fragments litter the surface (Figure 10-3).

10-3.   *Frost action. This slope is littered with angular rock fragments that have been dislodged from the main body of rock by repeated freezing and thawing of water in fractures.* [*Courtesy Wards National Science Establishment.*]

10-4. *An ice wedge exposed by gold mining operations near Livingood, Alaska. [Courtesy T. L. Péwé.]*

Frost action not only affects rock but also contributes to the further disintegration of loose material such as soil. A secondary effect occurs when the surface of moist soil freezes, expands, and bulges upward. As the surface begins to freeze, some water below migrates upward via the capillary sized openings between particles and feeds the freezing zone. The upward expansion heaves and fractures street pavement and causes wiser persons in the north country to build their foundations below the "frost zone."

At high lattitudes where the average annual air temperature is less than 0°C, the ground below a depth of a meter or so is permanently frozen, sometimes to depths of hundreds of meters. In unconsolidated material the presence of this layer of **permafrost** causes many problems in well drilling and foundation maintenance, whether of buildings or petroleum pipe lines. A feature characteristically associated with permafrost is **patterned ground**. There are various types of patterns depending upon the soil, rock fragment size, and topography, but all have some relation to frost action. During extreme winter cold, the ground surface may actually contract into a network of small fractures. Warmer weather permits melt water to enter and widen the cracks when it refreezes (Figure 10-4). Repetition of this process eventually causes the formation of a polygonal pattern of wedge shaped partings of the ground that are ice filled (Figure 10-5). In some areas where the surface environment is now quite mild, for example England, vestiges of such patterns are visible and give us a valuable clue to past climatic history.

In hot arid climates the growth of salt crystals is also effective in rock breakdown, especially in porous sandstones. Ground water with

10-5. *Patterned ground.* **Left.** *An example of pattern ground formed in a modern climate near Barrow, Alaska.* [*Photograph by J. Brown, courtesy T. L. Péwé.*] **Right.** *An example of vestigial patterned ground in England formed in a past climate.* [*Courtesy A. Baker.*]

10-6. *Formation of a shallow elongate cavern in sandstone by the action of salt crystal growth. Water moving downward to the lower sandstone acquires salt in solution and is evaporated at the cliff face. Salt crystals grow there, force apart and loosen sand grains, which may then be carried away by wind or water after rainstorms.*

10-7. *A shallow cavern formed in part by the action of salt crystal growth. [C. H. Dane, U.S. Geological Survey.]*

high salt concentrations evaporates and the crystal growth of the precipitated salt exerts pressure on the surrounding rock. Repetition of this in the pore spaces of sandstone eventually will cause separation of the sand grains. Some of the shallow caves where "cliff dwelling" Indians built their homes were probably caused by this process (Figures 10-6 and 10-7).

Organic Activity

The wedging apart of rocks by root growth or the burrowing action of some animals are relatively unimportant processes of physical weathering. In loose material, the physical activities of the lowly earthworm open the soil to aeration and water and thus promote chemical weathering; in this way earthworms play a very significant roll in soil formation.

Man as a rock breaker, with his explosives and bulldozers, probably occupies a place of importance, worldwide, between root growth and particle disintegration by running herds of hooved mammals. Locally, however, such activities as strip mining have contributed significantly to physical weathering, usually to the detriment of the local environment. Man's activity with axe and plow, however, has had a far more significant influence on weathering. The cutting of forests and the tilling of fields have mechanically opened up vast quantities of soil to further weathering and more rapid erosion. Man has probably more than doubled the rate of erosion for the entire world just since he invented the plow.

CHEMICAL WEATHERING: DECOMPOSITION

The processes of physical weathering we have just considered have one principal effect on solid rock: they break it into smaller pieces. Obviously this makes the transportation of particles by erosion and mass wasting easier, but the increase in surface area is the most important result. Chemical processes can only take place on the surface of a solid body, so any increase in surface area leads to an increase in the rate of a chemical reaction.

Chemists have formulated a law of mass action: the rate or speed of a chemical reaction is proportional to the concentration of each of the reactants. Conceptually, we can think of a mineral as a reactant and its surface area for any given volume as its "concentration." As an example, a cube of salt 1 m on each side has 6 m² of surface area available for chemical solution when immersed in water. If the cube is cut in half, the same 1 m³ has 2 m² additional of surface available to chemical solution. Going farther, if the original cube were granulated to the size of household salt the increase in surface area would be enormous. Therefore, it is easily appreciated that a body of granite with a well-developed joint system will chemically weather much faster than one with no joints at all (Figures 10-8 and 10-9).

Rock Composition and Chemical Weathering

Another factor that influences the rate of chemical weathering is the mineralogy of the rock. Many things are unstable when moved to an environment radically different from where they were born. A person in outer space requires a vast amount of paraphernalia to protect his physical and chemical stability. A plutonic igneous rock, stable in its birthplace deep in the earth's crust, becomes unstable when exposed to the environment at the crust's surface. Under lower pressures and temperatures and in the presence of high concentrations of free oxygen and water, the chemical breakdown of igneous material is a natural response. The products of the chemical decomposition of an igneous rock represent changes as the rock alters toward stability in its new environment.

In Chapter 6 you learned that, as a deep seated magma begins to migrate upward into regions of lower temperatures and pressures, there is a sequential order of potential mineral crystallization (Figure 6-5). Minerals such as olivine form first and others such as quartz later. Because olivine forms and is stable under conditions of higher temperature and pressure than quartz, it is less "at home" on the surface and should weather faster—and it does (Figure 10-10). It is said that on a sea coast where olivine-rich basalt is exposed, olivine is rare on the beach and is never found more than 100 m off shore; the olivine has been rapidly decomposed chemically. Yet on a granitic coast, quartz grains are apt to be very abundant even well offshore; quartz is simply more durable in our environment.

If we are to understand which processes of chemical weathering are most influential, it is desirable to know the relative amounts of various rock and mineral types exposed to those processes. These

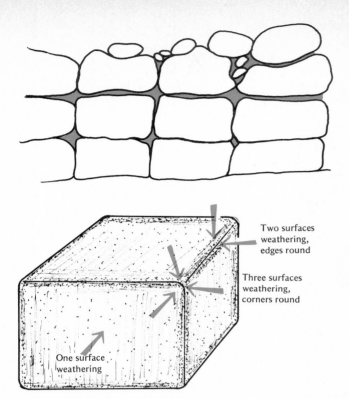

10-8. *The production of spheroidal blocks of rock by weathering along joints.*

Two surfaces weathering, edges round

Three surfaces weathering, corners round

One surface weathering

10-9. *Examples of igneous rock weathered along joints.* [*Courtesy R. Ojakangas.*]

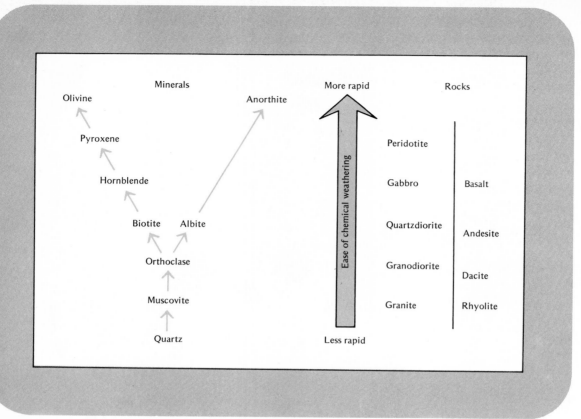

10-10. *Relative ease of chemical weathering of rocks and minerals at the earth's surface.*

data can only be estimated, for we do not know accurately the surface distribution of the various major rock types.

It has been estimated that the earth's upper crust is composed of about 8% sedimentary rocks, 27% metamorphic rocks and 65% igneous rocks by volume. However, the rocks at the crustal surface do not follow this quantitative pattern. About 75% of the land area has sedimentary rocks at the crustal surface. Igneous rocks make up the bulk of the remainder, perhaps 15% of granitic composition and 3–5% of basaltic-andesitic composition. Of the sedimentary strata, possibly 50% are shales, 30% sandstones, and 20% carbonates, mostly limestone.

The mineral composition of the sedimentary rocks varies widely; for example the percentage of clay minerals in shales and shalelike rocks may commonly range from less than one third to nearly two thirds. These variations, coupled with only rough approximations

Mineral	Abundance, %
Feldspars	29
Quartz	28
Clay minerals, & micas	15
Calcite-dolomite	14
All others	14

of the percentage of the major rock types exposed, make it difficult to determine accurately the percentages of minerals exposed to weathering. Figure 10-11 indicates an approximation of the exposure of major mineral types. It serves to illustrate that halite (salt), hornblende, and many other minerals well known to geologists are not quantitatively significant at the crustal surface.

Controls of Chemical Weathering

Climate. Moisture and warmth are the main climatic controls on chemical weathering; where rainfall is abundant and the temperature warm, weathering is most rapid. Hot arid regions generally receive enough intermittent rain to make chemical weathering important, whereas in polar regions chemical decomposition of rock is minimal. Under conditions of equal rainfall, heat is the most significant controlling factor in most chemical reactions; an increase of 10°C may greatly increase the reaction rate. Considering physical and chemical weathering together, weathering is slowest in a hot dry climate and most rapid in a hot wet climate. The startling preservation of mummies in Egypt is much less due to some secret process that was lost with the death of ancient embalmers than to the hot dry desert climate. Even the unembalmed bodies of common folk put into a hole and covered with sand are often very well preserved after centuries in any of the earth's hot arid climates.

Organisms. Plants and bacteria are the principal organisms that play a direct role in chemical weathering. The processes of growth and decay contribute carbon dioxide, oxygen, and organic acids to the soil, and these become involved in chemical reactions in the soil and on rock. A single rye plant has a total root and root-hair area of nearly 600 m^2, equal to the floor space in several average houses.

Plants remove certain nutrient ions directly from soil particles at the plant root, and this obviously must lead to a chemical alteration of that soil. The plant itself, therefore, is not just hydrocarbons formed from water, carbon dioxide, and oxygen. It also contains significant amounts of such metallic ions as potassium and phosphorus as well as lesser quantities of other metals. When these plants decay, the elements they remove return to the soil to be used again. Man, however, has altered the cycle—he harvests the plants.

When we ship plants, or beef that ate plants, away, often abroad, we remove many essential elements that the next crop must take by further chemical weathering. Soon the soil is depleted in these nutrients, and we must allow the land to lie fallow for a while or fertilize it. It is quite clear that, if we must mine minerals to put back into the soil, we have accelerated the normal processes of organic chemical weathering and must have caused at least a temporary depletion of our most important natural resource. It is no surprise that some scientists have termed chemical weathering the geological process most important to man. When one learns that the nutritional quality of wheat from the central United States has decreased over the last few decades, chemical weathering does indeed loom large as a process of great ecological impact.

If earthworms have a direct chemical weathering effect on rock particles, it is unknown. Their burrows and the circulation of fine particles through their digestive tracts undoubtedly make it easier for water and oxygen to carry out chemical weathering in soil. Charles Darwin collected worm castings and estimated that worms bring between 7 and 18 tons of material to the surface of 1 acre each year. It has since been estimated that in some tropical regions they move 100 tons/acre annually. Some tropical worms are over 2 m long; it is therefore possible that someday a geologist will feed and refeed them particles of granite and determine what effect, if any, their digestive processes have on rock.

Rates. Geologists have been able to simulate the chemical conditions found in nature and subject minerals to experimental weathering in the laboratory. Unfortunately, we cannot include a realistic time factor in the experiments. Most of the low temperature geochemical reactions occurring constantly at the earth's surface require many years to produce significant results. In addition, so many factors enter into chemical weathering that the time for a complete breakdown of a given mass of rock of a particular composition is a figure impractical to estimate.

In 1955, a granite exposure in Puerto Rico was still fresh looking enough to be used as a field illustration of that rock type. By 1972 the outcrop had been weathered to a depth of over 50 cm, and the residue could be cut with a knife. This is not to imply that a granite

10-12. *Weathering of marble headstones in a cemetery at Princeton, N. J. The stone at the lower left is dated 1828 and that at the lower right 1796. All were photographed in 1968. [Courtesy S. Judson.]*

monument that is in a tropical environment will be reduced to a pile of rubble in 20 years. The granite must have been weathering for centuries, but it seems that, once some level of chemical breakdown had been reached, physical breakdown became very rapid. Most weathering is a long term affair. Granites in the north of America and Europe that were laid bare and scratched by glaciers over 10,000 years ago may still show the fine striations; they have not been weathered appreciably.

One simple method of understanding something of weathering rates is to study tombstones. Most are quarried from relatively fresh rock, and their date of exposure to the elements can be found on the inscription. In the northeastern United States, marble headstones become illegible after about 175 years; those of slate, more resistant to weathering, still have a clear inscription after nearly 270 years (Figure 10-12).

Processes of Chemical Weathering

The chemical decomposition of a rock body in contact with a dry atmosphere occurs so slowly that is practically not measureable. The chemical processes that are of primary importance are largely dependent upon water. Water may act directly as a solvent or may carry gases and ions in solution to the minerals, which then react with these solutes. The reactions that will be discussed are simplified to avoid intermediate steps in the reactions and to emphasize the reactants and the products; it is more important to understand the concepts of the chemistry than the complexities.

Oxidation is a very common chemical process at the earth's surface and produces such familiar and often colorful compounds as rust. Minerals containing iron, manganese, sulfur, and copper are most commonly involved in oxidation, but these substances are relatively minor constituents of the crustal surface. Of them, iron compounds are most abundant, and man's use of iron makes it important to understand how it can be chemically removed from a mineral and deposited as an iron oxide, our most common type of iron ore. An example of oxidation (equation 10-7) has been included with the equations illustrating the more fundamental chemical weathering process known as hydrolysis.

Hydrolysis, as you might suspect from the name, involves water. Ideally, it involves the reaction between the hydrogen (H^+) and hydroxyl (OH^-) ions of water and some mineral. A reaction between pure water and a mineral can be expressed, but such an equation is

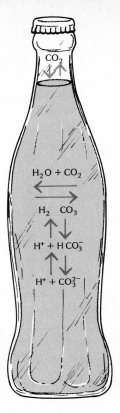

10-13. *Ionization in a carbon dioxide–water mixture, a bottle of "soda water."*

not realistic; in nature water is rarely "pure." Normally, the water involved in these reactions has come down in the form of rain, and in the atmosphere it combined with some carbon dioxide. As it percolated into the soil, the water took up even more carbon dioxide, which was released by soil bacteria involved in plant decay. So natural water is "charged".

$$H_2O + CO_2 \longrightarrow \underset{\text{carbonic acid}}{H_2CO_3} \qquad (10-1)$$

The weak acid, carbonic acid, itself ionizes, and the hydrogen ion content of carbon dioxide–charged water is much greater than that of pure water. A carbonated beverage provides a good example of the reactions. Figure 10-13 shows the equilibria conditions between carbon dioxide and water in a bottle of "soda." The additional hydrogen ions and the carbonate ions in natural water make it a very effective agent of chemical weathering.

Let us use an olivine mineral to show a simple form of the weathering reaction (equation 10-2).

$$Mg_2SiO_4 + H_2O + CO_2 \longrightarrow \quad Mg(HCO_3)_2 + H_4SiO_4 \quad (10\text{-}2)$$

magnesium   silicic acid
bicarbonate   in solution
in solution

However, olivine is not one of the most common minerals exposed at the crustal surface, and the hydrolysis of a common rock is a more meaningful process, although more complicated. Rock of granitic composition is the most common igneous rock exposed, and it is thought that the continental crust as a whole would approximate granodiorite if all rock types were combined. Assume that a grano-diorite composed of four crystalline minerals—sodium-calcium feldspar, potassium feldspar, quartz and biotite mica—is weathered. There is an equation (equations 10-3 through 10-6) for each mineral, and the insoluble products of each reaction are boxed.

$$NaAlSi_3O_8 \cdot CaAl_2Si_2O_8 + H_2O + CO_2 \longrightarrow$$
feldspar

$$NaHCO_3 + Ca(HCO_3)_2 + \boxed{Al_2Si_2O_5(OH)_4} \quad (10\text{-}3)$$
a clay mineral

$$KAlSi_3O_8 + H_2O + CO_2 \longrightarrow$$
feldspar

$$KHCO_3 + H_4SiO_4 + \boxed{Al_2Si_2O_5(OH)_4} \quad (10\text{-}4)$$
a clay mineral

$$SiO_2 + H_2O + CO_2 \longrightarrow \boxed{QUARTZ} + H_2O + CO_2 \quad (10\text{-}5)$$
quartz   relatively
unaffected

$$K(Mg,Fe)_3AlSi_3O_{10} + H_2O + CO_2 \longrightarrow$$
biotite mica

$$KHCO_3 + Mg(HCO_3)_2 + H_4SiO_4 + \boxed{Al_2Si_2O_5(OH)_4} + Fe(HCO_3)_2$$
a clay mineral

$$(10\text{-}6)$$

then by oxidation: $Fe(HCO_3)_2 + O_2 \longrightarrow$

$$\boxed{Fe_2O_3 \cdot nH_2O} + HCO_3^- \quad (10\text{-}7)$$

The reactions above represent the most significant concepts of chemical weathering, for even life as it now exists is directly or in-directly related to the hydrolysis of minerals. *The bulk of the sedi-ments, soil, and sedimentary rocks of the continents as well as the sediments and many of the dissolved ions in the sea had their funda-mental origin in the chemical decomposition of igneous rocks.*

As the granodiorite weathered, one new very common product was formed, clay; one common product was freed from the rock and remained relatively unchanged, quartz; and a number of elements were taken into solution. Their subsequent fates are related to erosion, deposition, and other processes yet to be discussed; however, to follow the concepts and importance of chemical weathering to at least a partial conclusion, it is worth our time to examine these products briefly.

Products of Chemical Weathering

1. **Clays** form a variety of mineral types containing aluminum and silicon, but the most common are derived from weathering. Having been produced at the earth's surface, clays tend to be relatively stable in most surficial environments. They make up a large and definitive part of shale, the most common sedimentary rock, and are extremely important in the development of soil and in the life processes of plants.

2. **Quartz**, as a mineral, is relatively resistant to chemical decomposition under conditions common at the earth's surface. It does go into solution, but only slowly. It is not known what percentage of the silica in solution as H_4SiO_4 was derived from crystalline quartz, but surely most of the H_4SiO_4 came from weathering of silicate minerals such as feldspar. The fate of much of the quartz freed from felsic igneous rocks is to be the common sand of the earth. Since quartz is quite at home at the earth's surface, it tends to stay around for a long time and it may be very old. The quartz sand on a beach may have gone through many cycles of deposition, lithification, weathering loose, transport, and so on, and the cumulative effects may be little more than a rounding and polishing of the grains.

3. **Solutes** have various and more complex routes to follow than do clays and quartz. Some of the silica in solution commonly precipitates around loose grains of quartz and may cement them together to form an all-silica sandstone. Small percentages of dissolved silica may precipitate from solution to form nodular lumps of such silica minerals as chert and flint in other rocks (Figure 10-14). One observation concerning dissolved silica is that the quantity in fresh water, in streams and rivers, is usually much higher than the 2–4 parts per million (ppm) common in sea water.

When the silica in solution reaches the sea, two processes appear to account for most of its removal. In the presence of sea water, some silica attaches to the omnipresent clays and deposits with them. This is probably a source of some of the silica in shale-type rocks. A second cause for the loss of silica in sea water is biological. Silica is taken up by some marine organisms, mainly the planktonic diatoms and radiolarians (Figure 6-12), and is used as construction material for their shells. Upon the death of the organisms, the shells may deposit in vast numbers. Under certain circumstances the protective organic coating on the shells may be lost and resolution of the silica may lead to the formation of beds of silica minerals such as chert.

The most common metallic ions taken into solution in the chemical weathering of igneous rocks are sodium, calcium, potassium, and magnesium ions. Potassium, calcium, and magnesium are essential elements of plant growth, and plants can absorb some directly from solution. Any of these common metallic ions may become attached to the surface of clay particles or even be taken into their crystal lattice. Plants commonly exchange hydrogen ions for these necessary metallic ions at their rootlets, and thus they further weather the clays, as we mentioned previously. For example, a year's crop of alfalfa from one acre (4 tons) takes up potassium equivalent to that contained in *2 tons* of granite!

Rivers carry those ions not taken into clay minerals or otherwise retained on land, to the sea, and they might be thought to stay there and continue to increase in quantity through time. To some extent this has occurred because, indeed, the sea is salty, and the sodium portion of the sodium chloride that makes it salty came mostly from chemical weathering of rock. Nonetheless, the sea is not a permanent home to the ions. It probably has not become any saltier for a very long time. Many things can happen to these ions. For example,

10-14.   *These nodules of chert are composed of silica that was precipitated from solution. They probably formed nearly contemporaneously with the surrounding limestone. This rock is also a good illustration of differential weathering; the chert is much more resistant than the limestone* [Courtesy K. Moran.]

	Ions added in 100 million years	Present amount of ions
Na	196	129
Mg	122	15
Ca	268	2.8
K	42	2.7

10-15. *Present amount of major metallic ions in sea water compared to estimated amounts added by rivers in 100 million years (in moles per square centimeter of total earth surface).*

some may be carried back inland with moisture-laden winds off the ocean, some will be traped with sediments on the bottom, and some will be carried down with oceanic crust and sediments at subduction zones.

The same fate does not necessarily overtake all the ions. Different things may happen to each of the common metallic ions after entering the sea. For example, there are more calcium ions in the river water entering the sea than any other. Yet the sea is salty principally with sodium, not calcium, potassium, or magnesium. Why is it that sodium can be retained in quantity while the other common metallic ions are depleted? Figure 10-15 indicates the relative amounts of the four principal metallic ions in sea water in contrast to their input over 100 million years.

Some of the potassium and magnesium has been incorporated by organisms, and much of that now lies on the sea floor as part of organic sediments. The major loss however, especially of potassium, results from an interaction between the metallic ions and suspended clay particles. There is a small attractive force between positive ions and the very small clay particles and the ions become attached, not chemically but physically, to the surface of the clay. The ions are said to be adsorbed onto the clays and are thus removed from the water and deposited with clay sediments. Through time much of this has been replaced on the continental mass as oceanic sediments are welded onto continental margins, or has been recycled by the subduction process at convergent plate boundaries. The residence time for potassium and magnesium ions in sea water is thus lower than that of sodium ions largely because of sedimentation.

The calcium ions have an even shorter residence time in sea water than potassium and magnesium ions. Although calcium ions are little adsorbed on clay, they are removed rapidly by biological activity. Many of the countless planktonic organisms, such as the Foraminifera (Figure 6-12), as well as larger organisms, such as clams and snails, build their shells of calcium carbonate, which they obtain by combining the calcium ions with the carbonate ions present in sea water. As the plankton die, their shells slowly drift down toward the ocean floor; some of the carbonate goes back into solution at depths greater than 4000 m, but above that depth much enters into the formation of limy sediment. Biologically produced calcium carbonate was the source of most of the marine limestone deposits that are now exposed so commonly on the continents. Magnesium may be further depleted from seawater when some substitutes for calcium in carbonate sediments and magnesium limestone or dolomite, $CaMg(CO_3)_2$, forms.

Sodium, once carried to the seas after being freed by the weathering of terrestrially exposed rock, has a very long residence time. Sodium ions are not much used by organisms for shell building, are not adsorbed on clays to the extent of potassium and magnesium ions, and thus simply stay in the water. Such depletion as does occur is mostly due to two factors. Some sodium, along with the other ions of course, is removed from circulation by entrapment of sea water in marine sediments and some is precipitated as salt deposits. The latter process occurs in shallow areas where the circulation of water is inhibited and a high evaporation rate prevails. As the solubility levels are exceeded, sodium chloride as well as other evaporites such as gypsum, $CaSO_4 \cdot 2H_2O$, form on the bottom.

It is not known whether the sea has always been sodium salty as it now is, but it does seem certain that the influx into the sea of solutes derived from rock weathering has had a fundamental impact on the origin and evolution of life on earth, since life is thought to have begun in the sea.

Chemical Weathering of Sediments

All of the three principal types of sedimentary rocks on land—shales, sandstones, and limestones—had their origin in weathering, but the process does not stop there. Sedimentary rocks themselves are all subject to further action by physical and chemical processes.

Limestone, in comparison to most sands and shales, is readily attacked by ground water. The reaction is comparable to hydrolysis; however, although hydrolysis theoretically involves *only* water but can be made more effective by the addition of carbon dioxide, the weathering of limestone is almost entirely dependent upon the presence of carbonic acid.

$$H_2O + CO_2 \rightleftharpoons H_2CO_3 \rightleftharpoons H^+ + HCO_3^- \rightleftharpoons H^+ + CO_3^{2-}$$

$$\text{hence: } \underset{\text{limestone}}{CaCO_3} + H_2CO_3 \longrightarrow \underset{\text{in solution}}{Ca(HCO_3)_2}$$

This reaction is called **carbonation** and CO_2-rich ground water moving through the limestone and along joints and bedding planes can be a very effective weathering agent (Figure 10-16). Along the route to decomposition, caverns may form in the limestone and sinkholes appear on the land surface when the cavern ceilings collapse. Once a cave system develops, the erosive forces of running water may modify it, but the initial cause of cavern formation is

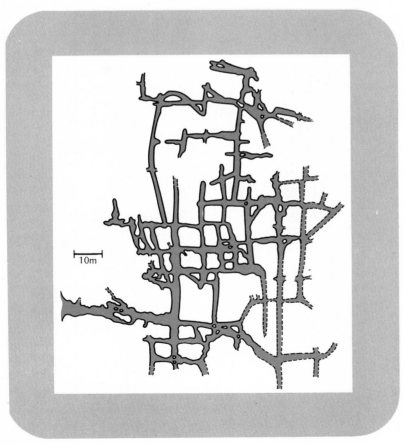

10-16. *Map of Hamilton Cave, W. Va., showing how chemical weathering developed the cavern system along joints. [Courtesy West Virginia Survey; after W. Davies.]*

10m

10-17. *This cavern was produced through the solution of beds of limestone by carbon dioxide—rich water. Flowing water later modified its shape. The stalactites formed as water with a high content of dissolved calcium carbonate evaporated as it came out on the ceiling surface and precipitated calcite to form the "dripstones." Some water dripped to the ground, evaporated there, and left the stalagmites which built up from the floor. The "tites" come down and the "mites" go up. [Courtesy Luray Caverns, Va.]*

the slow but incessant chemical action of fresh water. The process can be reversed when the water with dissolved limestone in the form of calcium bicarbonate, $Ca(HCO_3)_2$, evaporates or loses carbon dioxide, resulting in deposition of limestone. Both solution and

10-18. *Mature and immature sandstones.*
Upper photomicrograph. *A mature sandstone. The grains are all of quartz, they are well rounded and quite well sorted as to size. The different tones, black to white, are not the shades of the individual grains in normal light (they are clear) but are caused by their different crystalographic orientation to the polarized light used to make the grains stand out from the background.* **Lower photomicrograph.** *A immature sandstone. The grains are angular, of diverse mineralogy, and are poorly sorted. These grains have not been subjected to as much mechanical or chemical weathering as those in the upper illustration. They were probably deposited not far from the exposures of igneous rock from which they were derived.*

precipitation usually occur in caves, and the caverns formed by solution are often decorated with precipitates of calcite (Figure 10-17).

Sandstone is defined on the basis of the size of its grains, not on its composition. The composition depends not only on the parent rock, which may, of course, have been another sandstone, but upon the degree of chemical weathering to which the grains have been subjected. The two most important factors that determine the degree of weathering are time and climate. If the time is long and the climate warm and humid, the ultimate sandstone will be composed of quartz grains. Quartz is durable, and the rock has approached equilibrium with the surface environment.

Although erosion will be a topic of the next chapter, it must have become clear that, for accumulations of quartz grains to form, the clays and other products of the chemical weathering of an igneous rock must have been carried away. Normally over long periods of time, the free grains of quartz have been subjected to transport by water and abrasion by wave action on ancient shores. The result of all this is not only a mineralogical maturity but a morphological maturity as well—the grains become rounded.

The time factor in chemical weathering, in a restricted sense, is dependent upon the relief of an area. When plate interactions result in a mountain range, the high relief results in a rapid removal of rock material by flowing water. Even if the climate is warm and wet, the rapidity of transport and deposition will shorten the time during which surficial chemical weathering can act and an immature sand will result. The grains will not be predominantly quartz, and they will be more angular than rounded (Figure 10-18).

It is this rapid transport and the resultant short chemical weathering time that causes the graywackes deposited along arc-trench gaps to be immature. Their low quartz content (about 30%) and high feldspar and rock fragment contents indicate they have not been extensively chemically weathered. An analogous situation may exist on the continental side of a mountain range. The molasse-type deposits commonly contain large amounts of arkosic sandstones. The rapidity of their transport and low exposure time to weathering are reflected in their angular grains and high feldspar content. Simply by looking at the shape and composition of grains in a sedimentary rock, a geologist can get an idea about the ancient relief and even the ancient climate of an area.

Shale, because of its fine grained nature, is not very revealing of its mineralogy by normal visual inspection. The estimates of the minerals in the muds from which most shales were formed are usually the result of chemical or x-ray analyses. Typically, they contain 30 or 40% clay minerals and significant amounts of feldspars and quartz, predominantly of silt size, exist. Therefore, shale itself, with its diverse mineralogy, is open to further chemical weathering. The three or four common clay minerals that result from chemical weathering are related not only to the parent rock but to climate, and different types of clays can be produced from the same parent rock in different climatic zones. During and after transport, the clays may undergo compositional changes as ions are exchanged with or lost to their surroundings, but in most of the environments at the crustal surface they still remain clays.

Soil

Although the study of rocks would be greatly simplified by the absence of soil or weathered debris, and many field geologists have cursed it for covering up the answer to a geologic problem, soil is the most valuable product of weathering. The loose rock debris at the surface or on a former surface now buried, is called **regolith.** Through a combination of physical weathering, chemical weathering, and biological processes, the upper part of a regolith may be able to support rooted plants, in which case it is called soil. Soil can form either directly from the weathering material of the underlying rock or on sediment that was transported to and deposited at the site.

Once a soil has been initiated, the processes that created it continue and the soil, as it approaches equilibrium with its environment, develops a layered profile. Each layer of the profile is termed a **soil horizon;** a soil with a well-developed profile is mature, and one that lacks it is immature. Topography and vegetation do influence what type of soil forms, but climate and parent rock are the two overriding factors. In an immature soil, the parent rock is the dominant factor but ultimately it is the climate that determines what the mature soil will be. No matter what kind of parent rock, given enough time, very similar soils will develop in similar climates.

Because there are many climatic, rock, and topographic variations throughout the world, the detailed classification of soils requires a complex scheme. In the late 1800s a Russian geologist, V. V. Dokuchaiev, first noticed the climate-soil type relationship while traveling through Siberia and went on to develop the first widely accepted soil classification. Until little more than a decade ago, most soil scientists used some modification of the Russian system, and a farmer at work on the plains of South Dakota might have been told he was plowing a Chernozem (Russian for Black Earth).

Now, in the United States and several other countries, a new scheme is in official use and from part of its title, "the seventh approximation," you can get an idea of the problems in soil classification. The problem of trying to classify or pigeon-hole material that has infinite variability is common. In the opening remarks of the new system, the authors recognized this and stated, "Classifi-

10-19. *A soil profile of the A, B, and C horizons. A is dark and B is lighter due to an addition of silica from above and a lack of organic matter. [Courtesy U. S. Soil Conservation Service.]*

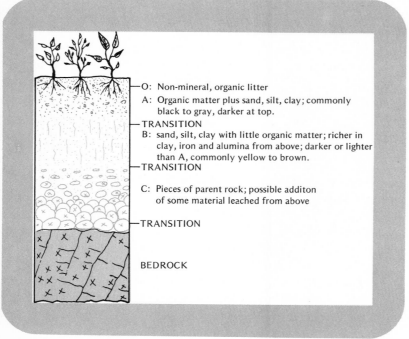

O: Non-mineral, organic litter

A: Organic matter plus sand, silt, clay; commonly black to gray, darker at top.

TRANSITION

B: sand, silt, clay with little organic matter; richer in clay, iron and alumina from above; darker or lighter than A, commonly yellow to brown.

TRANSITION

C: Pieces of parent rock; possible additon of some material leached from above

TRANSITION

BEDROCK

10-20. *Idealized soil profile.*

cations are contrivances made by men to suit their purpose. They are not themselves truths that can be discovered." This system has 10 major soil orders and many suborders; still, it basically reflects the influence of climate on major soil types. (It probably would not influence the style of plowing in South Dakota if that farmer realized that the name of his particular soil had been changed from the slavic

Chernozem to the modern Latin-derived Orthic Argaltoll variety of a Mollisol.)

Most well-developed soil profiles have three mineral horizons, A, B, and C. Below C is the unweathered parent rock (Figure 10-19). The maturity of a soil that has a distinct profile should not be considered to reflect its age in years directly. In the tropics a soil can reach maturity relatively rapidly, while in a cooler, dryer climate a soil may be exposed to surficial processes for a longer time and still be immature. The depth of the entire profile is variable, but A and B are usually confined to the top meter or so. C may be quite thin in a dry climate or approach a thickness of 100 m in a humid tropical one. The horizons develop as chemical and organic activity influence the upper zone and solutes and some clays move downward. The A horizon then becomes defined by its organic content whereas the B horizon develops through addition of material from above and its deficiency in organic residue. A generalized profile is shown in Figure 10-20, but it should be remembered that there are many variations on this theme.

ECONOMIC PRODUCTS OF WEATHERING

Soils are the most important product of weathering, and they are the most important natural resource of most countries. It is difficult to imagine any nation becoming a power or even an industrial nation without enough arable land to free its people from a struggle for food. In addition to this frequently overlooked natural resource, weathering products in one way or another are the source of most of the world's mineral wealth. The few products related to weathering that are listed below have obvious impact on financial activity in the world, yet the nonmonetary influences they have had, and will have, on our civilization is greater.

Petroleum. Petroleum itself is thought to have originated from the chemical alteration of decayed plant and animal material. Nearly all of the commercial deposits, or "pools," of petroleum are in sedimentary rocks. The commercial and cultural impacts of petroleum are remarkable; for example, in the United States nearly 99% of all transportation depends on petroleum products.

Iron. The majority of iron in commercial deposits probably was derived from the chemical weathering of rock. Most of these ores are sedimentary rocks, and many have had their percentage of iron increased when silica is removed by further weathering. The major iron ore bodies are of Precambrian age, and the iron occurs as oxides formed from the same oxidation process shown in equation (10-7) page 306. This age restriction implies that something occurred then that has not been repeated since on earth on a comparable scale. Because free oxygen is required for this process, the formation of oxides is thought to indicate that organisms capable of producing free oxygen photosynthetically had evolved and caused the deposition of the iron that had been accumulating in solution.

Nickel. Chemical weathering of ultramafic rocks may leave residual concentrations of silicates high enough in nickel to be commercial. This process has proceeded farthest under subtropical and tropical climates, and many nickel ores consist of a few meters of this weathered material atop an ultramafic parent. Although the major nickel production comes from Sudbury, Canada, and that ore body may be related to a meteorite that struck earth some 1.7 billion years ago, the weathered deposits in such places as Cuba, New Caledonia, and the Phillipine Islands have assumed increasing importance.

Aluminum. Until this century, Greenland was the principal source of a mineral from which the rare and expensive metal aluminum could be obtained. Pure aluminum was first produced in the 1820s, but even with "commercial" production it was so expensive that Napoleon III used it in preference to silver for cutlery at banquets. This very abundant metal of the earth's crust occurs predominantly in primary aluminosilicate minerals that, until recently, have defied attempts at commercial extraction. The mundane uses of aluminum today were made possible by low cost electrolytic extraction of the pure metal from chemically weathered products of clays and various rocks.

Under tropical conditions of high intermittent rainfall and high temperatures, silica may be removed from weathered material to the point where the aluminosilica structure breaks down and the aluminum is left as bauxite ore, in the form $Al_2O_3 \cdot nH_2O$. Deposits in Jamaica, Australia, Africa, and South America reflect the association of aluminum ore with the climate. In the United States, the lesser deposits in Georgia, Alabama, and Arkansas reflect a past subtropical climate during part of Tertiary time.

Clays. Virtually all clays originate from chemical weathering and their uses are broad, from cement and brick production to paper making and porcelain. The annual value of this homely fine grained material in the United States alone is hundreds of millions of dollars. The purest deposits are not normally those found in shales but are found as residual deposits, usually over a granitic parent. One common type of clay is kaolinite, which derives its name from *kau-ling*, Chinese for "high ridge." Clays were mined from a ridge near the town of King-te-chen, and a product was made which we have come to call "china."

1. What are joints, how can they be produced in nature and what general effect do they have on chemical weathering of rock?
2. What would be the general compositional type of igneous rock and the general climatic environment for chemical weathering to occur most rapidly?
3. How is most clay originally produced in nature?
4. Why is the sea salty?
5. What are the characteristics of a morphologically and compositionally immature sandstone and what does this immaturity indicate about its history?

Transport and Topography

eleven

Before delving into some of the specific processes of mass wasting and erosion, let us take a broad look at what effect they have, in conjunction with weathering, on the land masses of earth. All of the processes within these three categories are in continuous action and interaction, and their total effect is to lower the elevation of land. To lower a continent, material must be removed from it; because erosion is the prime mover of material, it is the prime leveler.

The principal agent of erosion is flowing water, and since water can only flow downhill there is a limit to its erosional capabilities. On a planetary scale the base of the continental "hill" is the sea. With this idea in mind a concept of **base level** has been put forth: base level is the lower limit of effective erosion of the land. Although the sea is the ultimate base level, it often is more practical to study erosion in terms of a **local base level,** which might be a lake where an entering stream ceases to be an effective agent of erosion.

A pertinent question for inhabitants of land is, "How long will it be before the processes of erosion reduce the land to sea level?" The continents now stand at an average elevation of about 875 m above the sea. Although it is very difficult to determine accurately the worldwide rate of erosion prior to the last thousand years or so when humans began to influence it, 2.5 cm/1000 years seems a reasonable average figure. A simple calculation indicates that in 35 million years, if only erosional processes took place, humanity would either swim or cease, for the land would be gone.

Of course, you now realize that plate tectonics act to add to continental areas, thus preventing complete reduction of the earth's land masses. Apparently there has always been land, and, if billions of years of geologic history are any guide, 35 million years from now much of our present continents still will be above sea level.

But what of the interiors of large continents, far from plate margin activities? Why aren't the interior plains of the United States, for example, at sea level? In fact, for long periods during the last few hundred million years the interior plains have been below sea level. Shallow seas have covered significant portions of all major continents. The reasons why the low lying portions of a continent may come to be above or below sea level are complex, and erosion is only part of the answer. The volume of water in the sea may change, major portions of the sea floor may rise or fall, or the continent itself may change in elevation.

The reaction of crustal material to the loss of mass from its surface is further uplift. In Chapters 4 and 8 you learned of this isostatic effect. However, the uplift is not a one-for-one reaction. If 500 m of land erodes away, 500 m of uplift does not necessarily occur as compensation. Thus, erosion eventually may reduce large areas of continents to relatively low levels. Land in Mississippi now over 400 km from the sea could be covered by sea water with nothing more geologically drastic occurring than the melting of the polar ice caps.

The whole process of uplift, weathering, and erosion can be viewed as part of a planetary energy cycle. As the internal energies of radiogenic heat and gravity result in the rise of crustal material, an energy imbalance is created. Gravitational energy exists in the rock mass that has been elevated above its surroundings. External energy, mostly solar, in conjunction with gravity, drives the erosive processes working to lower the elevated mass and thus restore the equilibrium or energy balance. If equilibrium were attained, we would have a smooth earth, the low areas filled by deposition of material eroded from the high areas and the whole covered by water. But, as far as we know, this has never occurred.

The internal processes leading to mountain building, isostatic uplift, outpourings of lava, and the creation of new continental crustal material are always creating imbalances of energy. Although the concept of energy imbalance is a useful overall view of worldwide erosion, it is somewhat sterile to think of a river roaring through some mountain gorge as merely a cog in a conceptual energy machine. Any agent of erosion with the capability of producing the Grand Canyon deserves to be studied in its own right, and must be so studied if its function in lowering the land is to be understood.

MASS WASTING

Mass wasting, as defined in Chapter 10, involves the downslope movement of material under the sole influence of gravity. A thorough study of all the processes of mass wasting would reveal a complexity not immediately evident from the definition. The processes may involve dry material or material that has been lubricated or made to lose its cohesion by the addition of water, and the movement of this material can range from the imperceptible to the awesome. For our purposes we will divide mass wasting into two categories: slow and rapid.

Slow Movements of Mass Wasting

Evidences of mass wasting are common in nearly any hilly terrain, and one of the most easily detected processes is one of the slowest, **soil creep.** The almost imperceptible downslope advance of soil or regolith will leave its mark in tilted telephone poles and fences or in the warped growth pattern of trees (Figure 11-1). The movement is usually measured in centimeters per year (just like sea floor spread-

11-1. *Soil creep: The downslope movement of soil has tended to tip the trees in the direction of movement. The trees' tendency toward upright growth has resulted in the curved trunks. [Courtesy R. Ojakangas.]*

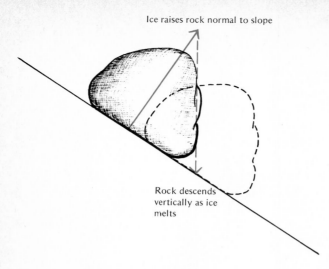

Ice raises rock normal to slope

Rock descends vertically as ice melts

11-2. *The downslope movement of rock by freeze-thaw action.*

ing) and, in general, the steeper the slope, the more rapid the creep.

Creep is most active where some disturbance of the soil particles occurs regularly, and very often this involves moisture. The alternate wetting and drying of the soil may cause expansion and contraction or cause minor volume changes in clay minerals, and together these factors create a small amount of free space for gravity to move the particles toward base level. Animals moving along slopes may contribute to creep, and an interaction exists because the animals may take preferred paths along the wrinkles on the surface that creep occasionally produces.

In cold climates, the expansion of freezing ground water will force grains apart and, upon thaw, allow some freedom for downslope movement. This action not only causes creep by particle disturbance but, at the surface, may move particles, even boulders, downslope in small intermittent "jumps." As the water around the base of a large or small particle freezes and expands, the particle is lifted up normal to the slope; as the ice melts the particle descends vertically and the net result is a downhill movement (Figure 11-2).

When soil on a slope becomes laden with moisture, a special type of creep may occur. Since normally the soil is bound to some degree by vegetation, it does not flow as mud. This type of creep is most common in cold climates where the surficial material has thawed but the still-frozen ground beneath prevents the melt water from migrating downward. The water saturated layer flows downhill as a unit. This **soil flow,** or solifluction (Figure 11-3), is seasonal and usually operates at rates of a few meters per year.

11-3. *Soil flow. Water saturated soil has slipped downslope forming tonguelike units known as solifluction lobes. (Courtesy T. L. Péwé.]*

Rapid Movements of Mass Wasting

Where the soil is not well anchored by plants, or when very heavy rainfall occurs, the surface material may become so liquified that it loses its cohesion and actually flows in a manner similar to water. The debris moving downhill may be as viscous as newly mixed

11-4. *A sandy mudflow. The lobate units are typical of the nature of most intermittent earth flowages.* [Courtesy C. L. Matsch.]

The angle of repose. Loose sand has assumed a 30° angle, about the maximum slope it can maintain. Earthquakes have triggered rapid mass wasting along this section of the Pan American Highway in Peru, and traffic has been blocked by a small slide. The sea is at the base of the slope. [Photo by D. Darby.]

concrete (Figure 11-4) or so soupy it is nearly very muddy water. These **mudflows** can be slow, but the more liquid types in mountainous regions can achieve velocities that would invalidate their classification as slow processes of mass wasting. Houses, trees and boulders may be taken as passengers in the more destructive mudflows.

When volcanos eject large quantities of ash that settles on their slopes, the material, with no vegetational anchor, is free to flow. The emission of water vapor and heat into the atmosphere makes thunderstorms a common associate of erupting volcanos, and the resultant mudflows of ash have buried nearby towns. One such flow covered over 100 villages in Java and killed over 5000 people; mudflows are not always harmless phenomena.

Most slopes in the world are gentle, only some 5° or less, and this is to be expected. Mass wasting and erosion are more active on steep slopes; thus, steep slopes do not endure for long periods of time. In the case of loose material, slopes of more than 30 or 35° are inherently unstable, and it does not require much action to cause a **landslide.** Dry sand, for example, when poured in a pile has a slope of about 30°; this is termed the "angle of repose." No matter how much sand you pile on, it spreads out at the base and tends to retain that angle.

Figure 11-5 shows a 30° slope of sand along a section of the Pan American highway in Peru. The traffic here is quite heavy, and, because the area is adjacent to a convergent plate boundary, earthquakes are common and a small one is enough to cause the sand to slide. When this occurs vehicles and their passengers may simply disappear from life and from the surface environment. Any rapid downhill movement of material is dangerous if one is personally involved; one falling boulder can be lethal to a person between the rock and its base level.

The most spectacular landslides are those that involve large quantities of rock that break off and move down steep slopes in mountainous areas. These **rockslides** usually occur when the strength of the rock or its limits of distortion under stress are exceeded, either by the force of some tectonic event or by the force of gravity. The reason why a rockslide occurs is not always known, but some are due to a river cutting into the base of a slope and steepening it or to the added weight of ice and snow atop a rock mass, which may cause instability. Earthquakes often are a trigger of massive rockslides.

Whatever the initiating factor, the rock fails, splits off along some joint, fault, or bedding plane, and comes down *en masse*. In 1963 large masses of rock broke from the flank of Mt. Rainier, Washington, fell, fragmented, and then roared across a glacier. At the end of the

11-6. *Talus cones along a mountain front. [Courtesy C. Craddock.]*

glacier the rock debris shot out into the air and came down onto the valley floor. The air beneath the hurtling mass was compressed and acted as a cushion and friction reducer. One rock mass went about 3 km beyond the glacier *after* passing over a small shack about 1.5 m high without damaging it! A later slide stopped short of the shack and the air expelled from beneath the rock blew the building hundreds of meters forward. Landslides cease to be only events of interest and become tragedies when people are in their path. Some have killed thousands of persons in a few minutes.

When large masses of snow and ice break off some mountain slope, the action is about the same as that of a rockslide, but this special case is called an **avalanche.** In many inhabited areas, small but potentially dangerous avalanches may occur as added snow creates an instability on the slope. Artillery shells are sometimes fired at the snow mass, bringing it down before its time but under controlled conditions. Occasionally a great mass of snow, ice, and rock debris weighing many millions of tons comes cascading down, destroying everything in its path.

Along the bottoms of steep cliffs, piles of loose rock often accumulate. Known as **talus** (Figure 11-6), the pieces are dislodged by weathering and fall or roll down forming a slope with an angle of repose of about 35°. Most of this rock and particle fall is channeled by small drainage ravines at the cliff's edge, and a series of conical piles, or talus cones, is a common feature near the base of very steep slopes.

Predictions of Slides and Avalanches

As mentioned, snow accumulations known to be approaching instability on a slope can be dealt with. Landslides and avalanches are known to be common in certain areas, and most people have enough common sense not to live at the base of a cliff. But, typically, the valleys in mountain areas have useful agricultural land and are inhabited. Since earthquakes, a common trigger for slides, are as yet unpredictable, what reasonable hope of early warning of mass wasting can scientists offer the inhabitants of mountain regions? Unfortunately, not much.

Good preliminary geology should be able to prevent dams and other engineering works from being built across faults or down-slope from zones where slides are likely to occur. Unfortunately, a conviction of the earth's permanence and stability is built into the minds of many. Until better predictive techniques are devised, the valley dweller may do well to pay attention to the lower animals. It is reasonably common for the survivors of earthquakes, landslides, and avalanches to remark about the odd predisaster behavior of dogs, cats, horses, and cattle.

Cattle may leave the slopes prior to slides, and other animals may become very nervous. Many people remarked about the barking dogs and the kicking and neighing fire horses prior to the San Francisco earthquake in 1906. Even in mines, rats have repeatedly squeaked warnings of ceiling collapse to the miners. Workers have been known to encourage the rats' presence by putting out food, and woe betide the new man who throws a rock at one of his potential saviors.

Deposits of Mass Wasting

Deposits of mass wasting are known by the general term **colluvium,** and it can be important to recognize these deposits in the rock record. A common characteristic of colluvium is poor sorting: large and small rock fragments and particles are mixed almost indiscriminantly. The most commonly exposed deposits that share this characteristic are of glacial origin, although material resembling colluvium can result from the downslope slumping of sediments on the ocean floor. Most sediments deposited by water or wind are easily distinguishable from colluvium because they are better sorted and have a layered nature. However, when the geologist is confronted with the necessity of explaining the origin of some poorly sorted ancient deposits, he may be unable to decide between a glacial or mass wasting origin. It makes a great difference to our understanding of the geologic history of an area whether glaciers or debris flows were active, but there are no hard and fast rules that always permit discrimination of their deposits.

Mass wasting processes normally do not cause material to be transported very far laterally. For the movement of a substance over a long distance, for example, a loose grain of sand from Missouri to the sea, water is the principal agent of transport. Mass wasting is an important geologic phenomenon, but in terms of the quantity of material moved and of its effect on the landscape it is completely outclassed by flowing water.

FLOWING WATER

Liquid water is the principal carrier of earth material and the principal sculptor of the land. How much material streams carry to the sea every year is a difficult fact to ascertain. Some areas today, for example, much of the western United States and about half of Australia, have no direct drainage to the sea, whereas the streams draining highly populated China, India, and Southeast Asia carry a much higher sediment load than is typical of other areas.

The high erosion rates of southern Asia are thought to be due to man's agricultural activities, although the erosion of the Himalaya Mountains in the background has a very strong influence as well. Mead, in 1969, calculated that the rivers of the eastern seaboard of the United States are carrying four to five times more sediment to the ocean today than they were before the Europeans arrived. The conversion of forest to crop land therefore has had a very real effect on the rate of erosion. Judson, in 1968, calculated that, prior to man's intervention, the rivers of the world were moving about 9.3 billion metric tons of sediment to the sea each year. Now they move about 24 billion tons!

If all this is true, will weathering and the creation of new soil fall behind man's erosional impact? If it is possible that the change of a continental-oceanic boundary from inactive to active is caused in part by the weight of large masses of offshore sediment initiating subduction, is mankind causing it to happen sooner? This type of speculation is interesting, but substantiating data are not yet avail-

11-7. *Outline of part of a small drainage basin* [*Courtesy U.S. Department of Agriculture Soil Conservation Service.*]

able. It seems certain, however, that man has caused acceleration of erosion, perhaps by more than two and one half times on a world-wide basis, and that this will affect his future. Very few study groups have yet been formed to look into the long range international effects man has on this planet, and estimates and ideas about things thousands of years in the future are subjects seldom sought by governments.

The Colorado River moves hundreds of thousands of tons of material daily through the Grand Canyon. If the river on its 2300-km journey to the Gulf of California were an average of 50 m wide and that load was the product of erosion *in* the river's bed, it would have to be cutting downward at over half a meter per year! At that rate it would be capable of producing a gorge as deep as the Grand Canyon along its entire length in about 3000 years!

Obviously the idea that a major river creates most of its own load is wrong. Most of a river's load is brought to it by large tributaries, by small tributaries, by water running down small gullies, and by water running not in channels but overland down some slope. The whole of this network of flow forms a **drainage basin,** that area which contributes water to a river channel (Figure 11-7).

A river and its tributary streams are not the principal producers of a wide well developed valley system. The erosive power of a river is confined to its channel, and in a given drainage basin very little of the area is channel. Rivers are mainly transport agents for the products of erosion; they are carriers more than they are eroders. True, stream erosion usually initiates cutting of the valley and is important in its development, but unchanneled water flowing down slopes after rain is the principal developer of the topography.

Most of the liquid fresh water of the earth is stored, temporarily, in the ground, a smaller amount is in lakes, and an even smaller amount is in channels. Most of this water is introduced to the land by rainfall, and that portion that does not infiltrate downward to form **ground water,** runs off. The runoff can be separated into **overland flow** and **channel flow** and in this chapter we will examine water existing in these conditions.

Ground Water

The ground water most familiar to us is that which came from rain and now fills or partially fills the pore spaces between particles of soil or rock. Of course, if the rain falls on granite, very little can enter the rock. Other than in fractures, or rarely in caves, a rock or soil can contain a fluid only in the pore spaces between particles. The amount of pore space relative to total rock volume is termed **porosity.** It is usually given as a percentage and is commonly expressed by the equation

$$\text{porosity} = \frac{\text{pore volume}}{\text{total volume}} \times 100$$

Porosity is dependent upon the size, shape, and sorting of the grains. Other than cavernous limestone, sandstones are the most porous rock; 30% is about the maximum porosity in sandstones and half that value is common. A bucket brim full of loose dry sand can usually have about 0.3 bucket of water added to it without overflowing. In this case the porosity is 30%. However, being porous does not guarantee that a rock will either absorb, or yield, a fluid. For that the pores must be interconnected. The ability of a rock to transmit a fluid is a function of pore space plus pore connection and is referred to as **permeability.** A shale may have 15% porosity, yet it is nearly **impermeable.**

A great deal of water in the subsurface that occurs in the pore spaces of rock did not infiltrate there from the surface. Every driller for oil is all too unhappily familiar with this water. Oil wells are drilled at a place where oil, not water, is expected to be in the pores of a permeable rock, usually at a depth of hundreds or thousands of meters. Unfortunately salty water is commonly found instead. Although it may have moved around quite a bit, much of this water was originally trapped among particles in a sediment at the time it was deposited in an ancient sea. It is called **connate** water. Our discussion will center on the more familiar, usually drinkable, rain derived variety—**meteoric** water.

A soil may rapidly take in rain water after a dry period, but as it becomes saturated its capacity for intake diminishes. Let us consider the ground water after the rain has ceased. Gravity will cause the downward movement of water through the soil and, normally, the water will come to a level where the forces acting on it are balanced. This level defines the **water table,** and its depth is easily measured in a normal situation: the level at which the water stands in a well.

The water table may shift its level upward with the next rain, or downward with the lack thereof, and its natural depth from the surface is a function principally of climate, soil, and topography. In a populated area the removal of ground water for drinking, washing, and so on, mostly during the day, causes fluctuations in the depth to the water table. In some populous industrial cities the high demands for water have resulted in a removal rate in excess of the rate of natural recharge, and water has to be imported. The consumption of water in the United States is expected to double or even triple within 30 years, and the ground water supply cannot support inefficient use of such amounts.

Counteracting the gravitational forces on ground water are the forces of surface tension and capillary action. These allow some of the water to cling to particles above the water table, or even to fill the pores completely some centimeters above it. This water is commonly used by plants, but there is no practical way for humans to use it since it will not flow. Remember the bucket full of dry sand that was capable of containing 0.3 bucket of water? If you poured water out you would not get 0.3 bucket back; some will cling to sand grains, defying gravity, as it were.

The water table defines the boundary between the **unsaturated zone** above and the **saturated zone** below. As you can see in Figure 11-8, the water table tends to accord, in a subdued way, with topography. It is normally closer to the surface in a valley than on a hill. In the figure, the water table intersects the surface at the stream, and, because the water in the saturated zone moves, it is adding to the stream. In an arid climate the local water table may be deeper and a

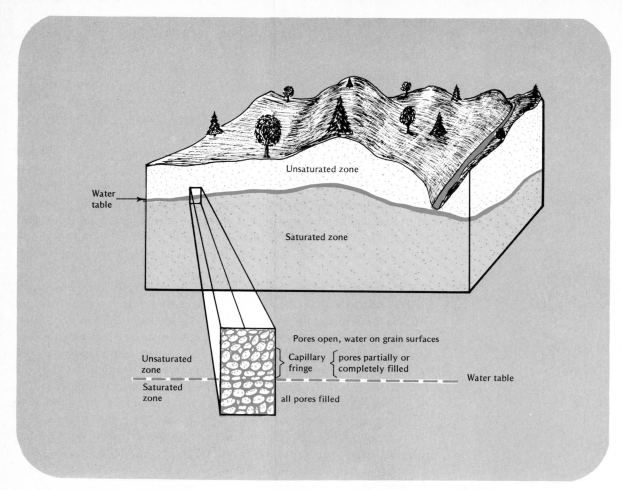

Water
table

Unsaturated zone

Saturated zone

Pores open, water on grain surfaces

Unsaturated zone

Capillary fringe

pores partially or completely filled

Water table

Saturated zone

all pores filled

11-8. *A typical ground water system.*

stream, such as the Nile, may release channel water into the underlying ground water system.

In some areas an impermeable **confining bed,** often a layer of
rock or clay through which water will not flow, will not permit an
unconfined ground water system to exist. Such things as an **artesian
system** or a **perched water table** can result. Figures 11-9 and 11-10
are examples of each.

Geysers and Hot Springs. When ground water comes into contact
with rocks heated by magma or hot gases at depth, it may come to
the surface nearly boiling. If a conduit exists, this heated water may

erupt as a geyser rather than forming a pool of hot water at the surface. The boiling point of water is normally about 100°C, but water will only begin to boil at a higher temperature when under increased pressure. Since water deep in a conduit is under pressure from the overlying water, it can be "superheated" to temperatures well above 100°C without boiling. When the lowest water finally "flashes" into steam, the rapid volume increase forces the heated water above to rise. When that water, also superheated, reaches a level where its previously attained temperature is the boiling point, it too vaporizes. The entire column turns to steam and there may be a violent eruption at the surface; ground water then infiltrates the partially

11-9.   *An artesian system. The drawing illustrates a permeable geologic formation, or aquifer, that has a confining bed overlying it. The pressure in the aquifer is the result of the water at higher elevation near the intake area. In a well tapping the aquifer, the pressure will force the water upward above the normal water table.*

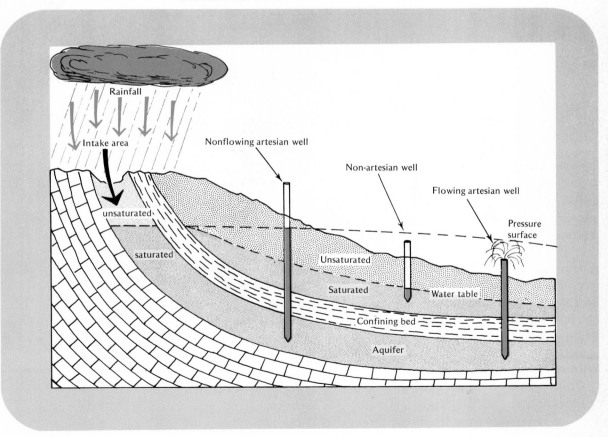

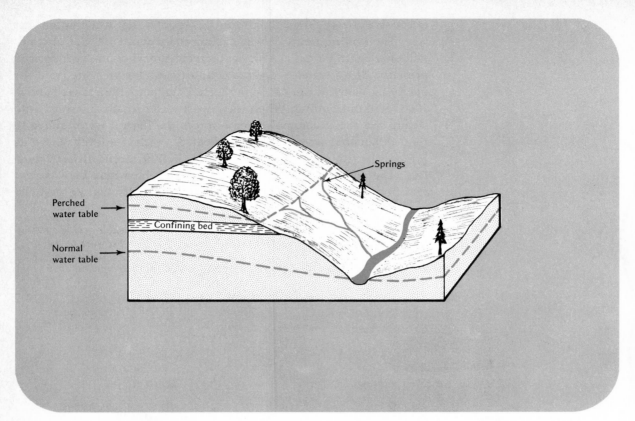

11-10. *A perched water table. An impermeable confining bed has prevented groundwater from reaching the normal water table below. A separate saturated zone lies perched above the normal saturated zone.*

emptied conduit and the process is repeated (Figure 11-11). Some geysers are quite old and faithful in their periodicity and have become tourist attractions.

Chemical Activity. Most caves are caused by the chemical weathering of carbonate rocks by ground water. (The carbonation reaction was discussed in Chapter 10.) Ground water solutions are generally the source of deposits of calcite, silica, and iron oxides that cement sediments or fill cracks and cavities. Buried bones, shells, and logs are often **petrified** by having their pores filled and material replaced with minerals from ground water.

Ground water used in households and industry may have too high a concentration of dissolved material, such as calcium and magnesium ions, to be used without treatment. These and other ions are responsible for "**hard water**" and will react with soap to form an

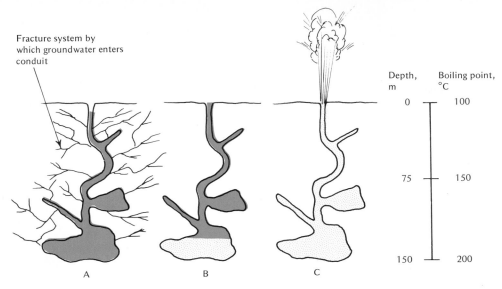

Depth, m	Boiling point, °C
0	100
75	150
150	200

A   B   C

Fracture system by which groundwater enters conduit

11-11. *Geyser action. A. An irregular conduit has nearly filled with groundwater. The boiling point at the bottom, 150 m below the surface, is 200° C because of the weight of the overlying water. B. The lowest water has turned to steam at 200° C and the increased volume has begun to cause the water above to rise in the conduit. Water rises into regions of less pressure, hence, lower boiling points, and it too turns to steam. C. The water in the entire column has passed through its boiling point and an eruption occurs at the surface.*

insoluble scum. This used to be very evident in the days prior to the popularity of showers in the form of a ring around the bathtub. To "soften" the water, these undesirable ions must be removed from solution. Synthetic detergents provided a cheaper answer to some problems than the installation of a water softening unit, but, if the detergents entering the waste disposal system cannot be broken down by organisms, they are not preferable to soap which is biodegradable. Extensive use of the older, nonbiodegradable, detergents in populous areas has had very adverse effects on the local water.

Overland Flow: Sheet wash

Overland flow describes water that moves downslope as more or less continuous films or sheets of water or as small trickles in rills across the surface rather than in definite channels such as a stream. This type of runoff begins when the rate of infiltration into a ground water system is exceeded by the rate of precipitation. On sparsely vegetated slopes, on overgrazed land, or on land where spaced crops such as corn are planted, these intermittently flowing sheets of water can move a surprisingly large amount of sediment (Figure 11-12). The usual term for this type of erosion is **sheet wash,** and although it is extremely common and operates worldwide, little quantitative data on it are available.

11-12. *Sheet wash erosion. This spaced crop has allowed much of the water after 3 in. of rain to run off the gentle slopes and erode the land. Soil from the background has been deposited in the foreground. The sediment has dried and contracted forming a system of polygonal "mud cracks." [Courtesy U.S. Department of Agriculture Soil Conservation Service.]*

In a drainage basin there is far more area in slope than in stream channel, so that sheet wash is probably the *major* aqueous erosive process on continents. Measurements by the United States Conservation Service on slopes of less than 5° in Missouri have shown that a grass cover permits only about 8% of the precipitation to act as slope wash, and the sediment loss is less than 20 g/m² each year. In contrast, a barren slope under otherwise equal conditions had over 25% of the precipitation run off, and the slope lost well over 5000 g/m² in a year! The erosion loss from a slope planted in corn was closer to that of a barren slope than to one covered in grass.

When we have this sort of data, it is easy to accept the previously mentioned opinion that humans are responsible for more than doubling the world rate of erosion. In some regions farmers plant spaced crops, such as potatoes or corn, in strips alternating with close growing crops such as grasses or alfalfa. The latter act as a groundcover to check the flow of runoff and preserve the soil.

Without soil conservation practices, rich farmlands have been destroyed in less than a lifetime. Uncontrolled soil erosion is not as dramatic as a war, but it can devastate the land quite as effectively (Figure 11-13 A, B, C). Rene Dubos once wrote: "Disease, warfare, and civil strife have certainly played important roles in the collapse of ancient civilizations; but the primary cause was probably the damage to the quality of the soil and to water supplies by poor ecological practices."

11-13. A. *Erosion in the early stages. The land has been deforested and the actions of sheet wash and gullying have begun to carry the soil downslope. B. Erosion progresses. Surface runoff has converged in this area of the overgrazed land and a gully has formed. The head of the gully eroded back 3 m in the 1 year before the photograph was taken. C. The bitter end. Both the soil and its tillers have migrated. [Courtesy U.S. Department of Agriculture, Soil Conservation Service.]*

A

B

C

Channel Flow: Streams and Rivers

The slopes of a typical drainage basin are normally dissected by numerous gullies that carry the intermittent rain water to a more permanently flowing body or stream. The stream, whether the smallest or largest member of a drainage system, is the principal carrier of the debris and dissolved material brought to it via slope wash, gully flow, or smaller streams. As material enters a stream channel the water transports this load in three components: as a **dissolved load**, a **bed load**, and a **suspended load.**

The Dissolved Load. The ions taken into solution by chemical weathering normally remain in solution when that water enters a stream system. It is reasonably evident that regional climate and rock types have strong influences on the quantity and type of material in solution in any stream. It is not as obvious that the concentration of dissolved material decreases as the quantity of water in the stream increases. A rainfall that adds water rapidly to a river cannot cause a proportionate increase in the rate of weathering, so the quantity of dissolved material is simply diluted as the flow rate goes up.

The percentage of the total load that is in the dissolved state varies; Leopold and others have reported ranges from about 1 to 64%, but 20% is probably an approximate average in the United States. The percentage is normally higher where the average yearly precipitation is higher. We have already looked into the fate of water borne solutes in Chapter 10, and, since they really do not affect stream flow, we need not discuss them further.

The Bed Load. Particles traveling along the bottom of the stream bed by rolling, sliding, or bouncing move downstream much more slowly than the water. In general, the bed load is formed of the larger particles in a stream, and the stream's ability to move them varies widely from place to place and time to time.

Measuring the bed load is not an easy task, but some general statements can be made. Once in motion larger grains will move faster than smaller ones and rounder particles more easily than those of flat or angular shape. The bouncing of some particles occurs when they are picked up from the bottom by turbulent currents in the stream, rise a small distance, and land farther downstream. The largest particle a stream is capable of moving along its bed is a measure of the stream's **competence.** Small swift streams may be very competent, whereas large rivers may have a very low competency.

In the case of liquid water, velocity is the main governing factor of competence. The impact of the flowing water on a particle may cause it to slide or roll, and, in general, the impact increases according to the square of the increase in stream velocity. Therefore, if a stream moving at 0.5 m/sec doubles in velocity to 1 m/sec, the impact force will increase four times. If the stream velocity triples to 1.5 m/sec, the impact on a particle will be nine times more than it was when the stream was flowing 0.5 m/sec. Entering a stream that does not seem to be flowing *too* much faster than the last one you waded across can be a mistake. A moment's consideration of the water's disporportionate increase in impact on your body compared to velocity might prevent an unpleasant or fatal experience.

Measurements under relatively ideal laboratory conditions indicate that quartz grains of 2 mm, 4 mm, and about 15 mm will begin to move when the flow velocity of the stream is about 40, 80, and 160 cm/sec, respectively. This suggests that, as the velocity doubles, a grain larger in diameter by the square of the largest grain in motion at the previous velocity will begin to move.

Streams, however, are more complex than water flowing in a pipe or other smooth conduit, and these mathematical relations are only approximations. On a stream's bed there is usually a mixture of different sized particles, one getting in the way of another, some grains have more adhesion to the bed than others, or have more cohesion with like-sized grains. Furthermore, if a device is lowered to the stream bed to measure velocities and particle movement near the water-bed interface, the mere presence of the device destroys the natural system. Some reasonably accurate measurements of the quantity of material moving along stream beds have been made. As you would suspect, the more gentle the flow the lower the quantity. Although there is great variation in nature, the average portion of the entire load (bed, suspended, and dissolved) being carried along the bed by rivers is probably about 10%.

Figure 11-14 is a graphical representation of the flow velocities of water necessary to initiate erosion of particles and to keep the particles in transport. There are several concepts to be gained by close study of the diagram:

1. The stream velocity necessary to initiate erosion of a particular size of particle is variable. The erosional velocity is shown as a zone rather than as a line. This is because the slope of the stream bed, the depth of the water, and the shape and density of the particle are all influential and variable.

2. The particle size that requires the least velocity to initiate

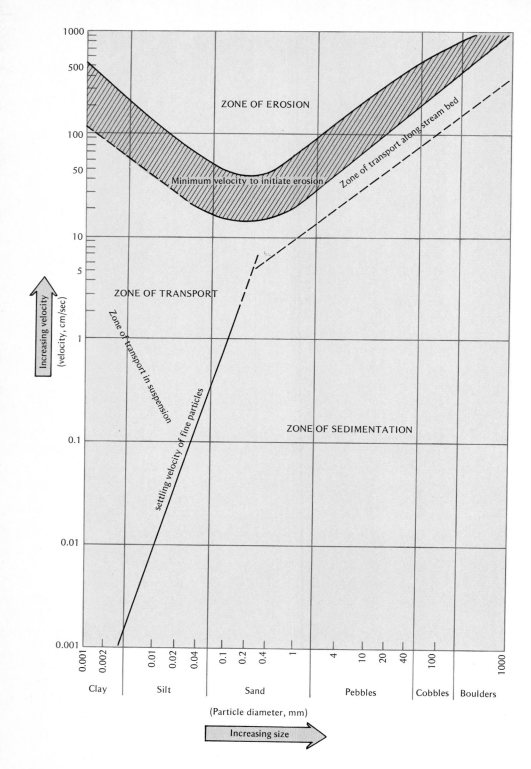

11-14. *The relationship between the flow velocity of water and erosion, transportation, and deposition of various sizes of sedimentary particles.*

movement is fine sand, not silt or clay. This does not mean that a filmy ooze of clay or silt lying on the bottom of a still pond will be undisturbed even if a current that would move fine sand grains is initiated. The ooze will move at a very low current velocity. The diagram indicates that when clay or silt is somewhat compacted, water flowing over it will not cause erosion until velocities are attained exceeding those which will erode fine sand. If the clay or silt grains are in contact, the forces of cohesion hold them together. Also the grains are so small they present very little surface area above the bed to receive the impact of the current.

3. Once set in motion, particles in a stream will remain in motion even if the current velocity decreases somewhat. The diagram indicates that larger particles, carried as bed load, will cease movement at velocities only slightly below their erosional velocities. This is to be expected for the impact of the water decreases greatly when velocity decreases. However, once clays and silts are put in motion, they are carried off the bottom and into suspension. The finer particles will not settle out even though the current velocity decreases markedly. For example, a silt-sized particle 0.01 mm in diameter will begin to be eroded only at a fluid velocity of about 50 cm/sec, yet it will not settle out until the currents drop to about 0.01 cm/sec!

The Suspended Load. Most streams are carrying the bulk of their load in suspension. Before examining the stream's action, let us examine the particle's action. Aristotle is blamed for the idea that, if an object weighs 10 times more than another object of the same composition and shape, it will fall 10 times as fast. This seems good sense and the idea was not seriously questioned until about 2000 years later when Galileo began experiments. These were carried out on inclined planes rather than from atop a tilted tower in Pisa, Italy, but the result was the same. No matter what their weight, bodies near the earth fall at the same speed (at least in a vacuum). A particle in water will also fall under the force of gravity, and we must consider two things about that particle. How fast will it fall and how far can it travel horizontally, for instance in a stream, before it hits bottom?

Consider this tried and true problem in elementary physics: If a bullet is shot horizontally out of a rifle over level terrain and, at the same instant, a similar bullet is dropped from beside the barrel, how long after the dropped bullet hits the ground will the fired bullet hit? Gravity is the only force acting to pull both bullets to ground, and they both hit at the same instant. Let us examine the fall of a grain of the finest sand after it has been introduced to a body of *still* water, analogous to the dropped bullet. The force of gravity causes it to fall, and two forces act to oppose that fall—frictional drag and the buoyancy of the water. Over 100 years ago the three force equations were combined, and an expression called **Stokes' law,** relating the terminal fall velocity of a small particle in a fluid emerged.

$$V = \frac{d^2 \, (d_p - d_f) \, g}{18u}$$

where V = the fall velocity in centimeters per second; d = the particle diameter in centimeters; d_p = the density of the particle; d_f = the density of the fluid; g = the acceleration due to gravity in centimeters per second squared; and u = the viscosity of the fluid in poises.

If we assume the water to be pure with a density of 1, the temperature to be 20°C so the viscosity of the water is 0.01 poise, and the small particle to be of quartz with a density of 2.65, we allow ourselves a short cut: the approximate terminal velocity, in centimeters per second, of a small particle (finer than fine sand) falling in water equals 9000 d^2.

So, for the finest sand of $^1/_{16}$ (0.0625) mm diameter, or 0.00625 cm, the fall velocity is about 0.35 cm/sec, or about 12.6 m/hr. For a typical size of clay particle, 0.002 mm diameter, the fall velocity is only about 0.013 m/hr. To get an idea of the difference, the finest sand would take 12 days to fall through pure water a distance equal to the average depth of the ocean, but clay would take over 30 years. These calculated fall velocities are only approximate guides to what may occur under natural conditions; for example, a flattened sand grain settles through water at different rate than a spherical one. Clay particles, entering the sea via a river, react with the ions in sea water and tend to clump together and fall faster.

The line on the diagram in Figure 11-14 labeled "settling velocity of fine particles" is a plot of Stokes' law. This defines the boundary on the left between the "zone of transportation" and the "zone of sedimentation." For particles larger than the size of fine sand, the density difference and viscosity of the fluid are relatively unimportant and another equation must be used. However, most of the suspended load is in the size range covered by Stokes' law, so we may ignore the other law and simply refer to the graphed line on the diagram that limits the upper boundary of the "zone of sedimentation."

Let us return to our theoretical stream. Imagine this same particle of the finest sand entering a stream whose depth is 1 m and whose velocity is 1 m/sec. Now the analogy is with the fired bullet. At the 12.6 m/hr fall rate, the sand should hit bottom in less than 5 min, never to rise again. While falling, the grain, like the fired bullet, would travel horizontally some distance. At 1 m/sec stream velocity the grain would travel 300 m in the 5 min allowed for its fall to bottom.

What conclusions might be derived from all this? One erroneous conclusion is that, to find a particle of the finest sand in suspension in a swift stream moving at 1 m/sec, one would have to catch it within 300 m from where it entered the stream. Why this is false will become clear when we look into the action of stream water. What one can *correctly* conclude is that, when a sand grain gets to a sea or lake where the currents are very slow, it cannot get too far from shore. The clay, on the other hand, can be carried far to sea

11-15. *Erosion on the steam bed. A pothole formed by the abrasive action of rock fragments. [Courtesy Swiss National Tourist Office.]*

by gentle currents before it reaches bottom. Therefore, under stable conditions, the particle size of grains deposited in the sea will diminish with the distance offshore.

This understanding of the action of particles in a fluid is of value in determining the geography and depth characteristics of ancient seas and their sediments. For example, if a widespread bed of rock originally deposited in the sea has more and more coarse particles in it as one examines the outcrops from south to north, the ancient shore was probably to the north.

If one samples the stream-borne sediments accumulated off a present coast and the particle size–distance offshore relationship does not hold true, there is a problem to be solved. Perhaps turbidity currents, discussed in Chapter 7 (page 179), carried larger particles into deeper water. Today, it is possible to dredge coarse sand grains in abundance from the sea bottom of the continental shelf many kilometers off the coast of some areas of eastern North America. There are no present currents capable of transporting them there, and there may be no evidence of turbidity currents. Our knowledge of the action of a particle in a fluid leads us to the possibility that sea level was lower in the not too distant past and what is now offshore was once the coast. Much can be made of a little knowledge and the somewhat uncommon human ability of wondering about the origin of a sand grain.

Stream Action

A stream has the capabilities of erosion, transportation of eroded material, and deposition. We have examined the fact that a stream can erode its bed by initiating the transport of bottom material, but this material in motion is also an erosive agent. The rolling and bouncing particles mechanically wear the other particles they hit. This **abrasion** includes impact and crushing action among particles and also the mechanical wear on the bedrock that may be exposed in the stream (Figure 11-15). The result is that larger particles, pebbles and boulders for example, are quickly rounded. The impact of sand and smaller particles is more "cushioned" by water and they are not as affected by abrasion.

Material in suspension has been dealt with from the particle standpoint, but what of the stream's action? The material in suspension makes up most of a typical stream's load, and if a stream had current motion *only* downslope little material could remain in suspension much past its point of introduction to the channel.

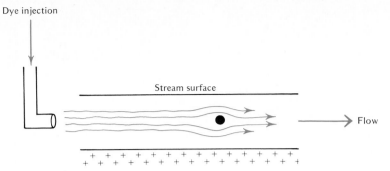

Dye injection

Stream surface

Flow

11-16. *Laminar flow. If water flows in a smooth straight channel at very low velocity, a small fraction of a centimeter per second, the molecular flow will be smooth and laminar even when deviating around a particle. This type of flow is not found in channels under natural conditions but is typical of ground water movement.*

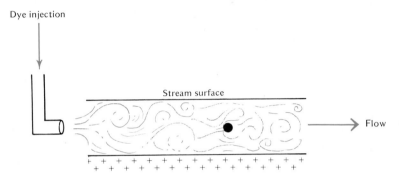

Dye injection

Stream surface

Flow

11-17. *Turbulent flow. Complex eddy currents moving in all directions are superimposed on the overall flow of the water. Turbulent flow is typical in streams and rivers.*

However, there are basically two styles of fluid flow: laminar and turbulent. Figures 11-16 and 11-17 are simple illustrations of each.

It is the complex multidirectional turbulent currents within the unidirectional overall flow of a stream that *keep* particulate matter in suspension. When some of the turbulent eddy currents in the stream move upward faster than the fall velocity of a particle, the particle will rise. It is the velocity of these *internal* currents, not simply the velocity of water movement in the downstream direction, that gives a stream the capability of carrying a large suspended load (Figure 11-18). The paths of clay, silt, and fine sand particles are complex and are similar to those shown by the paths of dye injected into turbulent water in Figure 11-17. Thus, the hypothetical grain of very fine sand that was introduced to the stream moving 1 m/sec would *not* come to rest 300 m downstream; in fact, it may stay in suspension many kilometers downstream from its point of introduction.

Energy losses due to friction and turbulence along the channel bottom and walls and at the air-water interface impede the flow of water in a stream. As a consequence the overall maximum velocity of a stream is usually less than one half the way down from the

11-18. *Deposition of the load. A. this dam is about 9 m high. When the flow velocity of the incoming water was diminished as the lake developed behind the dam, much of the suspended load and all of the bed load was deposited. B. Thirty-two years passed between construction and abandonment. [Courtesy U.S. Department of Agriculture Soil Conservation Service.]*

A

B

surface, and in a straight channel is midway between the banks (Figure 11-19). Figure 11-20 illustrates some channel shapes and their influence on flow velocity. Changes in channel shape, bed roughness, and curves in the channel all have effects on velocity and on turbulence. It is fair to state that the entire stream system is so complex that it is doubtful that the path of any one particle in it can ever be predicted.

When a stream is initiated on an inclined sediment-covered table under laboratory conditions, it begins by flowing in a relatively straight path. Within days its path becomes curved and finally sinuous. Why? One factor is that no stream in nature flows over absolutely uniform material. Small or large changes in the sediment or bedrock will alter its path; even the rotational forces of the earth act against straight flow.

Once a curve begins, the velocity maximum shifts from midstream toward the outside of the curve. This accentuates the tendency to curve because one side of the bed is then eroded more rapidly by the proximity of the higher velocity water. Figure 11-21 illustrates the development of a looplike bend, known as a **meander**, in a stream. When the erosion on the outside of the meander loop causes the river to meet itself, Figure 11-22, it will follow a straighter path once more and may abandon the water-filled meander, which then is known as an **oxbow lake**.

Meanders also come into existence because streams adjust to their slope, load, and flow. This adjustment may take the form of lengthening their route to base level, and meanders are a method of increasing length and thereby lessening the slope. When one meander is cut off, the length is shorter, but somewhere another meander is developing, so little overall change in the length of the river's path occurs. The principal concept to be gained from this discussion is not simply how meanders form but that streams can erode laterally as well as downward and that a meander functions as a method of adjustment.

After very heavy rainfall or when it receives melt water in spring, the velocity and quantity of water in a stream increases. Its capability for erosion also increases, and most of a stream's yearly erosion may be accomplished during a short period of time. The relation

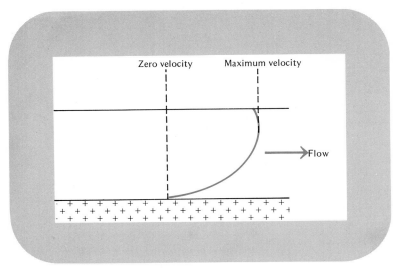

11-19. *Typical velocity profile in a stream. The maximum is at about one quarter of the depth. This may vary but the lowest velocity is near the bed since frictional forces and induced turbulence at the bed slow the flow.*

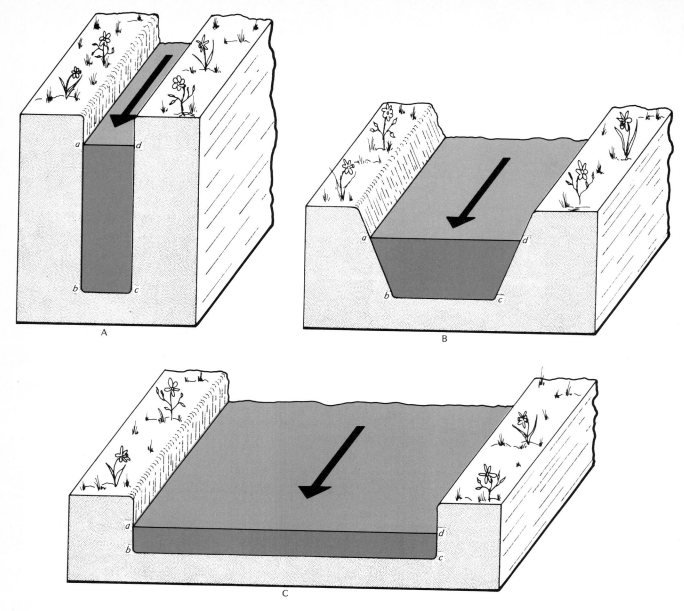

between quantity and velocity is expressed as a stream's **discharge** and is a measure of the quantity of water in the channel passing a point during a time interval. The discharge rate, Q, is often given in cubic feet per second (cusecs) but the equation is general and any volume-time relation may be derived from raw data.

$$Q = \text{surface width} \times \text{average depth} \times \text{average velocity}$$

11-20. *Stream channel shape versus velocity. The three diagramed channel shapes all have the same cross-sectional area of water. A and C have more water in contact with the channel along a,b,c,d than has B, hence A and C have more frictional loss. The water will flow more rapidly in channel shape B, all other things being equal.*

11-21. *Meanders. The hypothetical development of a meander is shown. The arrows indicate the location of the zone of highest flow velocity and tend toward the outside of curves. The river banks erode more rapidly near the higher velocity flow and some of the eroded material is deposited downstream on the inside of curves.*

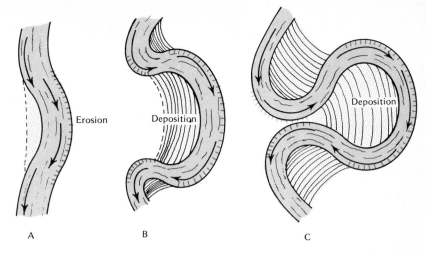

11-22. *Meanders and oxbow lakes. [Courtesy Royal Canadian Air Force.]*

When the discharge of a stream increases or decreases, some adjustment must occur in either its depth, width, velocity, or all three. The stream, then, tends to change in character according to the task at hand: less or more water, less or more load. When a tributary enters, the discharge increases and so does the depth and width. Does the velocity increase too? In general, there is a tendency toward a velocity increase downstream in many rivers.

It is common in introductory courses in geology to ask such a question as: "If a stream is 2 m wide, 3 m deep, and flowing at 0.5 m/sec, what is its discharge? " The correct answer of 3 m³/sec may get you two points, but do not allow that to be the limit of your thought process about discharge. There is more in that relationship than meets the eye.

Will a river adjust perfectly to the changes in slope, water content, and load content along its route? Since the river is part of an erosion system that decreases the elevation of land, the overall situation is only quasistable. Yet, given some relative stability, a perfect river is at least imaginable. Rivers can approach perfection more easily than humans, but, at the risk of anthropomorphism, one analogy with humans may help you to understand the concept.

A state of equilibrium can be thought to exist in a person whose balance between food intake and energy output is so perfect that no weight is gained or lost. Energy and material flow through the human, but all is balanced. In the case of our river flowing through Utopia, a small addition of water would cause it to alter and erode a small amount so that its sediment load increased and added work would be provided for the added water. On the other hand, a little added sediment would cause the river to deposit a small amount of its load to maintain its balance of forces when there was no extra water to deal with the new material. This is the concept of the **graded stream** which, since it neither erodes nor deposits, is perfect for its task.

Given a changing earth, this condition is never fully achieved, but the idea of this "goal" is a useful concept to help understand why rivers must react and why they erode or deposit when a change occurs. One concept related to all this is that humans cannot change a river without causing a reaction. The river will invariably come into adjustment with the change, and the way it adjusts may be unexpected. Let us examine a situation where a river reacted to human tampering.

To build a road down a mountain, one makes it longer by taking a circuitous route that gives the road a gentle slope in a steep area. Imagine that a major navigable river has adjusted to the land's slope toward base level by forming meanders, which serve to lessen the river's slope or gradient. Figure 11-23 illustrates the system. The water is flowing from A to B, a distance of 10 km, around the bend. The formula for slope is

$$\text{slope} = \frac{\text{difference in elevation}}{\text{horizontal distance}}$$

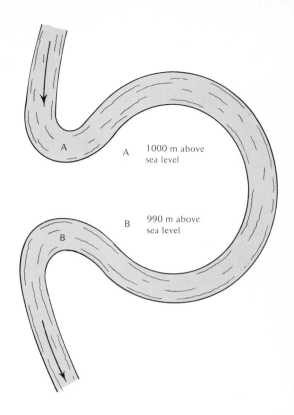

11-23. *A hypothetical meander prior to channelization between points A and B, which are at elevations of 1000 and 990 m, respectively.*

A 1000 m above sea level

B 990 m above sea level

Therefore, the river flows down a slope of 1 m/km (10 m/10 km). To decrease ship travel time, the river is shortened by eliminating the meander and dredging a channel from A to B, which are only 1 km apart; 9 km of travel has been saved but the slope is now 10 m/km, ten times as steep! The most immediate natural adjustment of the river is an increase in velocity down the steeper gradient. No more water is available to the system than prior to the channel change, so the discharge rate will come to equilibrium as other parameters change. $Q =$ width $\times$ depth $\times$ velocity so that, when velocity increases and no more water is added, the width and, especially, the depth decrease. The result is that the docks near this new channel are left too high above the river for the ships to use!

The Deposits and Landforms of Streams

The general term for river deposits is **alluvium**, and lithified alluvial sedimentary rocks are quite common. For example, as the

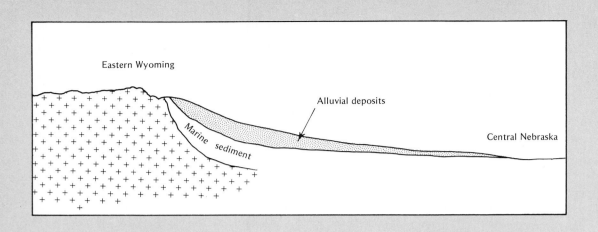

11-24. *Alluvial plain. Eastward flowing streams have deposited material shed off the Rocky Mountains. The deposits now form a gentle plain rising to the west from the central United States.*

Rocky Mountains began rising at the end of the Mesozoic Era, the streams bearing the weathering and erosion products off the highlands began to cover the lowlands to the east with deposits of alluvium. When the streams lost velocity as the gradients lessened eastward, they adjusted by depositing part of their load. Millions of years of erosion and deposition have produced a wedge shaped deposit of alluvium from the high Rockies eastward to the middle of the Great Plains.

The deposits have since been eroded in many places, and the Badlands of the Dakotas are examples of the more recent erosion. A person driving west from central Nebraska climbs this wedge and may go up about 1800 m in a distance of some 550 km. The wedge of alluvium, Figure 11-24, has such a gentle slope that for most of the journey the rise in elevation is imperceptible. This immense plain is the product of the deposits of many streams over tens of millions of years.

When a river begins in high country its path is relatively straight, and much of its erosive activity is directed toward deepening its bed. As it does, its valley will begin to widen as slope wash and the actions of tributaries progress. The result eventually is a valley in the shape of a V, Figure 11-25. At some later point in time or in some place downstream, the valley will begin to broaden as the gradient

11-25. [OPPOSITE] *A youthful stream valley in a mature topography. The V-shaped valley and lack of a well-developed floodplain are characteristic of a youthful stream. The area in view is nearly all slope, characteristic of a mature landscape. [Courtesy Oregon State Highway Commission.]*

11-26. *Lateral Erosion. As this stream began to erode laterally through its own floodplain deposits, rocks were dumped into the channel in an attempt to save the house. Not to be denied, the stream has penetrated weak spots in the armor and the house was lost. [Courtesy U.S. Department of Agriculture Soil Conservation Service.]*

lessens and the river erodes laterally, meandering and eroding bedrock or its own previous deposits.

Floodplains. The shifting river eventually produces a broad shaped valley, and the valley floor is defined as its **floodplain.** This plain, Figure 11-26, is not necessarily caused by floods, but when the river overflows it usually does so within the limits of that plain. The river's floodplain is not only fairly flat but the soil is often fertile due to deposition of fine material during flood stage; hence, floodplains are often areas of intense agriculture and population density (Figure 11-27). The floodplains of the Nile, the Ganges, and many other great rivers have been centers of civilization for thousands of years.

Terraces. Common features of many floodplains are the flat step-like zones, called terraces, above and roughly parallel to the stream in its channel. The stream has eroded downward through its own deposits (Figure 11-28). This increase in downcutting compared to lateral cutting has some cause and an increase in stream velocity and, therefore, in the ability to erode is one possible reason. Several factors can cause an increase in stream velocity, such as a climatic change involving increased rainfall, a tectonic rise of the land surface, or a lowering of base level.

Deltas: Alluvium at Base Level. Whether it empties into a lake or sea or even into another river, a stream effectively ends and, therefore, its ability to carry a suspended or bed load ends. This load may

11-27. *Flooding. The river has topped its banks and spread onto its flood-plain. These plains are usually flat and fertile and are commonly densely inhabited.* [*Courtesy* Courier-Journal *and* Louisville Times.]

be deposited in a fan shaped wedge, broadening outward, that often takes the shape of the Greek letter delta, $\triangle$, Figure 11-29. Material can be added to the delta as more and more particulate matter is deposited, and some deltas cover very large areas both above and below the surface of the water.

Alluvial Fans. Analogous to deltas, alluvial fans are formed when a stream reaches a local nonaqueous base level, usually the bottom of a steep slope in a relatively arid area, Figure 11-30. Much of its load is deposited at the sharp change in gradient as the stream leaves the confines of its channel.

Drainage Patterns. In any drainage basin the entire stream system forms a pattern. The pattern on a surface of relatively uniform material tends to resemble the trunk and branches of a tree or shrub and, for this reason, has been termed dendritic. Geologic structures, such as rock folds, domes, ridges of rock more resistant to weathering and erosion, joints and faults, and so on, may exert a control on the pattern, and the pattern may be a clue to the geology. Several important economic deposits have been found by coupling "creekology" with geologic insight. For example a salt dome, potentially with oil trapped in its overlying or flanking beds, may have risen near enough to the surface to cause a slight, almost imperceptible bulge. The drainage off a domal structure tends to radiate outward, Figure 11-31, and some oil fields were initially found by someone who noticed this pattern on a map.

The Erosional Evolution Model

Around the turn of the century a concept of an **erosional cycle** was put forth. In general terms the idea is that there is a natural sequence of both erosional landforms and stream character in an

11-28. *Terraces. A combination of lateral erosion and increased downcutting has left several flat terraces along the river's floodplain.*

area. The area and the streams develop from **youth** through **maturity** to **old age,** and the "age" of both the stream and the landscape in the sequence can be determined by their forms. The idea has come under criticism, some valuable, some not, and the concept has been modified and interpreted in many ways we will not consider. That the concept has endured this long is testimony to its value and to its relative simplicity in spite of great apparent complexity.

Two major factors enter and confuse any oversimplistic view of the concept. The first is that the earth is seldom, perhaps never, stable long enough to allow any major areas to progress from youth

11-29.   *The delta of the Nile. [Courtesy National Aeronautics and Space Administration.]*

11-30. *Alluvial fans. Streams reaching a sharp decrease in slope have decreased in velocity and deposited part of their loads in fan shaped patterns.* [*Courtesy U.S. Geological Survey.*]

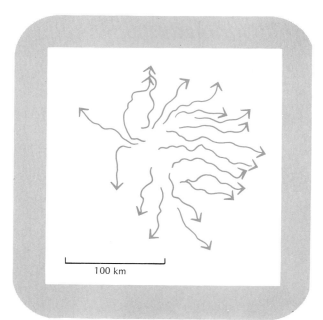

100 km

11-31. *Radial drainage. The drainage pattern shown is that of the Black Hills, a domal uplift in South Dakota and Wyoming.*

to old age without modification. Second, the streams and the landscapes may not coincide in their progress through the sequence. A stream may have the form of youth and the area through which it flows may have the form of maturity (see Figure 11-25 and then Figure 11-32 for a young landscape).

Figures 11-33A, B, and C illustrate the development of the ero-

11-32. *A young landscape. This area of Greenland has recently lost its cap of glacial ice and stream erosion has begun. The streams, which flow during summer, have cut narrow valleys and now form separate small drainage basins. As maturity is approached the valleys will connect and an integrated drainage pattern will develop. Slope will replace the relatively flat areas now between valleys as the principal topographic feature. [Courtesy Geodetic Institute, Copenhagen.]*

sional cycle as envisioned by its originator W. M. Davis. The final result of the erosional model, according to Davis, is that the once high surface is reduced to an erosional plain close to base level. Many geologists believe that tectonic activity and/or climatic changes never permitted such a near-plain, or peneplain as Davis termed it, to develop over any large area. Most plains are alluvial, are developed on flat lying sedimentary marine strata, or have had some marine or glacial erosional history that confuses the issue. Probably most geologists believe that at least remnants of peneplains exist, but arguments to the contrary are not hard to find.

An area may be uplifted during the latter portions of the cycle, or a more humid climate may develop; either event can restore youth or maturity to an area in old age. This rejuvenation may cause extensive terraces to develop or may cause a gentle meandering stream to erode downward rapidly enough to cut, or incise, its

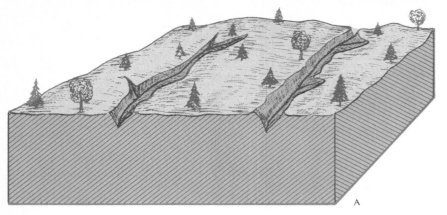

11-33.   The idealized geomorphic cycle.
A. Youth. The area has been uplifted and erosion has begun. The character of the landscape is one of moderate relief, most of the land is relatively flat and is between stream valleys. The streams are not close to being graded and are actively downcutting toward base level. They have a steep gradient and a narrow valley.
B. Maturity. Erosion has progressed to the point where nearly all of the area between the initial streams of youth has been streamcut. The land is mostly in slope and the topography has the maximum ruggedness. The principal streams have approached gradation and floodplains have broadened. The meanders that exist are about equal to floodplain width.
C. Old age. The area has been eroded close to base level and the land is nearly flat. Most of the surface is broad floodplains with rivers meandering along belts within them. All the streams have approached gradation, ox-bow lakes are prevalent, and the dominant stream process is lateral cut-and-fill rather than the downward erosion of youth.

meanders into the underlying rock (Figure 11-34). There are other signs of possible rejuvenation, and there are alternative explanations for any of them. Nevertheless, the idea of youth, maturity, and old age is a useful way to describe the morphology of an area. To speak of a mountainous region as having mature topography with youthful streams gives a mental picture with a minimum of words.

11-34. *Entrenched meanders. The apparent imposition of steep youthful stream valleys on a very well developed meander pattern suggests that rejuvenation has occurred. Perhaps the meanders developed in maturity and then the area underwent slow uplift causing the streams to renew the downcutting characteristic of youth.* [Courtesy L. J. Maher.]

GLACIERS AND GLACIATION

About 10% of the continental surface of earth is now covered by ice, and about 30% was covered not too long ago, geologically speaking. Although most of the present ice is in Antarctica, where no one lives permanently, and in Greenland, where few live, glaciers are common near many populated mountainous areas. Much of the present topography of many mountains, such as the Rockies of North America, where glaciation is not now very prevalent, is the result of glacial action.

A glacier is generally capable of moving and of carrying larger fragments of rock than flowing water. This high competence of ice is due to the nature of ice itself rather than to its velocity of flow, as is the competence of a stream. When water freezes, the molecules become linked together in crystalline arrangement much more structured than they had as a liquid. The crystalline nature of ice gives it a rigidity that enables it to transport very large rocks. The difference is easily appreciated if one stands on an ice-covered lake and attempts the same activity after the ice has melted. Although in one sense a glacier is flowing water, the structural form of that water has resulted in processes of erosion, transportation, and deposition of material quite different from those of streams.

Snow is the initial source of the ice that consitutes nearly all the mass of a glacier. The snow that accumulates in the upper regions of a glacier may be melted by solar energy or may be at least partially melted by the pressure from the overlying snow accumulation. The grains of snow are recrystallized and changes occur in their shape and size as they become transformed to ice. One of the most important mechanisms is **sublimation,** when molecules of the frozen snow enter into the vapor state without passing through a liquid state. The vapor can migrate and change back directly to a solid at another grain, so some grains grow at the expense of others.

Melting, sublimation, and recrystallization change the snow to subspherical particles, and frozen water in this state is known as **firn:** the ''corn snow'' of the skier. The name firn comes from German and refers to ''last year,'' implying that enough sublimation, and so on, has occurred since the previous winter to permit the granules to develop from flakes of snow. As more snow accumulates firn granules below become packed, grains are bonded together by continued sublimation, and recrystallization under pressure closes off the air spaces.

When most of the air has been forced out, the transformation of snow through firn to ice is completed. The ice is easily recognized by its bluish color. It is a crystalline solid, but relatively close to its melting point and therefore more easily deformed than most crystalline solids. When the pull of gravity is strong enough, ice ''flows''; its behavior, however, is more akin to deformation of a solid than the flow of a liquid.

Glaciers have two types of movement. One is within the glacier where the ice deforms by layers parallel to crystal planes gliding one over the other. In the other type of motion, the glacier itself slides over its bed. However, some glaciers are frozen to their bed and all their movement occurs apparently within the ice. Other glaciers may have liquid water at their base, perhaps from melting caused by geothermal heat flow, by pressure from the overlying ice, or as a result of the inflow of seasonal rain or melt waters. Some of these glaciers may move very rapidly. What occurs at the base of glaciers affects their ability to move and their ability to erode, but the base of a glacier is a very difficult place to study. Geologists are not of one mind about glacial flow.

A glacier is usually divisible into two sections, a **zone of accumulation** and a **zone of wastage.** In the accumulation zone there is a net annual gain of snow, hence firn, then ice, and in the wastage zone there is a net loss because not only is all of that year's snow lost but also some of the ice previously formed. This loss can occur from melting and subsequent runoff, from evaporation, sublimation, and wind removal of snow, or from ice breaking off into water and forming a berg (Figure 11-35). Since two or more of these processes may occur at the same time, a collective term for all forms of wastage is commonly used, **ablation.** If accumulation exceeds ablation over a number of years the glacier grows and advances, if the situation is reversed the glacier retreats, and if a balance is achieved the front of the ice remains in about the same place. One important concept is that, even when the glacier's front is stable or retreating, ice *within* is moving forward. The importance of this internal movement will become clear when we describe glacial deposits of rock debris.

Glaciers can be catagorized in more than one way. One method is to separate them on the basis of their internal temperature regimes, which often are reflections of the surrounding climate. By this method there are "cold" and "warm" glaciers. The former tend to be frozen at their base and the latter may have a film of liquid water there.

An easier way to classify glaciers is according to their morphology, and, although this method tells us little about the internal dynamics of the glacier, it is somewhat elegant in its simplicity. There are **valley glaciers** that flow downhill within the confines of a valley, and **ice sheets,** or "continental glaciers" which cover large

11-35. *About 90% of the ice forming this small berg is under water. The three ice breakers can push this fragment of the Antarctic ice cap only a few hundred meters in an hour. [Courtesy National Science Foundation.]*

areas of a land mass and flow in many directions. Of course, a valley glacier may advance beyond the confines of its valley and spread out somewhat, and when several join together over the plain at the mountain front the ice mass is called a **piedmont glacier.**

Valley Glaciers

In mountainous areas where yearly snow accumulations can build up in spite of summer melt, a glacier may form. At sufficient elevation this can occur no matter how low the latitude; there are glaciers in Africa and other equatorial lands with high mountains. Figure 11-36 illustrates the form of a simple valley glacier.

From the glacier's surface to depths of up to 60 m or so, the ice is not under sufficient pressure to permit plastic flow. In this more brittle zone cracks, or crevasses, develop (Figure 11-37). Crevasses are a danger to climbers and some unfortunates have provided empirical evidence of the internal flow of ice when their frozen bodies appeared at the glacier's front years after they fell into a crevasse far up the valley. Surficial flow is often studied by setting a series of stakes across the glacier's surface, but internal flow must

11-36. *Idealized longitudinal section of a valley glacier.*

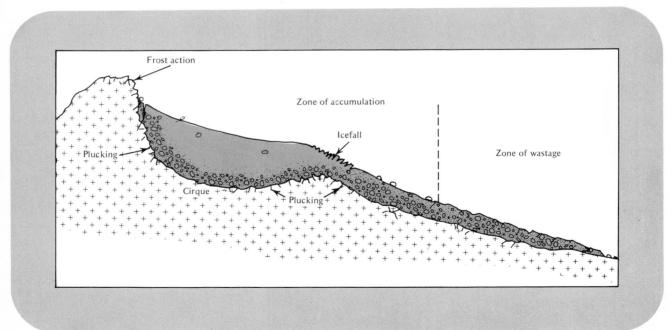

be studied by more sophisticated measurements, such as in boreholes. A typical velocity profile of ice flow is shown in Figure 11-38.

The rates of glacier movement are, of course, extremely variable, and variations even exist within a single glacier. Using the surface movement in the central area of a glacier as a guide, typical ice move-

11-37. *Crevasses. Ice movement in the more brittle upper layers of this large valley glacier has produced crevasses. Wind blown snow has drifted across the tops of the fractures.* [*Photo by D. Darby.*]

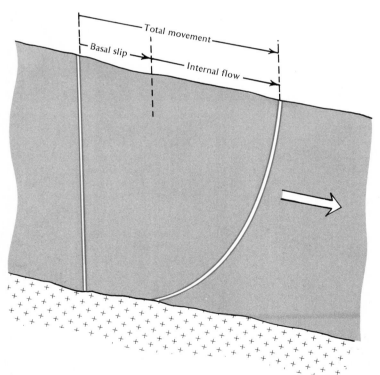

11-38. *Idealized velocity profile of a valley glacier. The original borehole at the left has been displaced down glacier to the right both by internal flow and slippage at the base. The hole has been deformed into a curve according to the different internal velocities.* [*Courtesy R. P. Sharp.*]

ment is less than 1 m/day. Some may move only a few centimeters per day and a few have been estimated to move 30 or 40 m/day for short periods. This kind of measurement is really of the ice velocity and should not be confused with the movement, if any, of the glacier's front. Normally if accumulation exceeds wasting, the front will progress down the valley, but that rate may be much less than the internal or surficial ice velocity since melting may be appreciable at the front. The front of an advancing glacier typically moves a few meters per year, but surges of movement up to kilometers per month are known. Why such surges occur is a question still awaiting an adequate explanation.

11-39. *Glacial polish. The passage of ice has rounded and polished the rock surface. [Courtesy U.S. Geological Survey.]*

11-40. *Glacial striations. This boulder, lying on a field in southern Africa, was faceted and striated when it was moved along the base of a late Paleozoic ice sheet. Bedrock often shows similar striations after an ice sheet has passed over. In the latter case the striations are clues to ice movement direction. [Courtesy J. C. Crowell.]*

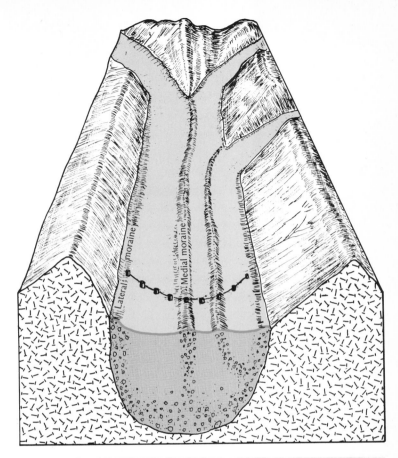

11-41. *Idealized section through a valley glacier. The line of stakes was originally straight across the glacier and the higher velocities in the central part of the ice have deformed the line. [After R. P. Sharp.]*

11-42. *A glacial valley. A good example of the U shape indicative of valley glaciation. Note the waterfall that begins at the side of the main valley where a small tributary glacier entered the main glacier. [Courtesy R. W. Ojakangas.]*

Erosion by Valley Glaciers. Rock fragments are a normal component of glaciers. Some of these fall onto the glacier from the valley walls, but much of the rock load of glaciers is derived from its own bed. Frost action seems to be the principal method of loosening bedrock. Melt water, either produced at the glacier's base or flowing in after summer melting in the area, freezes in joints and loosens rock. If this ice freezes to the glacier ice, the flow may cause the fragment to be plucked from the bedrock and carried downglacier. Once rock fragments are freed, they may become tools for the abrasion of the bedrock. The bedrock exposed by a retreating glacier is often rounded, polished, and grooved (Figure 11-39). Commonly both rock fragments and bedrock exhibit a linear pattern of scratches known as glacial striations, Figure 11-40 shows an example. The load of rock in a glacier typically contains some very fine "rock flour" thought to be a product of abrasion. Melt water streams at the glacier's front are sometimes milky colored from the suspension of this material.

The combination of **plucking** and **abrasion** by valley glaciers produces several distinctive features. One, of course, is the valley itself. As a mountain range began rising, it seems reasonable that in most climatic regions stream action began to form typical V shaped valleys. It also seems reasonable that when glaciers form they would use the same valleys. Since glaciers are very effective erosional agents and the ice fills the valley much deeper than water ever could, glaciers alter the shape of the valley from a V to a U. Even where no glacier now exists and a river has produced a floodplain across the valley bottom, the U shape is a giveaway to previous glaciation. Figures 11-41 and 11-42 are diagramatic and actual illustrations of a glacial valley.

In the area of the mountain crest, the glacial erosion commonly

11-43. *Cirques. A pair of amphitheater like depressions formed by plucking at the heads of Pleistocene valley glaciers.*

11-44. *A glacial horn. This type of peak is indicative of glacial erosion. [Courtesy Swiss National Tourist Office.]*

produces an amphitheaterlike basin called a **cirque** (Figure 11-43). If glaciers flow off the mountain in several directions the erosion may produce adjacent cirques and sculpt the mountain peak into a horn, for example the Matterhorn of Switzerland (Figure 11-44). Areas where valley glaciers were common, for example, much of the Rocky Mountains, are generally more rugged than they otherwise would be.

Ice Sheets

Ice sheets may flow over relatively flat terrain or they may be so thick that they simply cover terrain that has considerable relief. The large ice sheets today are confined to polar regions. The temperature of the large polar ice sheets is not constant throughout; it is warmest at the bottom and coldest near the surface. Measurements in a borehole of nearly 1400 m through the Greenland ice sheet indicated temperatures of −24°C at about 200 m and −13°C at the bottom.

The immense amount of pressure at the base of a thick ice sheet lowers the freezing point and may cause melting at temperatures a few degrees below 0°C. This pressure probably does not cause melting at the base of a polar ice sheet but is the significant factor in its internal flow. A thick ice sheet can flow radially outward from its center and advance over terrain almost independent of topography. Once an ice sheet extends into an area of warmer climate, pressure melting at the base may occur and contribute to the glacier movement and erosive action. If the ice sheet is frozen to its base it will not erode. The Antarctic ice sheet is over 4000 m thick in places, but most of the ice reaching the shores of the continent contains little rock debris. There, most of the ice is probably frozen to the base and ice movement is mostly internal.

In the past the ice sheet extended into more temperate climatic areas and extensive material was eroded, incorporated into the ice, and deposited as the glaciers retreated. A glacier, in relation to the surface over which it moves, is comparable to an elephant, capable of destruction or of gentleness; the elephant according to its mood, the glacier according to its temperature regime. Ice sheets have progressed over loose soil and upon retreat the soil profile was still preserved; in another part of the glacier where sliding at its base has occurred, rock material is eroded and the bed is gouged and striated.

The Deposits of Glaciers

As a glacier retreats, its load of rock is left behind. Even under equilibrium conditions deposition occurs at the front. Although ablation and accumulation may be in balance and the glacier front is not moving, the ice is *still flowing* to that front and bringing more debris. The material deposited directly from the ice is a nonhomogeneous mixture of sizes, but the melt waters may carry and sort things out. There are, therefore, two general types of glacial deposits, **unstratified** and **stratified.**

In the last century some people began to pay attention to the large boulders and unstratified deposits of debris common in parts of Europe. Although remote from mountains, some scientists familiar with glacial deposits in Scandinavia or in the Alps became convinced that ice had played a role in the history of these deposits. If at first someone had suggested that the fields of Germany once had been covered by a continental size ice sheet, perhaps kilometers thick, the roars of Teutonic laughter can be imagined.

It took many years for the European workers to build their case to a level of general acceptance. Even now, in spite of the overwhelming evidence, it is not easy to conceive that an enormous thickness of ice covered most of Canada and much of the northern United states. The early accepted theory for the deposits in Europe was that, during the great flood of Noah's time, icebergs drifted across Europe, melted, and let fall their load of rock. The deposits became known as **drift,** and the word is still used as a general term for any deposit of rock derived from a glacier.

It is necessary to maintain a mental distinction between the types, stratified or unstratified, of depositional material associated with glaciers and the landforms composed of these materials. Many depositional landforms contain more than one kind of drift.

Unstratified Deposits. Rock debris left by a retreating glacier may be a mixture of sizes from clay to boulders. The only consistent feature of this type of material is its inconsistency. Some unstratified drift may have immense boulders within the mass, Figures 11-45 and 11-46, whereas another deposit may have nothing larger than pebbles. **Till** is the term for drift that is unsorted and essentially unstratified, and the depositional landforms of till have their own terms. We will discuss only those varieties that are relatively common in glaciated areas. One of the most widespread glacial deposits is a plain of till left behind by the more-or-less uniform retreat of an ice sheet. This **ground moraine,** which may be either a thin veneer or a cover more than 100 m thick, is the base of much of the soil in Canada and parts of the northern United States.

Much of the till in a valley glacier is derived from frost action, mass wasting, and ice erosion along the valley walls, and this debris

11-45. *Modern till. This mound of till is nearly surrounded by glacial ice. The larger boulder to the man's left is termed an "erratic," and illustrates the competence of ice.* [*Photo by D. Darby.*]

11-46. *Pleistocene till. This shows the unsorted boulder-clay nature of till. Deposits similar to this are at or near the surface of much of Canada and the northern United States. [Courtesy R. W. Ojakangas.]*

11-47. *A modern valley glacier. The dark bands are till being carried by the ice. The till on the surface of the glacier is mostly from erosion along the valley walls. When two glaciers meet, as on the right, the lateral bands of till may merge and be near the middle of the down-glacier ice. [Courtesy N. Potter.]*

will flow with the ice downglacier forming elongate strings of till known as moraines, Figures 11-41 and 11-47. When the glacier retreats, these may be deposited as ridges, and depending whether they are on the sides or in midvalley they are called **lateral** or **medial moraines.**

11-48. *End moraine. A ridge of till deposited at the front of the glacier. [Wards Natural Science Establishment.]*

Any glacier that has rock material incorporated into the ice can form a moraine ridge along its front. This type of moraine has its axis more or less at right angles to the local direction of ice movement (Figure 11-48). Some of these **end moraines** deposited from large ice sheets form arcuate ridges that can be over 100 m high, kilometers in length, and may be the most prominent topographic features in a wide area. These end moraines are not the product of "bulldozing" by ice, although they can be modified by the readvance of a glacier. As we have already discussed, material is continuously brought to the front of the glacier by internal flow and freed from the ice by ablation at the margin. The system is analogous to a conveyor belt with a load of rock; ablation destroys ice, (the belt), at one end, and the debris is simply dumped on the ground.

Of course the melt waters affect the rock debris but, in general, the morainal mass is a jumble of material, clay-silt-sand-gravel and boulders. Figure 11-49 illustrates the formation of an end moraine and Figure 11-50 is a map showing some large end moraines in the area of the Great Lakes. As the ice retreats by excess ablation, changes in the accumulation-wastage ratio may occur, and minor advances and still-stands are marked by a series of end moraines progressively younger in the direction to which the ice retreated.

In some glaciated regions elongate smooth hills are a common topographic feature. Some of these are bedrock features that have been abraded by the moving glaciers, and their low smooth profile remind some of resting sheep. The "sheep rocks" have their gentle

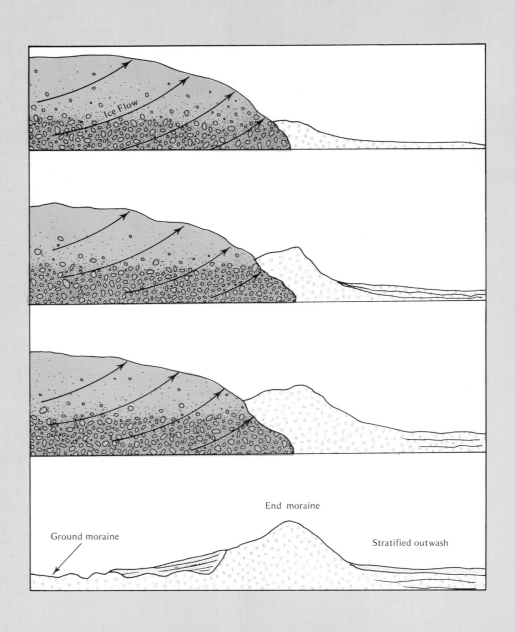

Ice Flow

End moraine

Ground moraine

Stratified outwash

11-49. *The formation of an end moraine. As the front of the glacier stabilizes for a time, the moving ice still carries debris to the front. A ridge of till is deposited, some of which will be carried away by melt water and be sorted and deposited as stratified outwash. When ablation exceeds accumulation for a period, the ice front will melt back at a more constant rate, depositing ground moraine.*

11-50. *End moraines in the Great Lakes area. [From: Glacial Map of North America, R. F. Flint et al.]*

slope in the direction from which the glacier came and a careful examination will often reveal striations parallel to the hill's long axis. A hill of similar shape may be composed of till. These hills too are streamlined and, because of this, they are thought to have formed beneath the ice in the zone of plastic flow. The moving ice has caused their gentle slope to be downglacier, the reverse of a sheep rock. These **drumlins** often occur in clusters and are commonly 30–50 m high and ½ km long (Figure 11-51).

Stratified Deposits. The inevitable melt waters accompanying most glacial activity, especially retreat, produce alluvial deposits

11-51.   *Drumlin. This elongate mound of till has its gentle slope in the direction if ice movement.* [*Wards Natural Science Establishment.*]

11-52.   *Esker. A sinuous ridge of stratified material thought to have formed beneath a glacier.* [*Courtesy D. J. Easterbrook.*]

11-54. *A Small kettle. A block of ice, broken from the glacier in the background, has melted. The circular depression in stratified outwash deposits will probably fill in as future melt waters deposit material in it.*

of various sorts. Some are little different than the deposits of any stream and may partially fill glaciated valleys or produce large alluvial **outwash plains** off the margins of ice sheets. Occasionally streams beneath the ice may form long sinuous stratified deposits known as **eskers** that are left as ridges after the ice is gone (Figure 11-52). The mechanism of subice depositions is not well understood. Near the ice front blocks of ice may separate from the main mass and their melt waters may create local deposits of stratified material that eventually are left as low, often conical hills known as **kames** (Figure 11-53). All of these deposits may be valuable sources for sand and gravel.

Lakes are a common product of glaciation, and they commonly form in cirques or in depressions gouged by the passage of an ice sheet. Moraines may block the local drainage and allow lakes to form behind them. Blocks of ice left behind in the till or outwash deposits will leave a depression, initially filled by melt waters, and later by the influx of ground water and runoff from rainfall (Figure 11-54). These **kettle lakes** are often circular and are very common today in Canada and the northern United States. Areas glaciated by ice sheets usually have many more than their share of lakes, and state or province nicknames, "Land of 10,000 Lakes" for example, are testimony to their glacial history.

The sediments that formed in these glacier associated lakes from particles entering with melt water streams may reflect the seasons. The silty material entering in summer results in a light colored deposit. When the lakes freeze over in winter, the finer clays can settle because wind is no longer able to create water turbulence to keep them in suspension. The clays are darker and organic matter, perhaps from decaying leaves as winter approaches, contributes to the color distinction between the two layers. The couplet, known as a **varve,** is usually a centimeter or so thick and represents 1 year (Figure 11-55); over 10,000 years may be represented in one deposit. Varves make an interesting subject of study, for they not only provide a chronology but also contain pollen grains that yield clues to the area's vegetational and climatic history.

Continental Glaciation

Large ice sheets have covered major land areas at various times in the past. The extent and nature of some of the more ancient ice sheets are still in doubt, but deposits of what appear to be cemented till, **tillite,** of Precambrian age have been found on many continents. Although rooted plants had not then evolved and erosion may have caused mud flows on a scale unknown in later times, the rocks and the associated striations have all the characteristics of glacial deposits.

The realization of glaciation about 700 million years ago and again in Ordovician-Silurian time over 400 million years ago is not simply an intellectual exercise for geologists. Large ice sheets take up enough water to lower sea level a hundred meters or more and may have caused the withdrawal of shallow seas which covered large parts of the continents. Such a significant lowering of base level must have affected the pattern and rates of erosion over the entire world, and of course the sediment input into the sea was affected. The marine organisms living in those shallow seas would have had their habitat reduced in area, and the increase in competition and rates

One varve (1 year)

11-55. Varves. These glacially derived clays and silts were deposited in Pleistocene lakes in the northeastern United States. Pins mark the annual boundaries of each pair of darker winter deposits and lighter summer deposits. [Courtesy of the American Museum of Natural History.]

11-56. *Tillite. Till cemented to form a sedimentary rock. This example is from a late Paleozoic glaciation of southern Africa. The ice came from the direction the man in facing. The deep Indian Ocean now occupies the source region. Similar deposits of similar age on other continents were one of the major criteria used to support early ideas of continental drift. [Courtesy J. C. Crowell.]*

of natural selection may have altered the entire course of organic evolution on this planet.

Geologists are very confident that a glaciation occurred some 270 million years ago in Pennsylvanian-Permian time, and the recognition and correlation of tillites and striations in South Africa, Australia, India, and South America was and is a major argument for ancient unification of those continents (Figure 11-56). Prior to the discovery of sea floor spreading, concurrent glaciations of now widely separated continents provided strong evidence that the continents had indeed moved.

If plate motions at any time were such that a continental mass entered a cold climatic region, the development of an ice sheet would hardly be surprising. It is at least imaginable that accumulation could exceed wastage by such a wide amount that this ice sheet could extend into lower latitudes of temperate climate. With this sensible outlook and our present knowledge that continent-bearing plates are capable of movement, one might think that the problems of explaining major glacial events in earth's history are solved. They are not.

For one thing, continental glaciation does not require that a landmass be brought into a position over a geographic pole. True, a continent situated over a pole may develop an ice cap; apparently Antarctica has had one for at least 10 million years. Yet, Antarctica probably has been over the South Pole much longer than that; why did glaciation not commence earlier? Today the Antarctic ice sheet does not seem to be growing and that continent, nearly the size of the United States and Europe combined, receives little snow. Thus a polar situation certainly does not appear to be an absolutely necessary criterion for the development of an ice sheet.

When the sea surrounding a continent is frozen most of the year, where is the continent to get its snow? A cold temperate climate seems to be preferable for the formation of a large ice sheet, and much of North America, Asia, and Europe have been at such latitudes long before the last ice age began. Also, during the long course of earth history, continents often must have been at latitudes where glaciation might be thought to occur but, with the numerically few exceptions mentioned, no such major glacial events are known.

Based on many lines of evidence, it seems clear that the climate in any given area of the world has undergone changes through time. The more ancient changes, hundreds of millions of years ago, are most commonly reflected in changes of plant and animal populations, now fossilized and in sedimentary deposits. For more recent times, changing ratios between certain isotopes of oxygen, archeological finds, and even weather records indicate the past changes.

What causes these changes is the major problem, and the occurrence of ice sheets is tied to the answers. The formation of major ice sheets apparently depends upon a combination of circumstances that occurs rarely. The most fundamental parts of this combination apparently involve changes in the amount or type of solar energy received by the earth, the positions of continents in relation to high latitude areas, and the elevation above sea level of these continents. The solar energy problem involves not only the sun but the earth's relation to it and the nature of our atmosphere.

R. F. Flint has grouped the more evident possible causative factors of ice sheet formation into six broad groups:

1. Changes in the nature and/or amount of solar emission.
2. Veils of cosmic dust.
3. Geometric variations in the earth's motions.
4. Variations in the earth's atmosphere, for example, changes in the quantities of ozone, carbon dioxide, or terrestially derived dust.
5. Lateral and vertical movements of the earth's crust.
6. Changes in the system of atmosphere/ocean circulation.

Chapters could be written concerning any of these factors. At present, and the idea may be discarded in the future, the most likely causative relationship seems to be between solar activity and movements of the earth's crust. The three conditions conducive to ice sheet formation might be

1. When a continent or continents are carried into geographical positions such that latitudinal, oceanic, and wind relations permit moisture to enter the atmosphere and be carried onto those continents for much of the year.
2. When climatic conditions are such that much or all of that moisture precipitates as snow and the level of annual snow accumulation is at a reasonably low elevation.
3. When the continents are at a sufficient elevation to have major portions above the level of snow accumulation.

The Pleistocene Ice Sheets

Glacial deposits are widespread over large areas of North America, Europe, and Asia. Based on well over a century of study of these deposits, it is evident that large ice sheets formed and advanced over those regions in relatively recent times. How old the oldest are and when the sheets disappeared are questions not always accurately answerable even with the advent of radiometric dating techniques. At least four major advances, with cycles of advance and retreat superimposed on them, have been documented. The ice sheets began to form approximately 2 million years ago and, with the exception of the Greenland ice cap, essentially disappeared from the Northern hemisphere less than 10,000 years ago.

This period of time is known as the Pleistocene, although the definitive factors that separate the Pleistocene from the Recent, or Holocene, epoch are debatable. During this time man as a biological and cultural being is thought to have come into existence, so Pleistocene glaciation has affected our history more directly than previous continental glaciations. The area of North America that was covered by ice at one time or another during the Pleistocene is shown in Figure 11-57. Note that the centers where the glaciers in Canada developed were not at the polar extremities, but to the south.

Since on land one glacial advance may destroy or alter the evidence of a previous one, the most complete undisturbed records are to be found in marine sediments of Pleistocene age. In recent years the United States has entered into a program of drilling the

11-57. *Approximate maximum extent of the Pleistocene ice sheets in North America. The arrows denote the general direction of glacial movement.*

ocean bottoms, and the results have already altered our previous ideas on the commencement of glaciation. One of several lines of evidence that is thought to mirror the climatic changes, at least in the temperature of oceanic water, is based on fossils. Certain Foraminiferal plankton, similar to the ones shown in Figure 6-12, prefer cold water to warm. Other similar types tend to coil clockwise in warm water and counterclockwise in cold.

Analyses of percentage ratios of these microfossils from deep sea cores in the Atlantic, combined with other temperature indicators, have resulted in curves similar to that shown in Figure 11-58. The curves indicate colder water, not necessarily a colder climate, for cold water may result from melting glaciers in a warmer climate. The water temperature trends, therefore, may not coincide precisely with advances and retreats of ice sheets. The last of the four major ice advances seems to have begun about 70,000 years ago. Since we are only 10,000 years or less into an interglacial period, when will a

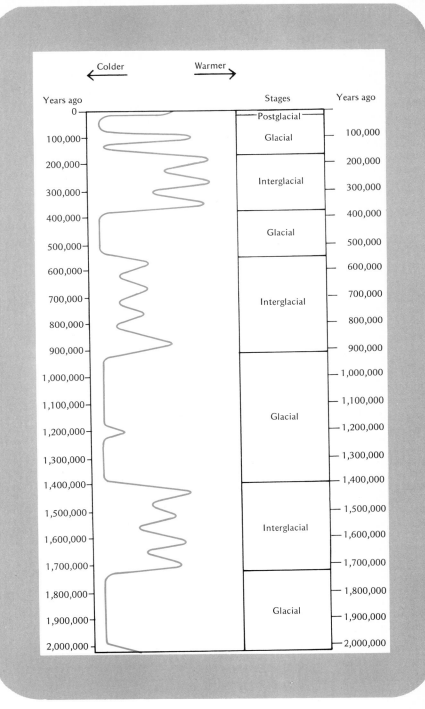

11-58. *Cooling trends of the Pleist-ocene. The cooling curves are based on faunal evidence from fossils found in cores of the sediments of the Atlantic and Caribbean. [After D. B. Ericson.]*

fifth advance begin? Or will there be one? Unanswerable questions for now.

In addition to deposits of sediment and an alteration of topography, the ice sheets of the Pleistocene left another, somewhat negative, legacy. Although ice is one third as dense as granitic crustal rock, the central parts of an ice sheet may exert a weight equivalent to 1000 m or more of rock. The present ice sheets covering Greenland and Antarctica have depressed the land, and if the ice were removed some areas would be below sea level. The reaction of a continental mass when unburdened of ice is the same as if erosion had removed

11-59. *Recent vertical movements, in millimeters per year, in central and eastern North America. The uplift is thought to be due to isostatic postglacial uplift.*

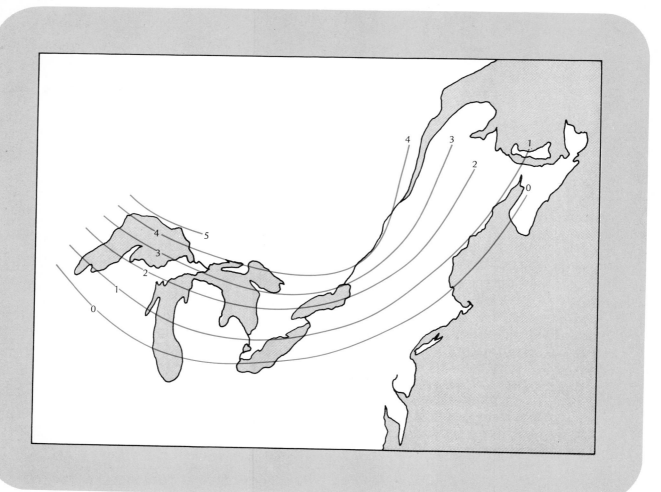

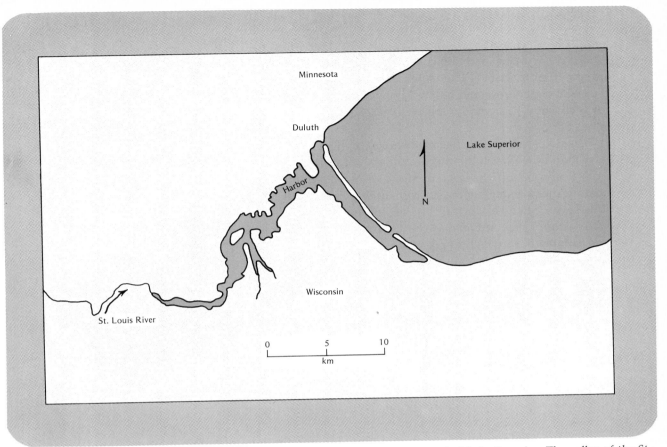

11-60. *The harbor at the western end of Lake Superior. The valley of the St. Louis River has been drowned as the land to the north and east rose in post-glacial time.*

large quantities of rock; it comes toward isostatic balance by rising to a higher elevation. This **isostatic rebound,** which began some 15,000 years ago, is still in progress. In North America, the central area covered by the thickest ice has risen nearly 140 m in the last 6000 years and continues to rise a few millimeters each year.

Figure 11-59 illustrates the yearly isostatic uplift in eastern North America. The water level in the Great Lakes was much higher at times when ice still prevented drainage into the Atlantic Ocean or into what is now Hudson Bay. As the lakes were freed of ice and developed their modern drainage pattern, the western ones, Lake Superior for example, had even lower water levels than today. But as isostatic rebound progressed, the eastern part of North America rose high enough to back up the waters and cause a rise in lake levels. Figure 11-59 shows that the northern and eastern areas of Lake Superior are rising faster than the southwestern part. One effect of this rebound has been to create a harbor at the southwest tip of Lake Superior (Figure 11-60). The valley of the St. Louis river was drowned as lake waters rose and entered it, and a port was provided through which moves a yearly tonnage exceeded by few other United States ports.

As the last major ice sheets began to melt, sea level was about 130 m lower than now. Recall the discussion about finding abundant coarse sand grains far off the eastern shore of North America and the speculation that a change in the position of the coast might be the answer. The lower sea level during the Pleistocene, as large quantities of water were taken up in ice sheets, provides a reasonable mechanism to explain the different positions of the coast line.

Any significant rise of sea level in the future would cause many of the world's most populous cities to be flooded. If the ice now covering 98% of the large continent of Antarctica were to melt, sea level would rise over 50 m. That does not, however, seem to be a probability in the immediate future because the continent would have to shift away from the south polar region, or a major climatic change would have to occur.

THE WORK OF THE WIND

Of the three agents of erosion active on the land's surface, wind is the least competent. Normally wind is an effective transporter of material only in areas where vegetation is sparse or lacking. Man has contributed to wind erosion when, through poor farming practices, he has allowed the wind to blow away topsoil and create "dust bowl" conditions over land not otherwise open to extensive wind erosion. Other than this, the areas of the world where wind, as well as water, is a prime mover of material are deserts.

"What are deserts?" is a question usually answered in terms of hot, dry, barren, and sandy. Arid regions are often defined as receiving 25 cm or less average annual precipitation, but much of the arctic and the vast ice covered continent of Antarctica fall into this category. Thus, there are polar deserts as well as hot deserts, and, although ice, not wind, is the principal erosive agent on the Antarctic continents, some of the sand dune forms found in the

11-61. *Butte. An erosional remnant of once extensive horizontal strata.*

Sahara have their analogs in the wind-blown snow of polar regions. Excluding polar deserts, arid and semiarid regions make up about 25% of the continental surface. Vegetation is normally sparse in such regions, which are often hot during the day. More area in deserts is covered by rock than sand, however, and water plays a very important part in the erosion of many desert areas.

The precipitation in hot deserts is often of short duration, and the water evaporates quickly. Many of the stream channels resulting from the infrequent cloudbursts are short and commonly they terminate in alluvial fans or in shallow dry basins. The channels, locally known by various terms—wadis, draws, arroyos, and so on—are often steep sided and dangerous places to camp because, when it rains hard, the runoff is high and they quickly fill with torrents of water.

Many landforms in desert regions are produced by the same processes as in more humid areas, but intermittently or more slowly. Most of the prominent rock features of dry deserts are remnants of stream erosion. Flat lying sedimentary strata may be eroded to form mesas and buttes as the streams cut down around them (Figure 11-61). Rock towers, badland topography, canyons and caves are mostly derived from the actions of water, not wind. Desert topography tends to be characterized by steeper slopes and, in general, the relief is sharper; chemical weathering does not round off the angles very rapidly (Figure 11-62).

The action of wind is similar to that of water, but wind is not as competent an agent of transport as water. Pebbles and larger rock fragments are not generally moved much by wind, and, if the sand

11-62. *The intermittent rains in a dry climate, where chemical weathering is slow, lead to an erosional landscape with marked angularity. [Courtesy U.S. Department of the Interior, Geological Survey.]*

being supplied to an area keeps being blown away, a "lag" deposit of coarse material will be left behind on the surface. This "desert pavement" is common in many regions; more of Arabia, for example, is floored with loose rock than with sand. The critical erosional velocity curve for wind is similar in shape to that of water, but the velocities required for movement are much higher because air is much less dense.

In general, the particles eroded by wind move in two ways, saltation and suspension. Under most wind conditions these serve to separate geographically the two most common small particles found in dry deserts, sand and silt. (You may at this point wish to make a statement to yourself as to why clay is not abundant in a dry desert environment.)

Saltation

When the wind reaches some critical velocity, a grain of sand begins to move. As you may recollect, a quartz grain is only 2.65 times as dense as water, but it is 2000 times as dense as air; therefore, when the moving grain strikes its neighbor, there is not much cushioning effect. Upon impact the neighboring grain, the original grain, or both may be ejected a short distance upward into the air stream. Air very close to the ground surface has little velocity because of friction so that the airborne grain has little forward movement. As it rises it begins to achieve a velocity close to that of the

11-63.   *Saltation. The diagram illustrates the paths of bouncing wind-blown sand grains over a surface of pebbles and loose sand.*

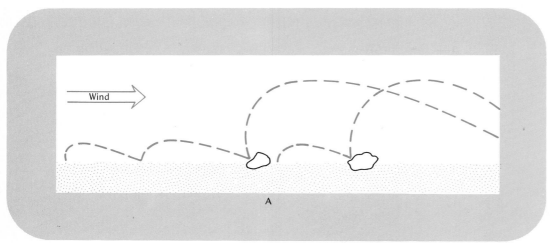

air. It takes an arcuate path then strikes other grains upon impact and the system continues. The grains descent at about a 10 or 15° angle, and the struck grain is impelled a few hundred grain-diameters into the air and begins its own flight. This jumping movement of wind driven sand is given the name saltation, not after salt but from the Latin word "to jump," *saltare*. Figure 11-63 shows some typical paths of saltating sand grains.

Once the critical erosional wind velocity is reached, the grains may continue to move when the wind velocity diminishes somewhat. Just as in the case of water, it requires less velocity to keep particles moving than it does to start them. The impact of one grain transmits enough energy to another to start its jump, so the system once activated continues downwind an indefinite distance. Under high wind velocities most sand will attain maximum heights of about 2 m, but the average height is much less, around 10 cm. The maximum heights are not reached on sand deserts but in areas where much of the surface is pebble covered. The sand grains transmit much less of their energy to an immobile pebble upon impact than they would to another sand grain. So the sand sized grain rebounds much higher and gets up into even faster air where it takes on more velocity, strikes the surface again with even more energy, and so on. Eventually heights of 2 m are achieved. Being peppered by saltating sand grains can be very unpleasant, but climbing aboard a camel elevates one above the zone of maximum discomfort.

Sand grains moving by saltation can abrade or sandblast outcrops of rock or fragments on the surface, Figure 11-64. It is a common misconception that the natural bridges, shallow caves, and other desert features are produced by sand abrasion. These features may be modified by sand but, for the most part, the infrequent streams of water, or the streams of a more humid climate in the past, were the agents of production.

11-64. *This boulder has been faceted by erosion caused by wind blown particles. Rocks with similar shapes, known as ventifacts, are often found in both hot and polar deserts.*

Suspension

When the turbulent eddy currents within the moving air mass have a greater velocity than the velocity of fall of the particle, the grains remain off the ground, a situation quite analogous to suspension in a stream. Some very fine sand may remain in suspension, but most of the suspended material is of silt size. What is often referred to as a sandstorm, is, in reality, a siltstorm. These storms may carry fine particles to heights of thousands of meters and move the dust long distances (Figure 11-65).

11-65.   *Weeding operations have finely pulverized the soil of this fallow field. The lack of vegetation has made the soil open to wind erosion. If a strong windstorm were to occur during a dry period, much of the topsoil would be removed.* [Courtesy U.S. Department of Agriculture Soil Conservation Service.]

Wind Deposits

Wind deposited sediments are collectively known as **eolian** deposits, perhaps after Aeolus, a Greek god of winds. The two types of wind transport yield two types of deposits: Loess and dunes.

Loess is formed when the silt transported in suspension is deposited. These relatively unstratified blankets of silt are common in many areas. The thick and extensive loess deposits in China are thought to have been derived from the Gobi desert. There are ex-

11-66. *Loess. This deposit of wind blown, silt-sized particles in Iowa retains a steep slope. [Courtesy Wm. Burns, Iowa State Hwy. Commission.]*

tensive loess deposits in Iowa, Indiana, and other areas of the United States that have no apparent connection with deserts. In the latter case, glaciers, as they melted back from the northern United States, left extensive deposits of rock debris, and prior to the growth of vegetation the wind carried and deposited immense quantities of silt. A characteristic of loess is that it is more stable on a nearly vertical slope than on a 30° slope where rains can cause erosion. Highway engineers, upon recognizing loess, make the roadcuts and sides of drainage ditches nearly vertical (Figure 11-66).

Dunes are the product of the wind's bed load. They too, of course, may occur in nondesert regions, for example along sandy coasts. Yet it is in desert areas that dunes reach the size and perfection that inspired R. A. Bagnold, a modern authority on desert sand to write: "In places vast accumulations of sand weighing millions of tons move inexorably, in regular formation, over the surface of the country, growing, retaining their shape, even breeding, in a manner which, by its grotesque imitation of life, is vaguely disturbing to an imaginative mind."

Depending upon the sand supply, the nature of the wind, and the amount of vegetation, various types of sand dunes can form. Some of the types common in desert areas are shown in Figure 11-67. No type of dune afforts a better artistic expression of nature than a large symmetrical barchan all alone on a stony surfaced desert. With little or no new sand being supplied, these dunes can move across the desert a few meters each year (Figure 11-68). The grains moving over the top and around the sides are covered by succeeding sand grains, and after a time the grains in the lee became the new supply at the windward edge. Figure 11-69 indicates the paths of sand particles over the top and, as the dune progresses to the right, through the dune.

11-67. *Types of sand dunes. The various shapes and orientation are products of variations in sand supply, constancy of wind direction and the presence or absence of vegetation (scales unequal).*

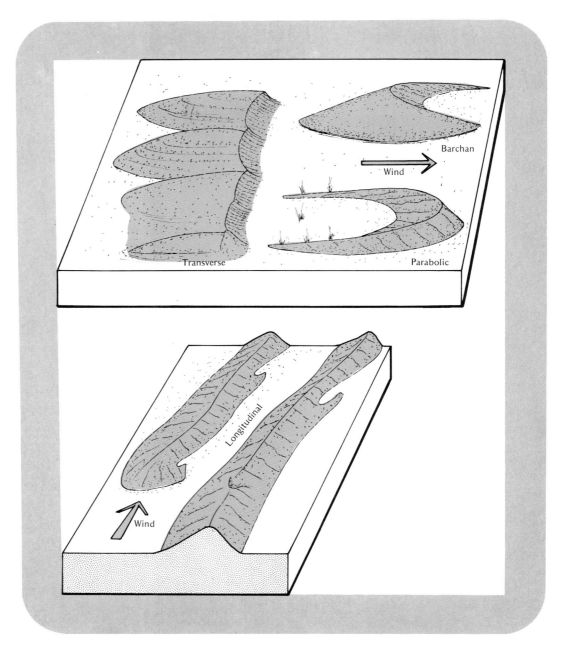

11-68. *A barchan shaped sand dune. [Courtesy Wards National Science Establishment.]*

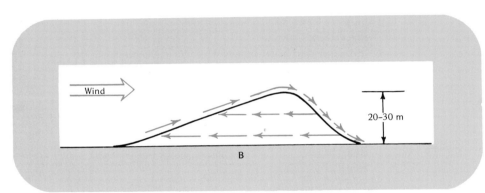

11-69. *The paths of sand grains over and circulating through a large migrating barchan dune.*

Wind

20–30 m

B

PLATE TECTONICS AND EROSION

In the first part of this chapter some data on the worldwide rates of erosion were given. The erosion processes were examined, and you should have perceived that the whole is the sum of very complex parts. Earth history is not a subject of this text, but what has passed has governed what is, at least to some degree. Were it not for sea floor spreading and the conservation of area by subducted oceanic plates, perhaps a major part of the earth's erosional record would exist somewhere on the sea bottom. The oldest rocks yet known are on the continental crust and are about 3.8 billion years old, but a sharp contrast exists between the age of the ocean floor and that of much of the continental crustal rock.

The oldest known crustal rocks from the present ocean basins are less than 200 million years in age. This fact is a major foundation for the idea of sea floor spreading. A corroboration of the age of the ocean floor should be available if one can estimate the amounts of past erosion and determine how thick the deep marine sediment should be. Given erosional rates less than half those thought to be occurring at present, thus eliminating most of the human factor, the several hundred meters of sediment commonly found in the deep ocean can be accounted for in less than 200 million years.

Prior to the origin of the sea floor spreading concept, these data were known, and the need to account for this anomaly was a very real problem. Some proposed the possibility that older marine sediments were interlayered with the basaltic oceanic crust and hidden, others considered the startling possibility that there was almost no deep marine sedimentation throughout geologic time. Now, of course, a more reasonable answer is that the deep sea sediments deposited prior to about 200 million years ago have either been carried beneath the crust in subduction zones or incorporated into a continental mass, as outlined in previous chapters. The modern theory of plate tectonics is reinforced not only by magnetic, seismic, thermal, and geochemical data but by the lowly products of streams, wind, and ice.

1. You have been given the responsibility of determining potential home sites along a ridge located in a region with a humid climate. The north-south trending ridge is composed of shale with a dip of 30° east. Would you recommend building along the eastern or western slopes? Why?
2. What are some criteria that might serve to distinguish deposits of colluvium, alluvium and glacial till?
3. Would you expect a river in a cold temperate climate to be capable of carrying more sediment in suspension when the water is warm or cold? Consider this in relation to the variable factors in Stoke's equation.
4. Why do glacial crevasses usually end at a depth of about 60 m. or less?
5. What are the two types of ice movement that may occur in a glacier and what are two postglacial features that might indicate the the direction of glacier movement?

twelve

Geology and Man

Man is profoundly affected by many geologic processes. The science of geology is not merely the outgrowth of man's intellectual curiosity about the earth but a subject closely related to his present and future well-being. Man, through various of the activities in which he engages, has become a significant agent of geologic change. For example, human agricultural activities disrupt natural surface environments and alter drainage patterns. Natural geologic processes such as earthquakes, landslides, or floods can have adverse (even fatal) effects on people and their property.

Furthermore, man's utilization of the energy and mineral resources of this planet forms the very foundation stone of modern civilization. From his humble beginning as a maker of stone tools, virtually every step of man's development to his present level of technological competence has been accompanied, in one way or another, by increased use and dependence on materials obtained from the earth's crust. This heavy dependence on limited earth resources, coupled with recent trends in world population figures and per capita consumption, has caused many knowledgable earth scientists to voice grave misgivings about the future well-being of the human race.

MAN AS A GEOLOGIC AGENT

Soil Erosion

Many of man's activities have measurable effects on geologic processes and environments, and one of the most obvious of these is the tilling of soil for agricultural purposes. It has been estimated, for example, that since approximately 3000 B.C., when man first engaged in agriculture, the average sediment load of the world's rivers has doubled. The bulk of this increased sediment load represents soil lost from agricultural areas.

As we saw in Chapter 10 soil represents an intermediate stage between unweathered rocks and minerals and their eventual removal by surficial processes. It functions both as an important reservoir of nutrients and as a medium for nutrient exchange for the growth of plants. In view of the importance of soil, both in terms of quantity and quality, to agricultural productivity, any net loss of this material is a serious problem in a world where food production must keep pace with expanding populations (Figure 12-1).

In many areas the balance between soil and vegetation cover and the runoff of rainwater is delicate, and heedless human interference with this natural balance can have disastrous consequences (Figure

12-1. *Plot to illustrate growth of human population with time. Note sharp inflection of curve at about 1700 A.D. This coincides with the Industrial Revolution and the birth of modern medical techniques.*

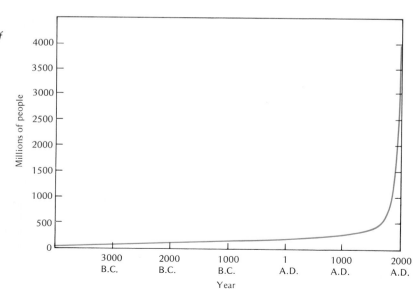

12-2. *Total destruction of farmland by soil erosion is shown by this photograph of a homestead in Bolivia. [Photo by Paul Almasy. Camera Press (RBO) London, No. 7177-2.]*

12-2). In addition to physical disruption of soils, in many places man has induced chemical changes in soils (for example, application of chemical fertilizers) that tend to have beneficial short term effects on agricultural productivity but adverse long term effects on soil quality. Furthermore, a certain proportion of these chemicals inevitably are washed into drainage systems, where they contaminate water in streams, rivers, and lakes.

Manipulation of Surface Water

River environments have repeatedly been the target of attempts to control natural processes. The damming of rivers, mainly to con-

12-3. A. *The Nile drainage system in its original configuration, showing waterfalls, rapids, and swamp areas. B. Lake Nasser, a long, immense body of water backed up by the Aswan Dam. The Red Sea is on the horizon in background.* [*Photo No. 71-HC-310, courtesy of NASA.*]

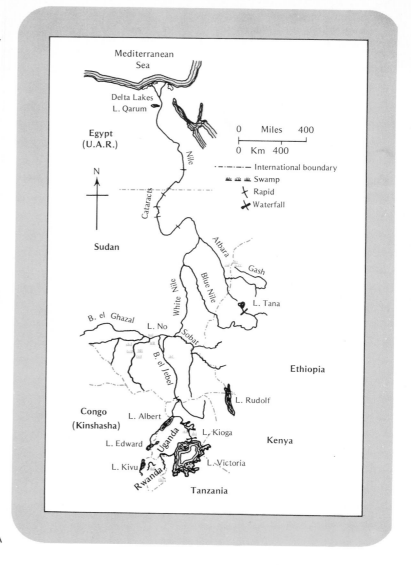

A

trol flood waters and to facilitate the generation of hydroelectric power, has a long history and is still very much in vogue. All too often the beneficial aspects of dam building are to some extent offset by unforseen readjustments of the natural system to manmade alteration. Probably the most interesting case history that can be cited in this regard is that of the recently completed Aswan High Dam project on the Nile.

The Nile is one of the major rivers of the world (Figure 12-3) and the water it carries northward from equatorial Africa and the

B

Ethiopian Highlands has aptly been described as Egypt's lifeblood. Attempts to manipulate the Nile drainage system have a long history. In 1881 a series of spillways were built in the Delta area to aid irrigation projects. A dam at Aswan was completed in 1902, and finally in 1967 the new Aswan High Dam was closed. This last project represents one of man's most ambitious attempts to control natural phenomena.

The rationale behind the construction of the Aswan High Dam included flood control, electric power generation, irrigation development, and a fishing industry in the newly created Lake Nasser. All of these developments will undoubtedly benefit the Egyptian nation, but they are being partially offset by other changes taking place in the Nile drainage system.

As a result of the Aswan High Dam the heavy load of silt that accompanied the annual flooding of the Nile now deposits in Lake Nasser and, therefore, is no longer delivered to the lower reaches of the river. The absence of this silt and the nutrients it contains has had a series of adverse effects on the river system and on some related human activities.

1. Marine fisheries in the Mediterranean Sea just off the Nile Delta and the Levant Coast have declined severely. The same is true of fisheries based in several large Delta lakes.

2. The outward building of the Nile Delta has ceased, and in fact shoreline erosion is resulting in encroachment of the sea. Consequently, valuable agricultural land may be lost in the Delta area, and the Delta lakes, which are protected from the sea only by fragile sandbars, are threatened.

3. The beneficial silt and its nutrients, which for thousands of years have enriched the agricultural land of the Nile Valley, are being replaced by chemical fertilizers.

4. The lower Nile has commenced an accelerated cycle of erosion of its bed and banks.

5. Water-related parasitic diseases that affect livestock and man (for example, bilharzia or schistosomiasis) have spread rapidly along irrigation networks, and pose a serious problem in the area.

Lake Nasser, when completely full, will have a surface area of 5860 km^2, and it is calculated that 12% of the annual flow of the Nile will be lost by surface evaporation from this body of water. If aquatic weeds, like the water hyacinth, establish themselves in the lake (a distinct possibility) water losses to the atmosphere will be considerably greater. Such has been the case in the Kariba Lake in central Africa, presently the largest artificial reservoir yet created by man. This lake, which displaced 57,000 persons in the Zambesi River valley when it filled, was rapidly infested with aquatic water ferns (at one stage covering 10% of the lake surface), causing considerable hardship to local fishing communities.

Many of man's activities have a negative impact on the quality of surface waters, even where direct modification of drainage systems, such as dam building, is not involved. The strip mining of high-sulfur coal in parts of Pennsylvania, West Virginia, and Kentucky in the eastern United States has had disastrous effects on the streams and surface environments in this area (Figure 12-4). It is estimated that the strip mining of coal in the eastern United States has led to the near devastation of over 2000 km² of land. The iron sulfide minerals associated with the coal deposits oxidize rapidly when exposed to the atmosphere by mining activities, and this leads to the production of sulfuric acid (H_2SO_4). The acid enters

12-4. *Aerial photograph showing effect of strip mining on wooded hills in Kentucky.* [*Photo by Billy Davis, III. Used with permission.*]

drainage systems in concentrations that kill fish and other aquatic life and render the water unfit for human consumption for tens of kilometers downstream.

Natural lakes are sources of water, provide man with important recreational facilities, and in some instances support commercial fishing operations. In geologic terms the lifetime of lakes is very short and most now existing were formed during the Pleistocene by glacial ice, volcanism, or structural movements. Lakes eventually are either drained by erosional lowering of their outlets or filled up by sediments. Man's activities can substantially accelerate these processes, especially the latter.

More serious perhaps, the nitrogen- and phosphorus-rich wastes generated by man and his activities within lake drainage basins lead to an overenrichment of these nutrients in the waters of many lakes. This, in turn, allows an almost explosive increase in the quantity of microscopic plant life (plankton) present in the lake waters and a marked decrease in the quality of the water.

In terms of total volume, the largest body of fresh water in the world is Lake Baikal in the Soviet Union. This lake contains (or contained until recently) water of great purity and a unique aquatic fauna. Today large volumes of raw effluent are being dumped into the lake from pulp and paper mills and other industrial sites along its shoreline. Lake Erie, one of the Great Lakes of North America, has been so abused by man that it is now seriously overfertilized and could only be restored to its original state at a great cost.

Land Subsidence

Man's activities have caused land subsidence in some areas. This phenomenon results from excessive pumping of ground water and has been particularly bothersome in the southern portion of the Great Valley of California (Figure 12-5). Here, the withdrawal of large volumes of ground water for irrigation purposes has led to land subsidence over broad areas (7800 km^2). Careful surveys have shown that in places nearly 3 m of subsidence has taken place, with consequent damage to ditches, canals, roads, and pipelines. In addition, near-surface subsidence of as much as 8 m has taken place over more localized areas due to the compaction of geologically recent fan shaped stream deposits (Figure 12-6). Apparently water from irrigation projects percolating downward through the originally dry underlying material causes it to soften, weaken, and compact under the weight of overlying material.

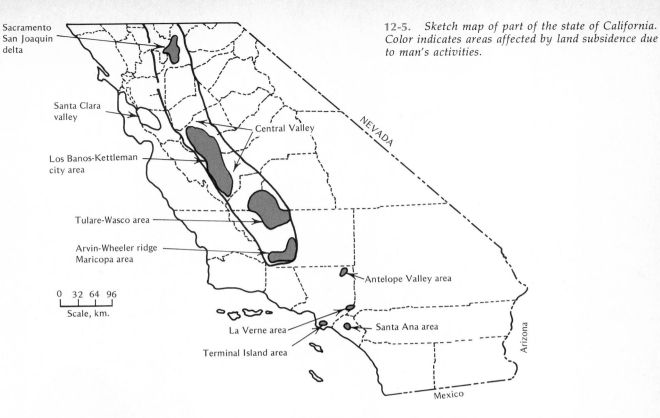

12-5. *Sketch map of part of the state of California. Color indicates areas affected by land subsidence due to man's activities.*

Sacramento San Joaquin delta

Santa Clara valley

Central Valley

Los Banos-Kettleman city area

Tulare-Wasco area

Arvin-Wheeler ridge Maricopa area

0 32 64 96
Scale, km.

Antelope Valley area

La Verne area

Terminal Island area

Santa Ana area

NEVADA

Arizona

Mexico

12-6. *Example of land subsidence caused by wetting of underlying dry materials by irrigation and artificial surface water storage. [Courtesy U.S. Department of the Interior Geological Survey.]*

Mexico City has been seriously affected by land subsidence problems. The older parts of the city are built upon a sequence of fine grained, soft lake sediments, and as a result of building and changes in ground water condition the surface of the ground has subsided as much as 7 m locally. This in turn has caused many problems in water transport, drainage, and construction in the heart of the city.

Large scale withdrawal of subsurface fluids in areas of petroleum and natural gas production has caused problems in many areas. In the Po Delta area, on the Adriatic Coast of northeastern Italy, extraction of methane from the poorly consolidated delta sediments resulted in subsidence rates of as much as 30 cm/year locally. The problem has become so severe that a cessation of the valuable natural gas industry has been suggested.

To what extent the natural gas extraction has affected the subsidence of Venice is not clear, but this historic city has been slowly sinking into the soft sediments on which it was built. The Italian government has yet to act on measures to save Venice, but any solution, whatever the engineering details, will be extremely costly.

The pumping of petroleum from the Wilmington oil field that underlies the Long Beach Harbor area of southern California resulted in up to 9 m of subsidence between the years 1928 and 1962 (Figure 12-7). In an attempt to reduce the problem in this heavily urbanized area, repressuring of the oil zones by water injection was undertaken. Fortunately, partial rebound of the ground surface occurred and oil recovery from the field was actually improved. Combating

12-7. *Photograph of Long Beach, Calif., showing contours of equal subsidence (in feet). Subsidence of land surface is due to the withdrawal of oil and gas from the underlying Wilmington oil field.* [*Courtesy City of Long Beach, Calif.*]

this artificially induced subsidence was nevertheless an expensive undertaking, for the repressuring installations cost an estimated 30 million dollars.

Solid Waste Disposal

The seemingly simple task of the disposal of man's solid wastes has become a matter of deep concern in the last few decades. In the United States, for example, the amount of solid waste produced each day averages 3–4 kg/person. If such were true for every inhabitant on the earth, the daily waste generation would total approximately 12 million tons, or 4.4 billion tons/year.

The cost of solid waste disposal in the United States is at least $10 per ton, so it can be appreciated that waste disposal is a matter of major economic importance in many places. As living standards and population levels rise around the world, disposal of waste will become more and more problematic and increasingly expensive.

At the present time, the two main procedures used for waste disposal are shallow burial in landfill sites and dumping in the oceans. Serious objections can be raised to both these procedures. Shallow landfill operations involve the danger of pollution of ground water supplies, and waste disposal in the oceans can, at least locally, disturb the delicate balance of marine life. It has even been suggested that the best possible site for dumping man's wastes would be active oceanic trenches, where the material might eventually undergo subduction and effectively be removed from the earth's surface environment. The cost of hauling large tonnages of solid waste from distant cities to the nearest subduction zone makes this intriguing suggestion impracticable.

GEOLOGIC HAZARDS AND CATASTROPHIES

Earthquakes

In some instances man finds himself subjected to geologic processes entirely beyond his control. The phenomena associated with earthquakes and volcanic eruptions are the most hazardous of these processes. It has been estimated that, during the past 1000 years, earthquakes and volcanos have resulted in the loss of 3–5 million lives. Discussions of earthquakes, their locations, and causes are

12-8. *Relationship between Modified Mercali Earthquake Intensity Scale and Richter Magnitude Scale.*

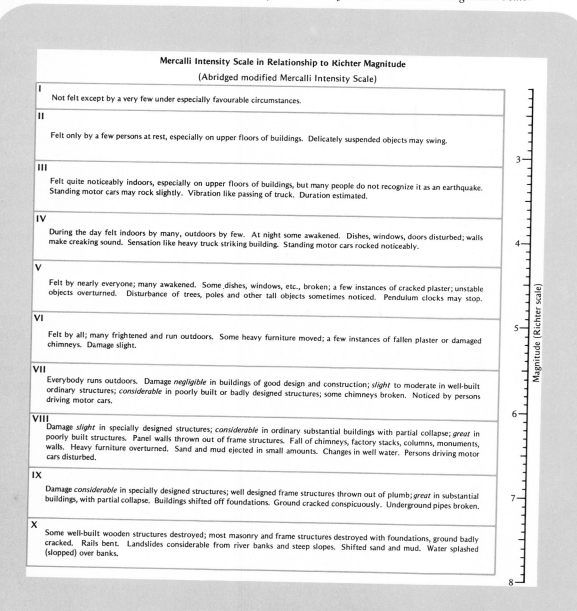

Mercalli Intensity Scale in Relationship to Richter Magnitude
(Abridged modified Mercalli Intensity Scale)

I
Not felt except by a very few under especially favourable circumstances.

II
Felt only by a few persons at rest, especially on upper floors of buildings. Delicately suspended objects may swing.

III
Felt quite noticeably indoors, especially on upper floors of buildings, but many people do not recognize it as an earthquake. Standing motor cars may rock slightly. Vibration like passing of truck. Duration estimated.

IV
During the day felt indoors by many, outdoors by few. At night some awakened. Dishes, windows, doors disturbed; walls make creaking sound. Sensation like heavy truck striking building. Standing motor cars rocked noticeably.

V
Felt by nearly everyone; many awakened. Some dishes, windows, etc., broken; a few instances of cracked plaster; unstable objects overturned. Disturbance of trees, poles and other tall objects sometimes noticed. Pendulum clocks may stop.

VI
Felt by all; many frightened and run outdoors. Some heavy furniture moved; a few instances of fallen plaster or damaged chimneys. Damage slight.

VII
Everybody runs outdoors. Damage *negligible* in buildings of good design and construction; *slight* to moderate in well-built ordinary structures; *considerable* in poorly built or badly designed structures; some chimneys broken. Noticed by persons driving motor cars.

VIII
Damage *slight* in specially designed structures; *considerable* in ordinary substantial buildings with partial collapse; *great* in poorly built structures. Panel walls thrown out of frame structures. Fall of chimneys, factory stacks, columns, monuments, walls. Heavy furniture overturned. Sand and mud ejected in small amounts. Changes in well water. Persons driving motor cars disturbed.

IX
Damage *considerable* in specially designed structures; well designed frame structures thrown out of plumb; *great* in substantial buildings, with partial collapse. Buildings shifted off foundations. Ground cracked conspicuously. Underground pipes broken.

X
Some well-built wooden structures destroyed; most masonry and frame structures destroyed with foundations, ground badly cracked. Rails bent. Landslides considerable from river banks and steep slopes. Shifted sand and mud. Water splashed (slopped) over banks.

Magnitude (Richter scale)

3 — 4 — 5 — 6 — 7 — 8

included in earlier chapters of this book; here, then, we need to focus on their effects on man and his works. Although major loss of life is associated with some intense earthquakes, it is not the earthquakes themselves that result in fatalities but the secondary effects of the release of earthquake energy. This can be readily illustrated by consideration of three major earthquakes that have occurred during this century.

In 1906 a major earthquake (magnitude ~8.3 on the Richter Scale, Figure 12-8) resulted from a sudden movement along the northern sector of the San Andreas fault (see Figure 8-37). Strong ground waves accompanied the earthquake along a 300-km belt, and in some instances permanent horizontal displacements of the ground of up to 6 m occurred along the fault trace. Damage to manmade structures in the San Francisco area was extensive and spectacular (Figure 12-9), both as a result of the earth movements and of a major fire that swept through the city shortly thereafter.

In all, more than 600 persons perished and property damage was assessed at $400 million. Since 1906 no major movement along the San Andreas fault has taken place in the San Francisco area, although this structure is known to be the main boundary between the Pacific and American Plates, which are moving relative to one another at

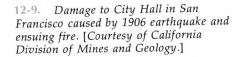

12-9. *Damage to City Hall in San Francisco caused by 1906 earthquake and ensuing fire.* [*Courtesy of California Division of Mines and Geology.*]

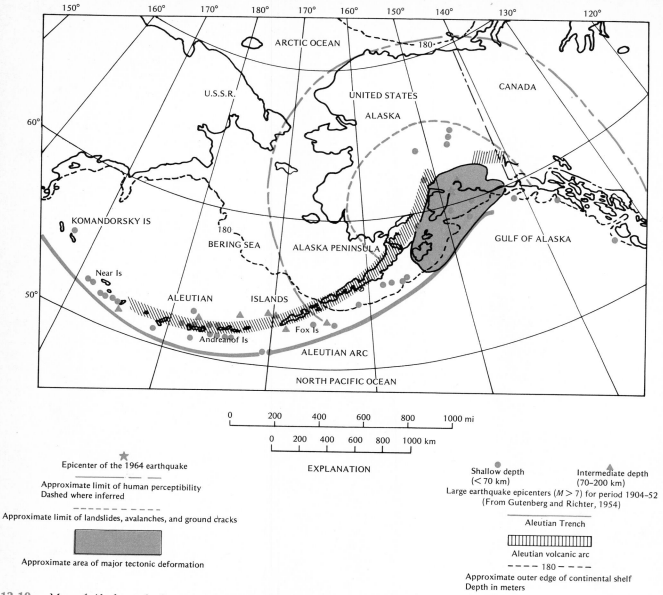

12-10. *Map of Alaska and adjacent areas showing the location of the 1964 earthquake, the area affected by the earthquake, epicenters of previous major earthquakes, belts of active volcanism, and the Aleutian Trench. Note that area contains all the typical elements of a convergent plate boundary.* [*After U.S. Department of the Interior Geological Survey.*]

6 cm/year. There is thus a very real possibility that another destructive earthquake will strike the area in the future.

An earthquake of magnitude 8.6 rocked south central Alaska in

12-11. *Damage in suburbs of Anchorage that resulted from the 1964 Alaskan Earthquake. [Courtesy U.S. Army.]*

the vicinity of Anchorage in 1964 (Figure 12-10). The Great Alaskan Earthquake, as this seismic event is now called, and its effects on buildings, ports, transportation networks, and natural environments have been subjected to considerable study and documentation. Shocks from the earthquake were felt over an area of more than 1.3 million km², and major damage occurred over an area of 130,000 km².

Although loss of life in this relatively sparsely populated region only amounted to 115 persons, property damage exceeded $300 million. During the earthquake strong ground waves moved across the area causing snowslides, rockfalls, and landslides on steep slopes. In Anchorage itself, severe tilting and fracturing of the ground surface occurred with considerable resultant damage to buildings (Figure 12-11). Vertical movements of the earth's surface over hundreds of square kilometers could be readily demonstrated locally by the exposure of marine rock dwelling organisms such as barnacles

12-12. *Photograph to illustrate uplift of shorelines in vicinity of epicenter of 1964 Alaskan earthquake. Boundary between dark upper zone and underlying light grey zone represents change from encrusting lichens to lighter colored barnacles. Actual uplift in this area was slightly in excess of 1 m. [Photograph courtesy G. Plafker, U. S. Department of the Interior Geological Survey.]*

and seaweeds above sea level (Figure 12-12). The uplift amounted to nearly 10 m in some places.

A series of seismic sea waves, or **tsunami,** generated by the earthquakes swept across the Pacific, causing damage along coastlines as far distant as Oregon and California, and eventually spent themselves against the shores of Antarctica. Seismic sea waves are generated by sudden movements of the sea bottom or large scale slumping of unconsolidated marine sediments and pose a periodic threat to life and property along the margins of the Pacific and the Hawaiian Islands.

In the open ocean these waves, which travel at speeds of up to 800 km/hr, are only about 1 m high and 160 km from crest to crest; thus, they may go unnoticed by ships. However, when these speedy wave forms encounter shallow water, the enormous energy they possess is greatly concentrated; they curl into giant crests which smash their way inland across shorelines (Figure 12-13). The largest seismic sea wave on record slammed against Cape Lopatka at the southern tip of Kamchatka in 1739, breaking at a height of 65 m!

Because of the ever present danger along the Pacific shorelines of destructive tsunami, a sophisticated early warning for the detection of these waves has been set up with its nerve center in Honolulu. Whenever a strong earthquake is recorded, the epicenter is located within 1 hr with the cooperation of other seismographic stations. Special tide gauges near the event are then checked by remote control for the presence of seismic sea waves, and, if necessary, warnings are disseminated to areas in their path. This system, installed in 1948, has in all probability already saved a great many lives.

The seismic and related events that occurred on May 31, 1970, in west central Peru (Figure 12-14) have been described as the greatest natural disaster on record in the Western Hemisphere. More than 50,000 lives were lost and roughly 186,000 buildings were rendered uninhabitable. The effects of the earthquake were typical of those associated with strong seismic events: strong ground vibrations leading to collapse of structures, subsidence of ground, and landslides.

12-13. *Damage caused by seismic sea waves along the shore of Kodiak Island after the 1964 Alaska earthquake* [*Courtesy U.S. Navy.*]

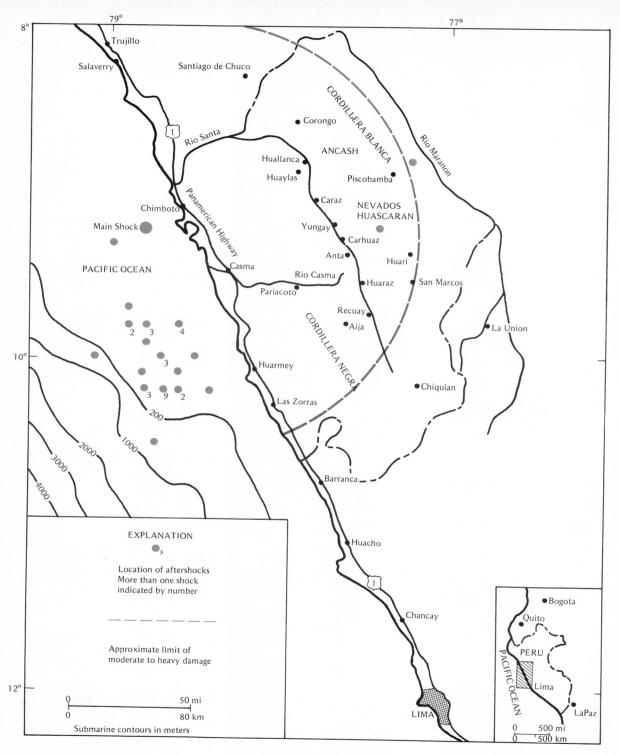

Main Shock

PACIFIC OCEAN

EXPLANATION

●₃

Location of aftershocks
More than one shock
indicated by number

— — — — —

Approximate limit of
moderate to heavy damage

0   50 mi
├────────────────────┤
0   80 km

Submarine contours in meters

414   THE EVOLVING EARTH

12-14.[OPPOSITE] *Map of central western Peru showing areas affected by the 1970 earthquake. [After U.S. Department of the Interior Geological Survey.]*

In addition, an avalanche of unprecedented proportions occurred. An estimated 25 million m³ of rock and ice broke away from the very steep north side of Huascaran Peak (Figure 12-15), the highest mountain in Peru, and plunged into the valley below with a deafening roar. The avalanche reached a velocity of about 400 km/hr and transported individual boulders of up to 14,000 tons in weight. It completely obliterated everything in its path, including two towns and nearly all their inhabitants (17,000 dead). If the epicenter of this earthquake had been located 200 km to the southeast, near the large city of Lima, the loss of life may have been even greater.

Earthquake Prediction and Control. The major hazard represented by seismic events along the great earthquake belts of the world (see Chapter 5) has now been recognized, and a great deal of research money and effort is being directed toward earthquake prediction and possible control. Active fault systems, such as the San Andreas fault in California, are carefully monitored for observable changes that could indicate an impending earthquake. Although the nature

12-15. *Oblique aerial photograph to show path of landslide that plunged down from steep upper flank of Huascaran Peak into the Santa Valley below. [From G. Plafker, G. Ericksen and F. Concha,* Seismological Soc. Amer., Bull. 61, No. 3, *Fig. 4 (1971). Photograph courtesy G. Plafker, U.S. Department of the Interior Geological Survey.]*

of earthquakes and earthquake belts are relatively well understood, the possibility of accurately predicting major earthquakes, much less controlling them, still appears remote.

Probably the most intriguing development in this field relates to the Rocky Mountain Arsenal earthquakes near Denver, Colo. In 1962 injection of fluid wastes into the Precambrian basement rocks nearly 4000 m below the ground surface was followed by a series of small earthquakes. Studies of this phenomenon have indicated that the injected fluid lubricated a fault zone under stress and thereby promoted release of seismic energy. Experimentation at the site is still going on, but the connection between the fluid injection and earthquake activity was quickly realized. Accordingly, the suggestion has been made that potentially destructive earthquakes could be converted into numerous nondestructive minor earthquakes by using fluid injection techniques. Thus, although man has no hope whatever of controlling the major earth movements that lead to earthquakes, he may in some cases be able to influence the manner in which earthquake energy is dissipated.

Volcanos

Volcanos have a long history of sudden, sometimes disastrous, eruptions into human affairs. In earlier chapters some of these events have been documented. Fortunately, in certain areas (for example, Hawaii) it is now possible to predict volcanic eruptions with some accuracy. Prior to the onset of an eruption the active volcanos in Hawaii swell slightly, and small earth tremors are recorded, no doubt in response to the new body of magma moving upward. This subtle distortion of the shape of the ground can be monitored by sensitive tiltmeters and seismographs and provides advance warning of an impending eruption. It has also been found that very slight but distinctive variations in the local magnetic field occur prior to eruptions in Hawaii and New Zealand. The eventual widespread use of these techniques in areas of volcanic activity should reduce the danger of sudden eruptions to human life.

Landslides

Catastrophic landslides or mudflows such as the one associated with the 1970 Peru earthquake and those that followed the 1959

12-16. [*Left and right*] *Photographs of Madison Slide and Earthquake Lake that resulted from the Hegben earthquake in Montana.* [*Courtesy U.S. Department of the Interior, Geological Survey.*]

Hebgen Lake earthquake in Montana, U.S.A. (Figure 12-16), are extreme examples of a normal geologic process. Landsliding, due to the force of gravity acting on earth materials on natural slopes, is an important agent in shaping the earth's surface. Downslope movement of earth materials occurs everywhere, but generally at imperceptibly slow rates. In cases where a discrete unit of materials moves downslope more rapidly than adjacent material, such movements are called landslides. Earthquake-induced landslides are obviously the most hazardous. In 1920 in the Kansu Province of China, over 100,000 persons were killed by massive landslides of unconsolidated sediments mobilized by an earthquake. At the other extreme are slow moving slides involving very local downslope movement of soil and weathered rock at rates of less than 1 m/year (Figure 12-17).

Whether catastrophic or slow moving, landslides represent a major geologic hazard to any artificial structure that stands in their way. In California, for example, uncontrolled development of hillsides underlain by poorly consolidated geologic materials has probably resulted in hundreds of millions of dollars damage due to

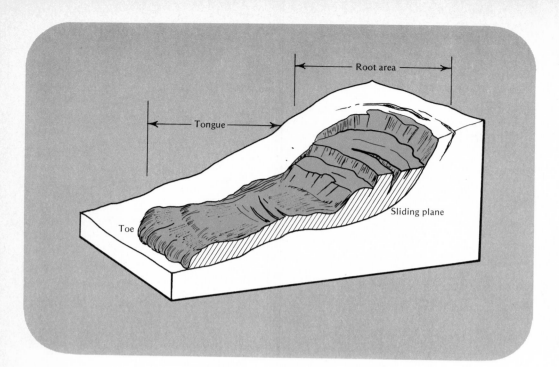

12-17. *Diagram to illustrate the typical features of a slow-moving landslide.*

Root area

Tongue

Sliding plane

Toe

12-18. *Landslide on the north side of the Palos Verdes Hills in southeastern Los Angeles County. [Courtesy of California Division of Mines and Geology.]*

12-19. *The Pacific Palisades slide near Los Angeles, California. The principal motion of the slide is along bedding planes that dip gently seaward. [Photograph courtesy of K. Emery.]*

landslides (Figure 12-18). A very small fraction of this sum spent on careful geological investigation of the sites could have prevented virtually all of these losses. A special case of landsliding is the erosion of sea cliffs by wave action and subsequent periodic collapse of material along the cliff face. The coastline of Southern California, where poorly consolidated sediments are exposed to the constant action of wave erosion, has been particularly subject to this hazard (Figure 12-19).

Floods

Periodic flooding of streams and rivers due to temporary excessive runoff within catchment areas are natural events. In fact the configurations of stream and river valleys are strongly influenced by these events (see Chapter 11). The flat lying floodplains that border many rivers contain rich soils, and thus tend to be extensively settled by man. Because these floodplains are created by the deposition of river borne mud and silt during flood-stage events, human occupation of floodplains involves the risk of flood damage. In addition to this, man's activities within a drainage basin (for example, deforestation) can lead to a sharp increase in the percentage of water lost by surface runoff. This merely serves to aggravate the potential hazard of flooding.

Low lying deltaic floodplains are particularly unsuitable areas for human settlement, for a combination of flooding and storm induced high tides can result in the inundation of extensive areas. It was just such a combination of events that in 1970 resulted in the death of an estimated 500,000 people in the fertile areas of the lower Ganges Delta.

The construction of dams is a widely used measure to control flooding. Large dams impound considerable volumes of water; therefore, very careful geologic investigation of potential dam sites is necessary if dam failures and catastrophic flooding of the valleys below them are to be avoided. Unfortunately, this type of study has not always been carried out.

12-20. *Collapse of Malpasset Dam, France. The dashed line indicates the original elevation of the dam. [Photograph courtesy of C. Fairhurst.]*

In 1959 the Malpasset Dam in France collapsed, sending its impounded waters raging down the valley below. Approximately 400 lives were lost and extensive damage resulted. Apparently the failure was related to seepage, under the pressure of the water in the dam, into cracks present in one of the abutments at the side of the dam wall. This led to the weakening and eventual collapse of the structure (Figure 12-20).

The events that occurred at the Vaiont Reservoir in Italy on October 9, 1963, provide another example of the necessity of thorough geologic surveys prior to the construction of a dam. A great mass of rock (more than 240 million m^3) suddenly slid into the reservoir sending a great wall of water 100 m high surging over the dam and down into the valley below. Some 2117 persons were killed in a matter of 7 min. During the planning and construction of the dam the instability of the area on the south side of the reservoir was not heeded. The rocks consisted of fractured and jointed limestones that dipped steeply towards the reservoir. The rise of the water level in the reservoir led to weakening and slow downhill movement of this material. Eventually the rock mass broke loose on the day of the catastrophe and plunged into the reservoir.

EARTH RESOURCES

General Considerations. Few people realize the extent to which we all depend on earth resources, quite aside from the fact that every bite of food we eat and, probably, the air we breathe at one time came from the earth. This book contains earth materials in addition to paper products, and if you are reading it with the aid of artificial light, many other earth materials, such as tungsten, copper, iron, and mineral fuel resources, have been used to bring that light to you. In fact as you look around (presuming you are indoors), virtually everything you can see represents material obtained from the earth.

Right at the beginning of this discussion of earth resources, several points need to be made clear.

1. The supply of exploitable mineral and mineral fuel resources is finite, and, with the exception of ground water, nonrenewable.

2. Increasing population and increasing per capita consumption of earth resources is making us ever more dependent on an uninterrupted supply of these materials.
3. Earth resources are unevenly distributed over land areas of our planet, and no one country is self-sufficient with respect to all its mineral and mineral fuel requirements.
4. The large scale extraction and consumption of earth resources inevitably involves damage to natural environments.

All of these factors bring the subject of earth resources into the arena of social, economic, and political concern. Many well informed earth scientists have expressed concern about our high rate of earth resource consumption and the problem of long term supplies.

At least since the time the Conquistadores came lusting after the gold of the western hemisphere, earth resources have been an important factor in world politics. In the future, natural resources can be expected to play a larger and larger role in influencing international affairs.

Ground Water Resources

Ground water can be regarded as the most critical and valuable of all earth resources. Although ground water is involved in the hydrologic cycle (Figure 12-21) and is therefore a renewable resource, ground water resources require protection against man's tendency to deplete and/or pollute them. Some examples of man's tendency to affect surface waters adversely have been pointed out earlier in this chapter. Surface waters are more accessible for use than ground waters, but they represent less than 5% of the fresh water available to man. The remainder is almost all ground water (Figure 12-22). The quantity and quality of ground water resources are a vital factor with respect to man's utilization of many of the arid areas of the earth.

Ground or **subsurface water** represents water held in a temporary reservoir that is available to man. All near-surface rocks and geologic materials contain in their cracks and pores some ground water, which tends to move through the surrounding medium towards the sea. This movement is in general considerably slower than that of surface water, and the residence time of water in the ground may range from hours to thousands of years. Several factors control residence times of ground water, but the two most fundamental are the

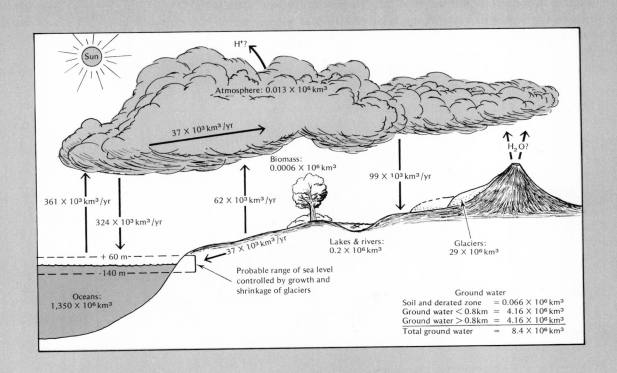

12-21. *The hydrologic and the hydrologic cycle. After A. L. Bloom:* The Surface of the Earth. [*Prentice-Hall, Inc., Englewood Cliffs, N.J., (1969, p. 13).*]

local surface relief and the permeability of the surrounding geologic materials. These largely control the configuration of the local water table, (see Chapter 11, page 321). If impermeable layers (for example, shale beds) are not present, the ground water table is typically a relatively flat or gently sloping surface which is a subdued reflection of the surface topography (Figure 12-23). In situations where impermeable layers are present the ground water table has more complex configurations. (See Figures 11-8, 11-9, 11-10.)

Although the amount of ground water is several thousand times the amount of water present at any moment in rivers, the utilization of this resource presents difficulties in some cases. The porosity and the permeability of diverse geologic materials are highly variable. As we outlined in Chapter 11 the porosity of a rock determines the quantity of water it can hold when saturated, but the permeability determines the rate at which the contained water can be extracted. Geologic materials that lack permeability are simply not suitable for water extraction, because the rate at which water can be withdrawn from them is too low. Thus shales, some of which have high porosity, are unsuitable materials for ground water extraction.

12-22. *Distribution of the water in the hydrosphere.* [*After U. S. Department of the Interior Geological Survey.*]

	Location	Water volume	Total water, %
Surface water			
Fresh-water lakes	1.25×10^5 km³	0.009	
Saline lakes and inland seas	1.04×10^5 km³	0.008	
Average in stream channels	1.1×10^3 km³	0.0001	
Subsurface water			
Vadose water (includes soil moisture)	6.66×10^4 km³	0.005	
Ground water within depth of half a mile	4.16×10^6 km³	0.31	
Ground water—deep lying	4.16×10^6 km³	0.31	
Other water locations			
Icecaps and glaciers	2.91×10^7 km³	2.15	
Atmosphere	1.29×10^5 km³	0.001	
World ocean	1.32×10^9 km³	97.2	

Labels within the diagram:
Rain cloud
Precipitation
Evaporation while falling
Cloud formation
Evaporation
Evaporation and transpiration
From streams
From soil
From plants
From ocean
Surface
Soil
Infiltration
water
Capillary fringe
Intermediate vadose zone
Runoff
River or lake
Runoff
Water table
Percolation
Fresh ground water
Salt water
Ocean
Confining beds

12-23. *Diagram to illustrate the movement of ground water and the water table, and recycling of water among oceans, atmosphere, and the earth's surface and subsurface.*

The most suitable geologic materials for adequate recovery of ground water are beds with high permeability and a porosity (pore-space volume) in excess of 10%. Such units are called **aquifers.** They can be of two types: unconfined aquifers and confined aquifers. In the latter case the water-producing unit is confined between beds of low permeability; thus, the ground water of a confined aquifer is less prone to contamination from the surface. The flow direction of water in confined aquifers is determined by the distribution of **head** (water pressure) within the aquifer system. In some instances this head is sufficiently great to push the water in the aquifer up to the surface (see Figure 11-9), giving rise to free flowing **(artesian)** wells.

The rate of recharge or replenishment of aquifers is an important factor in ground water utilization. Recharge rates increase with the amount of rainfall in the catchment area but are reduced by the amounts of water lost by surface runoff, transpiration by plants, and evaporation. In the arid southwestern United States, serious depletion of some ground water reserves by man has already taken place. The problem of ground water depletion has actually become so accute locally that certain ground water basins have had to be closed to further development of ground water resources. Such action obviously places a severe limit on additional growth prospects in an area.

Finally, the quality of ground water is an important factor in its utilization. Water that contains dissolved salts in concentrations in excess of 0.05% is unsuitable for human consumption, and concentrations of salts in excess of 0.2% render water unsuitable for almost all uses. The major demand for water in many arid areas is for irrigation of crops. Most plants have a low tolerance for salts of various kinds; therefore, ground water used for irrigation must be relatively pure or the buildup of salts in the soils will rapidly poison them for agricultural purposes.

Wise management and use of both ground and surface water resources is a matter of major importance in all parts of the world. Water consumption rates tend to reach high levels in industrialized nations. In the United States, for example, water usage reached levels of 370 billion gal/day in 1970. This is equivalent to an average per capita consumption of 1800 gal daily! Although most of this water is available for reuse, nearly 25% is lost by evaporation or rendered unfit for reuse by pollution and other causes. By 1980 water consumption in the United States is projected to total 400 billion gal/day. It is hardly surprising then that problems of water supply of varying degrees of seriousness have already affected most parts of the United States and are very much in evidence in most other nations.

Efficient and well informed management of water resources is clearly one of the major priorities on which mankind's future will depend.

Energy Resources: Fossil Fuels

Man's technological progress is intimately related to his increasing consumption of energy. At the present time well over 90% of the energy utilized by man is provided by the irreplaceable fossil fuels: petroleum, natural gas, and coal. Fossil fuels represent solar

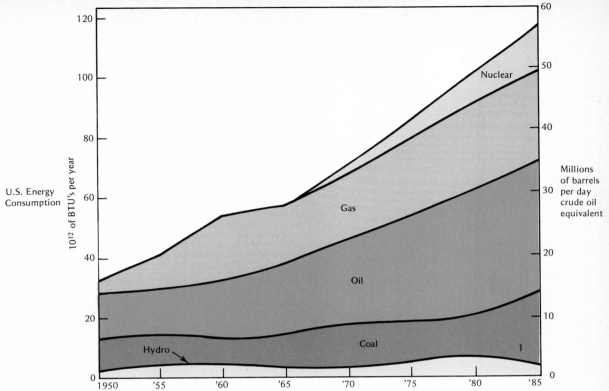

12-24. *Plot to illustrate sources of energy that are used in the United States and how they varied in the past and are projected to vary in the future. BTU = British Thermal Units and are a measure of the amount of heat necessary to raise 1 lb of water 1° F in temperature. [Courtesy Humble Oil and Refining Co., and Oil and Gas Journal.]*

energy that was converted during past geologic ages by biological processes into carbon rich substances. Upon combustion, these substances release their stored energy.

The United States of America, which has approximately 6% of the world's population, currently accounts for approximately 30% of world energy consumption. Close to 75% of this energy is supplied by petroleum and natural gas (Figure 12-24).

Petroleum Consumption. Let us concentrate our attention for the moment on some of the available data regarding petroleum consumption and availability. In 1970 world petroleum consumption reached a level of close to 50 million barrels per day (bbl/day). (1 bbl = 42 U.S. gal = 158,970 cm³). This figure may seem surprisingly high, but of far greater concern is the fact that the consumption of

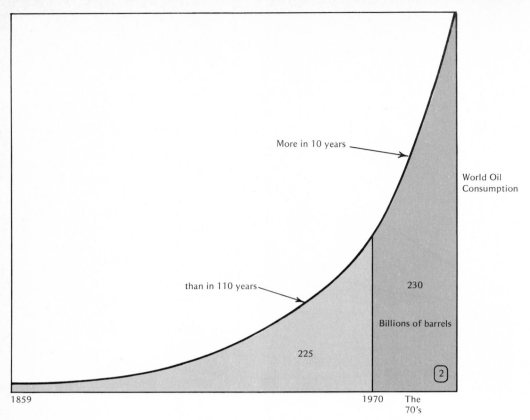

12-25. *Diagrammatic plot to illustrate rapidly accelerating trend of world petroleum consumption. Courtesy of Humble Oil and Refining Co., and* Oil and Gas Journal.]

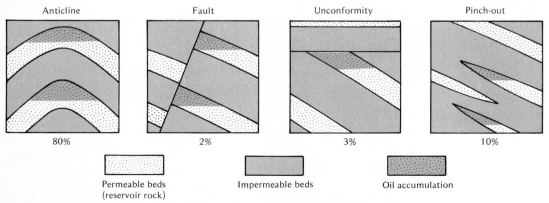

12-26. *Series of diagrams to illustrate various types of oil traps. Percentage figures indicate relative importance of each type with respect to world oil production.*

petroleum has been increasing during the last two decades at an average rate of $7\frac{1}{2}\%$/year worldwide. The full impact of this statistic is perhaps not easily grasped, but it indicates a doubling of oil consumption level every 10 years. In fact the consumption of petroleum by man between 1960 and 1970 was approximately equal to total cumulative consumption of petroleum prior to the year 1960 (Figure 12-25). The growth rate of oil consumption during the coming years will probably slow somewhat, but it seems certain that man's petroleum needs will exceed 85 million bbl/day by 1980. The crucial question we must face is, "How long can these high consumption levels of petroleum be maintained before world reserves run out?"

Formation and Accumulation of Petroleum. Petroleum liquids and natural gas originate from the decay of marine organisms trapped in sea bottom sediments. The decay processes lead to the production of **hydrocarbons,** molecules composed primarily of carbon and hydrogen combined in various ways. The simplest such compound is methane (CH_4). This compound is a gas, and **natural gas** consists largely of methane and other simple hydrocarbon compounds that contain four or less carbon atoms per molecule. Hydrocarbon compounds that contain larger numbers of carbon atoms per molecule form liquids, and naturally occurring petroleum consists of complex mixtures of various types of these hydrocarbons.

In order that petroleum liquids and/or natural gas accumulate into extractable "pools" within rocks, three basic conditions must prevail. Initially petroleum must be generated by the decay of organisms within fine grained sedimentary rocks **(source beds).** Subsequently the petroleum must be able to move out of the source beds during their compaction into nearby relatively porous and permeable coarser grained sedimentary rocks **(reservoir beds).** Finally the petroleum must migrate through the reservoir beds to eventually encounter some barrier to its continued migration (a **trap),** where it can accumulate.

Traps can be of many different kinds (Figure 12-26), but structural traps formed by anticlines are by far the most important. Although petroleum has been found in rocks of Paleozoic age, over 90% of the world's petroleum has come from sedimentary basins of Mesozoic and Cenozoic age. In fact young sedimentary basins tend to be more productive in terms of petroleum than older basins of comparable volume. This could indicate that the total amount of potential petroleum producing organisms has been increasing with geologic time, but it probably merely signifies that petroleum accumulations eventually leak out of most traps.

Commercial quantities of petroleum and natural gas can accumulate in sedimentary rocks deposited in any marine environment, but in plate tectonic terms three environments of preeminent importance can be singled out. These are inactive continental margins, molasse basins formed behind active continental margins and adjacent to zones of continental collision, and within the thicker marine sequences of continental cover rocks deposited in intraplate environments.

Initially all petroleum was obtained from land areas, but in recent years more and more petroleum production has come from offshore oil fields. One result of this has been an increase in the amount of oil inadvertently spilled into the sea, where it eventually fouls beaches and takes a heavy toll of marine life and sea birds. At the present time exploration for new fields is being heavily concentrated in offshore areas along continental shelves. Although the technology for recovery of petroleum from below water depths of about 200 m is now being made available, the current price of oil makes production costs prohibitive. Nevertheless, oilmen are pushing their search for petroleum into deeper and deeper water.

Petroleum Reserves in the Future. In view of ever increasing demand, some oil geologists have attempted to calculate the total theoretical potential of world petroleum reserves. Given current trends of petroleum consumption the presently known world reserves, approximately 500 billion bbl, will not last much more than 15 years. Projections of the total world potential for petroleum are difficult to make with certainty, but on the basis of the known volume of Mesozoic and Cenozoic marine sedimentary rocks it can be estimated that 4000 billion bbl of petroleum could eventually be produced. If present trends continue, even this amount will be consumed prior to the year 2030, assuming of course that geologists are equal to the formidable task of finding all this oil before consumption overtakes production capability.

Two facts seem to emerge clearly from the presently available data. First, all of us can expect to pay more, perhaps a good deal more, for the petroleum products we use (for example, gasoline) in the future. Second, we will eventually have to turn to alternate sources for our future energy requirements. In the United States clear evidence of depletion of petroleum resources can be seen. Excluding recent major oil discoveries in Alaska, the petroleum reserves of the United States have been declining for several years, and already this country is compelled to rely on imports for nearly one third of its petroleum requirements.

12-27. A. *Teapot Rock, an outcrop of thinly bedded oil shale of the Green River Formation.* [*Courtesy U.S. Department of the Interior Geological Survey.* B. *Varves (annual bands) in organic-rich shale from Green Formation* × 4. [*Courtesy U. S. Department of the Interior Geological Survey.*]

A

B

Oil shale and Tar Sand. Three other types of fossil fuel are available to meet man's energy needs: oil shale, tar sand, and coal. Oil shale (Figure 12-27) and tar sand are rocks in which hydrocarbons occur as solids rather than liquids. In order to obtain the hydrocarbons from such rocks they must be mined and the constituent hydrocarbons must be distilled out of them by heat. At the present time (1972) these sources of petroleum are too costly to complete to any significant degree with liquid petroleum. It is only a matter of time, however, before decreasing availability and increasing cost of liquid petroleum provide sufficient impetus for the large scale exploitation of oil shales and tar sands.

A large fraction of world oil shale and tar sand reserves occur in North America (see Figure 12-28). Unfortunately the development of these resources will inevitably create environmental problems. To hold down extraction costs the oil shales of the western United States will have to be mined by stripping away the petroleum rich beds and their cover rocks. As we have seen earlier, these mining techniques can have a disastrous effect on local environments. Nevertheless, the eventual utilization of oil shales and tar sands should add more than 1000 billion bbl to world petroleum reserves.

12-28. A. *Location map of Athabasca tar sands in Alberta, Canada. Commercial exploitation of these sands has already begun near Fort McMurray, where 45,000 barrels of oil are produced from 100,000 tons of sand mined each day. [After B. Skinner:* Earth Resources, *Prentice-Hall, Inc., Englewood Cliffs, N. J. (1970, p. 120).]* B. *Location map of the oil shales of Colorado, Wyoming, and Utah, western United States. These organic rich shales were deposited in an extensive fresh water lake during Eocene time. The darkly shaded areas represent shale of sufficient thickness and oil content to represent a potential future source of petroleum liquids. [After U.S. Department of the Interior Geological Survey.]*

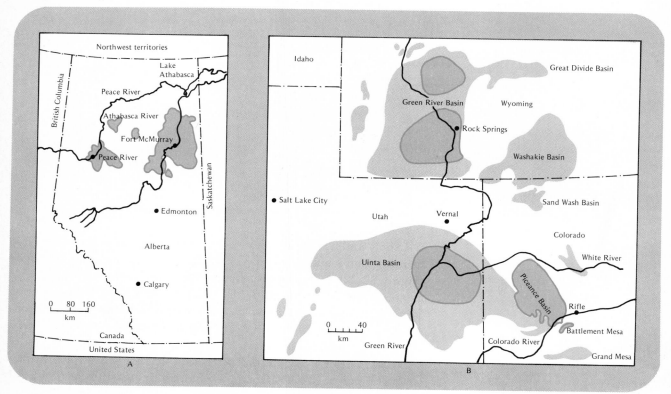

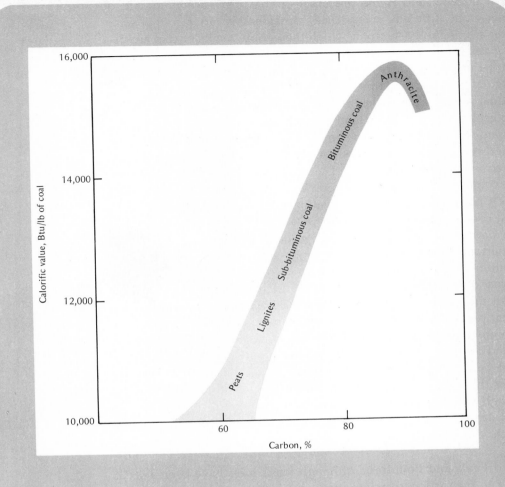

12-29. *Plot to show relationship between carbon content and calorific value (heat producing capacity) of various ranks of coal. Btu = British thermal units and are a measure of the amount of heat necessary to raise 1 lb of water 1°F in temperature. [After B. Skinner:* Earth Resources, *Prentice-Hall, Inc., Englewood Cliffs, N.J. (1970, p. 111).]*

Coal. The other important fossil fuel available to man is coal. Coal represents fossil plant material (mostly from ancient swamps) that has been buried and changed into coal in place. Potential coal forming environments appear to be rare at the present time, but the Dismal Swamp in Virginia and North Carolina is probably a good example. Here a layer of **peat** with an average thickness of 2 m has been formed over an area of 5700 km². Peat is carbon-rich decayed plant material and is the precursor of coal. The carbon content of coals from different areas varies considerably. In fact this characteristic is the basis upon which the various ranks of coal are distinguished (Figure 12-29). The **rank** of coal is determined principally by the extent to which pressure and temperature have affected the original peat since its time of burial.

Coal reserves are widely distributed throughout the latter part of the geologic record. Once land plants had evolved and become widespread (by mid-Devonian times) coal deposits tended to form wherever local environments permitted the accumulation and preservation of plant material. The most favorable sites, however, were molasse and intracontinental nonmarine sedimentary basins (see Chapter 8). The Upper Devonian rocks of the Canadian Arctic area contain the earliest coal deposits of significant size, but it was the Carboniferous* and Permian Periods that represented the most important interval of coal formation in earth history. In fact the Carboniferous Period was so named in recognition of the abundance of coal deposits formed during that time in Europe.

All the major coal basins of the world are thought to have been already discovered, but only a small fraction of the world's coal inventory has as yet been utilized by man. In fact it has been estimated that the world's reserves of recoverable coal are in excess of 8000 billion tons. In terms of usable energy this is equivalent to more than 34,000 billion bbl of petroleum. It is thus clear that, although world petroleum stocks may become exhausted during the next century, fossil fuel supplies in the form of coal will last very much longer.

Fossil Fuels and Pollution. Although fossil fuels have been a great boon to man in his search for sources of energy, their recovery and use have had some serious side effects. The air pollution that affects virtually every large city in the world today and endangers the health of its occupants is largely the result of the combustion of fossil fuels and their derivatives. One result of our dependence on fossil fuels for energy is the spewing, by automobile exhausts and fossil fuel burning power plants, of at least 5 million tons of carbon gases into the earth's atmosphere every day!

No one is sure just what long term effect this may have on global weather patterns, but some scientists fear that the carbon gases will act to trap increased amounts of the solar energy received by the

*Carboniferous is a term used by European geologists. In North America the Mississippian plus Pennsylvanian Periods are used to designate this portion of geologic time.

earth. This so-called **greenhouse effect** could accelerate the melting of the polar ice caps, with a consequent rise in sea level. Offshore oil operations have led to some serious oil spills, as have accidents involving ships carrying oil. Finally one must consider the ill effects of the strip mining of coal that have been discussed earlier in this chapter.

Alternate Energy Resources

It is clear that, before too long, man's heavy reliance on fossil fuels as an energy source will have to be greatly curtailed. Other cleaner sources of energy, less prone to depletion, will have to be developed. Several alternate sources of energy have been explored, but the most promising appears to be nuclear energy.

Nuclear Energy. The key element in the development of nuclear power by **fission processes** is uranium. Only 0.7% of all natural uranium consists of the unstable fissionable isotope uranium-235 (^{235}U). When bombarded with neutrons the ^{235}U nucleus splits to form lighter elements with production of additional neutrons. If these processes occur under controlled conditions, a sustained chain reaction takes place that produces large quantities of heat. This heat can then be used to generate electricity via steam turbines.

In the United States in 1972 approximately 1% of the total energy used was generated by nuclear power plants, but this figure is projected to rise rapidly during the next 10 years. The amount of ^{235}U available from uranium ore deposits is limited however, and the cost of isolating it from the more abundant ^{238}U is high. Fortunately ^{238}U and thorium-232 (^{232}Th) can be converted into fissionable isotopes by a process known as **breeding** (Figure 12-30).

Breeder reactors have yet to be perfected, but, if and when this technological step is taken by man, the long term availability of nuclear energy will be assured. This is true for two reasons. The use of ^{238}U will mean a hundredfold increase in the amount of nuclear fuel available, and various very large, but presently subeconomic, deposits of uranium will be worth working for the energy potential of the uranium they contain. Certain organic rich shales and phosphate rich sedimentary rocks, for example, contain enormous low grade reserves of uranium (Figure 12-31).

Uranium deposits that can be mined under current economic conditions occur mainly in Canada, the United States, and South Africa. In Canada and South Africa the principal deposits occur in certain Precambrian conglomerates in which the mineral uraninite (UO_2) occurs, apparently as detrital grains. In the Colorado Plateau area of the western United States, economic uranium deposits have been formed by the deposition of uranium bearing minerals in certain sandstones rich in organic matter. The process of ore for-

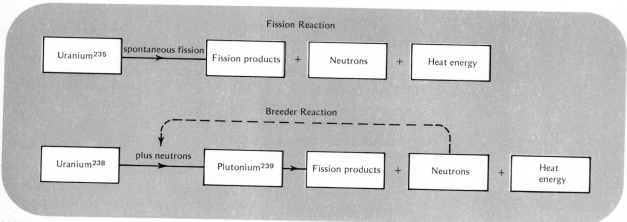

12-30. *Schematic illustration of fission and breeder reactions.*

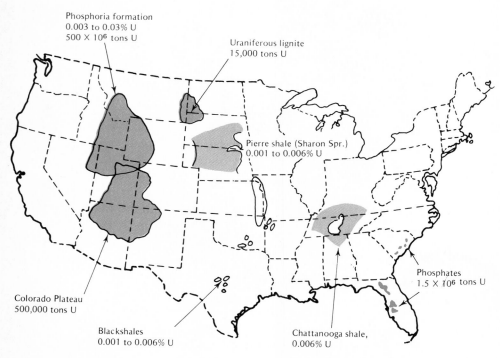

Phosphoria formation
0.003 to 0.03% U
500 X 10^6 tons U

Uraniferous lignite
15,000 tons U

Pierre shale (Sharon Spr.)
0.001 to 0.006% U

Colorado Plateau
500,000 tons U

Blackshales
0.001 to 0.006% U

Chattanooga shale,
0.006% U

Phosphates
1.5 X 10^6 tons U

12-31. *Map of the United States showing location, quantity, and grade of low grade uranium deposits. If the breeder reactor becomes a reality, much of this low grade material could eventually be tapped to help supply man's future energy requirements. [After M. K. Hubbert:* Energy Resources, *Publ. 1000-D. Committee of Natural Resources, National Academy of Sciences—National Research Council, Washington, D.C., 1962.]*

mation involves the reaction of oxygen rich ground waters that contain small amounts of uranium in solution with the organic materials or iron sulfides in the sandstones. As a result the uranium in the ground waters is precipitated.

The generation of energy in nuclear power plants is a clean process in terms of air pollution, but other environmental problems are encountered. Large amounts of heat have to be dissipated into the local environment. This thermal pollution can have serious effects, especially on the delicate balance of local fresh water or marine ecological systems. Furthermore, in addition to the ever-present hazard of accidents in nuclear power stations, there is the problem of disposal of radioactive wastes. Finding a suitable dumping site for these dangerous materials, so that there is no possibility that they enter the biologic cycle, is a major headache. Suggestions have even been made that they be shot out into space or dumped into a subduction zone. One of the more practical disposal techniques now

12-32. *Map showing geothermal areas of the world. Note that geothermal energy is already being used in a number of areas. [After U. S. Department of the Interior, Geological Survey.]*

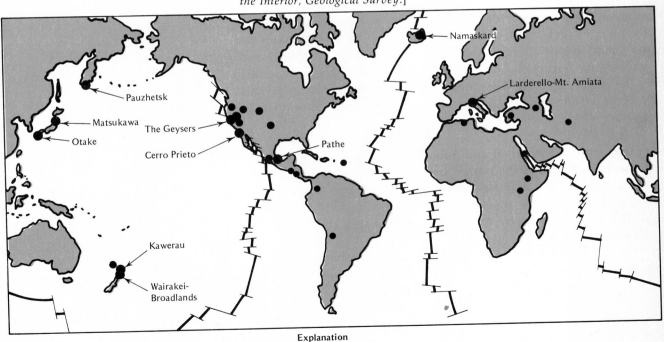

Explanation

● Producing geothermal fields

● Promising geothermal areas

Midocean rifts; centers of ocean floor spreading along which heat flow is high

receiving attention is storing these wastes in solution cavities specially created for the purpose in deep lying salt deposits.

Fusion reactions provide the energy for the sun's radiation, and also for hydrogen bombs. Although man now has a fairly complete understanding of fusion reactions the technology is simply not available for harnessing the enormous amounts of energy released by the fusion process. Nevertheless, considerable research into this problem is now underway, for its solution would allow man to generate almost unlimited amounts of energy.

Hydroelectric, Geothermal and Solar Energy. Another clean source of energy is hydroelectric power. Unfortunately, its development is limited to countries with suitable river systems, and in most cases hydroelectric projects involve tampering with natural drainage systems. As we have seen, this can have unforeseen and unwelcome side effects.

In some areas (Figure 12-32) there is considerable potential for harnessing the earth's internal heat energy (geothermal power). This is already being accomplished locally (Figure 12-33), but the potential of geothermal power in certain regions is very large and is being closely studied.

Finally, man may learn to utilize effectively some of the enormous quantity of energy that falls upon the earth as solar radiation. In isolated cases limited use of solar radiation is already being made, and research workers are busy designing systems to expand usage of this possible energy source.

Mineral Resources

The mineral resources man obtains from the earth encompass a wide spectrum of materials; from sand, gravel and rock for building purposes, to rare metals such as mercury, silver, gold, and platinum. Surprisingly, the value of the sand and gravel obtained from the earth during any year exceeds the combined value of all the metals obtained during that same period. Mineral resources, excluding fuels and water, can be divided into three distinctive categories: nonmetals, abundant metals, and scarce metals (Table 12−1). One example from each category will be used for illustration.

Phosphate Deposits. One of the most critical nonmetallic minerals, in terms of man's well-being, is **apatite** $[Ca_5(PO_4)_3OH]$, the most important phosphorus-bearing mineral. Phosphorus is a vital element in the physiological processes of all vertebrates, including man, and is essential for plant growth. It is necessary to apply phosphates to crop land subjected to intensive cultivation in order to replenish the original supply. Apatite is present in trace amounts in most rocks, igneous, metamorphic, and sedimentary. The most important economic deposits of apatite rich rock, however, are found in marine sedimentary sequences, and over 90% of world production comes from such sources (Table 12−2).

The precise origin of marine phosphate deposits is still uncertain, but it is probably related to the upwelling of cold phosphate rich

Table 12−1 Classification of Earth Materials of Potential Value (not all-inclusive)

*Numbers in brackets after individual metals represent their average continental crustal abundance in percent.

**Rare metals are designated as those with average crustal abundance less than 0.01% (100 parts per million).

Nonmetals	Abundant Metals	Rare Metals**
Sand and gravel	Aluminum (8.0)*	Copper (0.0058)
Building stone	Iron (5.80)	Nickel (0.0072)
Asbestos	Manganese (0.10)	Zinc (0.0082)
Clays	Chromium (0.01)	Lead (0.0010)
Gypsum	Titanium (0.86)	Molybdenum (0.00012)
Sulfur		Tin (0.0015)
Potassium salts		Tungsten (0.00010)
Phosphate		Mercury (0.000002)
Fluorite (CaF_2)		Silver (0.000008)
Barite $(BaSO_4)$		Gold (0.0000002)
Diamond		

**Table 12−2 Major World Producers
of Phosphate Rock (1971)***

*From U.S. Bureau of Mines, January 1973, Commodity Data
Summaries.

Area	Quantity, Thousands of Short Tons
United States	38,886
Morrocco	13,237
Tunisia	3,402
Other free world	10,402
Communist countries	26,845
World Total	96,449

ocean water. Subsequent precipitation of apatite occurs when
these waters are carried into shallow marine basins. This is thought
to be the origin of the phosphate rich sediments that form the
extensive Phosphoria Formation of Permian age in the western
United States (Figure 12-34).

Another source of phosphate is found in the nodular bodies of
apatite that occur in some limestones formed in shallow marine
environments. Major concentrations of these nodules are in some
instances found in surface environments where large scale solution
and removal of the limestone host rocks has taken place. The extensive "landpebble" phosphate deposits of Florida (Figure 12-35)
were formed in this manner.

The populous nations of the Far East have no known major phosphate deposits. Consequently, they are forced to import large
quantities of phosphates to support their agricultural industries as
well as using animal and human wastes to maintain the necessary
phosphorus levels in their soils.

Iron Deposits. Iron, which accounts for 95% of all metals that have
been mined from the earth's crust, can be considered one of the
foundations of modern civilization. Many other metals, manganese,
nickel, cobalt, chromium, tungsten, and vanadium, are exploited
mainly for use as additives to iron in various steel making processes. The discovery of methods by which natural iron oxides could
be converted into metallic iron, before the year 1000 B.C., represented
a major step along man's path to his present type of civilization.
Today (1973) world iron and steel production has reached $>500 \times 10^6$
tons/year and is rising at the rate of ~2% per year.

Average continental crust contains nearly 6% iron, making iron
the second most abundant metal after aluminum (average crustal
abundance ~8%). Most economic deposits of iron **ore** contain be-

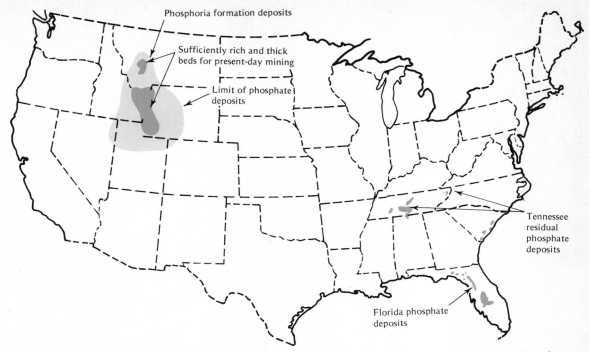

12-34. *Map of United States showing important areas in which phosphate deposits occur. The extensive Phosphoria Formation deposits are capable of supplying the needs of the United States for many centuries. [After U. S. Department of the Interior, Geological Survey.]*

tween 30 and 60% iron. Thus a concentration of iron of five to ten times average is required to form ore grade material. Iron is a constituent of a great many minerals, but only four are important as economic sources of iron (Table 12–3). Important iron ore deposits fall into three classes:

1. Deposits associated with igneous rocks.
2. Sedimentary deposits.
3. Residual deposits (these mainly represent surficial enrichments of low grade sedimentary deposits).

Table 12–3 **Important Minerals of Iron in Major Iron Deposits**

*Magnetite or hematite are by far the most abundant iron minerals in virtually all very large iron deposits.

Name	Composition	Iron Content, %
Magnetite*	Fe_3O_4	72.4
Hematite*	Fe_2O_3	70.0
Geothite	$FeO(OH)$	62.9
Siderite	$FeCO_3$	48.2

12-35. *Photograph of thirty bonded Precambrian iron formation. Light layers are iron oxide, darker layers are chert. [Photo by M. Kaiper.]*

12-36. *Location of major Precambrian banded iron formation deposits. The enormous size of these deposits ensure an adequate supply of iron ore for several hundred years into the future.*

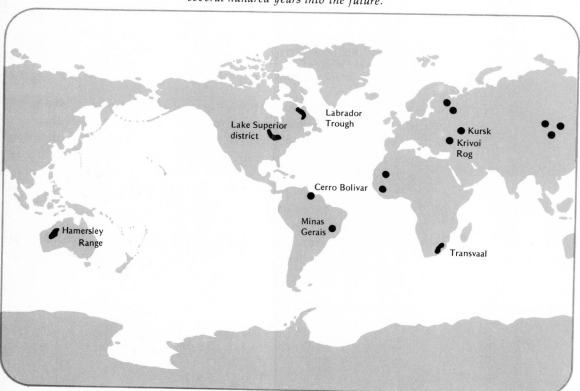

Labrador Trough

Lake Superior district

Kursk

Krivoi Rog

Cerro Bolivar

Minas Gerais

Transvaal

Hamersley Range

In the past, iron ore deposits associated with igneous rocks were an important source of iron, but today the large bulk of iron ore is mined from sedimentary deposits and residually enriched iron rich sedimentary rocks. Especially important, particularly in terms of future iron resources, are **banded iron formations** (Figure 12-35) that are almost exclusively of Precambrian age. Banded iron formations consist of chemical precipitates containing alternating thin bands (often less than 1 mm thick) of iron-rich and silica-rich material. Enormous amounts of iron ore are present in some banded iron formations, and a recent United Nations survey of world iron ore resources estimated that 10^{12} tons of economically minable iron ore are available for exploitation by man. Clearly adequate supplies of this metal are at hand to supply man's needs for many centuries to come.

The imposing size of many banded iron formations and their widespread occurrence in Precambrian sedimentary and volcanic terranes (Figure 12-26) has led to a lively controversy regarding their origin. Whatever the source of the iron in these iron formations may have been—volcanic exhalations, iron leached from preexisting rocks, or a combination of both—it is generally agreed that the composition of Precambrian atmospheres (see Chapter 9) must have been an important factor in the transport of iron in surficial environments and its solubility in sea water. It is of interest to note that most of the largest units of banded iron formation were formed in the time interval from 2.2 to 1.8 billion years ago. After this time large scale deposition of banded iron formation essentially ceased.

Copper Deposits. Copper, unlike iron, occurs in the continental crust in very small amounts. In fact its average concentration in continental crust is only 60 parts per million (ppm). Copper deposits are unevenly distributed around the world and seven nations ac-

Table 12-4 Major World Producers of Copper (1971)*

*From U.S. Bureau of Mines, January 1973, Commodity Data Summaries.

Area	Quantity, Tons
United States	1522
Canada	720
Chile	791
Peru	235
Zaire	449
Zambia	718
Other free world	1231
Communist countries	999
World Total	6665

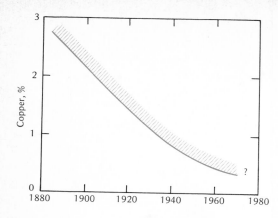

12-37. *Plot to illustrate the steady decline in the minimum grade of copper ore that can be worked at a profit. This is largely the consequence of new techniques for the surface mining of large low grade deposits. [After U.S. Bureau of Mines.]*

12-38. *A. Aerial view of the Bingham Canyon open-pit copper mine, Utah. Over two million tons of rock are moved every week in this operation which represents the largest copper mine in the world. B. Large powershovel used for loading are in the Bingham Canyon Mine. The trucks have a load capacity in excess of 150 tons. [Photographs courtesy of Kennecott Copper Corporation, Salt Lake City, Utah.]*

A

count for more than 80% of the world copper production (Table 12–4). Like iron, copper has had a long history of use by man. Economic deposits of copper ore represent concentrations of copper 100—200 times higher than that of average crustal material. Advances in the technology of mining and processing of copper ores have resulted in a steady decrease in the minimum grade of exploitable copper ore (Figure 12-37). As a result, the lowest grade of copper ore that can be mined profitably has now reached $1/2\%$. The discovery of many large near-surface deposits of low grade copper ore spurred the mining industry to develop special techniques and equipment for the large-volume low cost open pit mining of such deposits (Figure 12-38).

B

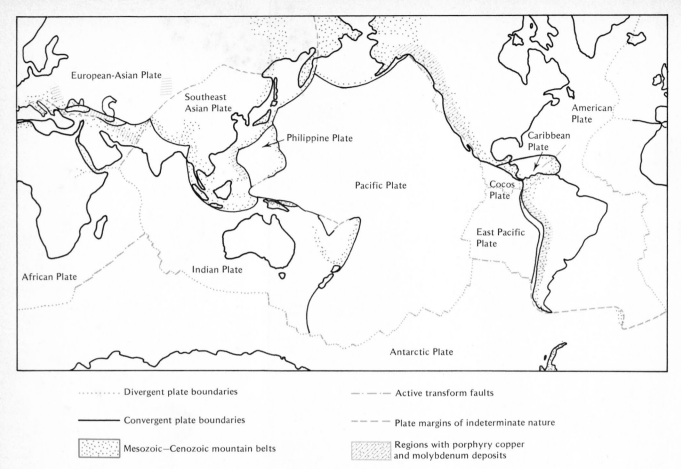

European-Asian Plate

Southeast
Asian Plate

Philippine Plate

Pacific Plate

African Plate

Indian Plate

Antarctic Plate

American
Plate

Caribbean
Plate

Cocos
Plate

East Pacific
Plate

.... Divergent plate boundaries

——— Convergent plate boundaries

Mesozoic—Cenozoic mountain belts

—·—·— Active transform faults

— — — Plate margins of indeterminate nature

Regions with porphyry copper
and molybdenum deposits

12-39. *World distribution of porphyry copper deposits showing their restriction to narrow belts in the vicinity of convergent plate boundaries. [From R. Sillitoe: A plate tectonic model for the origin of porphyry copper deposits.* Economic Geology, 67: *p. 187, (1972).]*

These deposits are called **porphyry coppers** because of their invariable association with **porphyritic** (see Chapter 2) intermediate to felsic intrusives. The major copper bearing mineral in porphyry coppers is chalcopyrite ($CuFeS_2$), which occurs more or less evenly disseminated, as small blebs or in veinlets, throughout large volumes of rock. The largest porphyry copper depoists contain up to 3 billion tons of such copper enriched material. At the present time well over 50% of the world's copper is produced from porphyry copper deposits, and a large number of them are now known (Figure 12-39).

One particularly interesting aspect of porphyry copper deposits is their apparent relationship to plate tectonics. All known examples

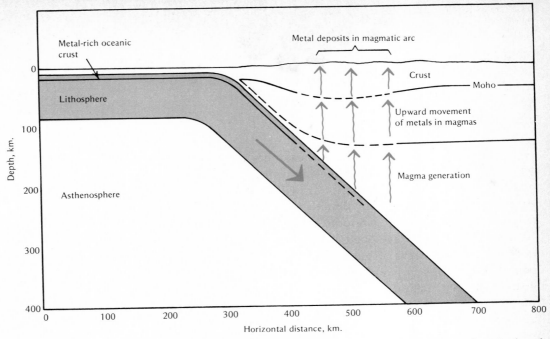

12-40. *Diagrammatic illustration of concept that metals in ore deposits found in volcano-plutonic arcs largely represent metals originally present in oceanic crust.*

of these deposits occur on the inner side of present or past convergent plate boundaries and lie within, or immediately adjacent to, subduction-related intrusive igneous rocks. In fact, it has been suggested that the metals in porphyry copper, and many other metal deposits in volcano-plutonic arcs, are derived from the metals originally present in oceanic crust. The metals are returned, after subduction of oceanic crust, to upper crustal environments by uprise of magma formed by partial melting along Benioff zones (Figure 12-40). Final emplacement of the metals in concentrations of ore grade is effected by hydrothermal solutions that move the metals from the cooling magma to their point of deposition. Metal deposits of this type are largely confined to young (Mesozoic and Cenozoic) arc systems because they form at shallow levels within the crust and, in older arc systems, tend to be removed by erosion.

Plate tectonic theory promises to be an important key to understanding the space-time distribution of, and source of, metals in many types of metal deposits. This concept is of major importance, because availability of many of the rare metals will eventually be threatened by depletion of known ore reserves. For example, world demand for silver is already well in excess of primary production. Fortunately, unlike fossil fuels, most of the metals won from the earth are not destroyed by use. They can be recovered by recycling techniques and used again. Growing emphasis must be placed on the importance of recycling metals.

Earth Resources and Politics

Finally, the political aspects of earth resources must be mentioned. The combination of sharply increasing world population figures and increasing per capita consumption will inevitably lead to heavy demands on the resources man obtains from the earth. It is unlikely, for instance, that adequate amounts of the necessary earth resources could be found and developed to support the inhabitants of all nations at the standards of living now prevailing in the highly industrialized nations. If world populations reach a figure of 7 billion by the year 2000, adequate resources would be impossible without truly major advances in technology. It can be argued that earth resources are the natural heritage of all mankind, but at the present time we are indeed a long way from that situation.

Competition among the industrial nations inevitably develops for earth resources that fall into short supply. As a result, individual powerful nations succumb to the temptation to influence the political development of developing resource rich nations. In fact, these considerations are already a growing factor in international politics, as nations and power blocks vie for the control of resources. This type of situation obviously has a negative effect on world stability and the efforts of developing countries to achieve economic and political maturity. Since it is the geologist who must find the untapped resources of our planet and take inventory of its known and potential resources, he plays a key role in these matters.

CONCLUSIONS

The central conclusion that can be drawn from this chapter on Geology and Man is the need for man to live in harmony with his natural environment and to develop and use the earth resources available to him wisely and prudently. This goal will be increasingly difficult to achieve as world population pressures mount. It is also clear that many of the technological solutions used by man to achieve a more complete mastery over natural environments can produce serious, often unforseen, side effects, which can largely negate the originally planned benefits.

Fortunately, in recent years an awareness of the science of **ecology** has grown substantially. Ecology teaches us that natural phenomena, especially biological and geological systems, are complexly interrelated and that they adjust to man-induced changes in ways that cannot always be predicted. It is none too soon for man's adverse impact on natural environments to become a matter of national and international concern.

QUESTIONS

1. List the adverse results of the building of the Aswan High Dam on the Nile River in Egypt.
2. What is the major cause of man-induced land subsidence? Cite three examples of areas where such land subsidence has caused problems.
3. List as many of the effects of the 1964 Alaska earthquake as you can.
4. Explain how seismic sea waves are generated, how they move across the oceans and what happens when they encounter shallow water.
5. By what techniques can volcanic eruptions be predicted.
6. Illustrate with a diagram the main features of a slow moving landslide.
7. Explain why low-lying deltaic floodplains are hazardous places for human settlement.
8. Indicate the various factors that influence groundwater recharge rates.
9. Why is the quality of groundwater an important factor in its utilization?
10. Trace the progressive steps by which marine organisms can form petroleum liquids and eventually accumulate into extractable 'pools" within rocks.
11. In what types of geologic environments do coal deposits tend to form?

12. List the energy resources other than fossil fuels that are available to man.
13. Explain in general terms how marine and landpebble phosphate deposits form.
14. What are banded iron formations and during what geologic time period were large volumes of this material formed?
15. Explain the relationship between plate tectonics and porphyry copper deposits.
16. Why are earth resources an increasingly important factor in world politics.

Appendix 1
Powers of Ten and Conversion Factors

The science of geology is concerned with both very large and very small numbers. We need to be able to express simply and calculate easily with numbers as large as the size and age of the earth, and as small as the size and mass of an atom. Using powers of ten or *scientific notation* makes writing and calculating with large and small numbers quite simple. Prefixes on the names of units of measurement are often used to denote powers of ten.

Number	Scientific Notation	Prefix
One billion = 1,000,000,000	10^9	giga-
One million = 1,000,000	10^6	mega-
One thousand = 1,000	10^3	kilo-
One hundred = 100	10^2	hecta-
Ten = 10	10^1	deca-
One = 1	10^0	
One tenth = 0.1	10^{-1}	deci-
One hundredth = 0.01	10^{-2}	centi-
One thousandth = 0.001	10^{-3}	milli-
One millionth = 0.000001	10^{-6}	micro-
One billionth = 0.000000001	10^{-9}	nano-

Thus one thousand meters = 10^3 meters = 1 kilometer, and one thousandth of a meter = 10^{-3} meter = 1 millimeter.

Powers of ten are multiplied by adding the **exponents,** the superscript numbers at the upper right hand corner of the 10. For example, $1000 \times 1,000,000$ can be written as $10^3 \times 10^6 = 10^9$. In division, the exponent of the divisor is subtracted from the exponent of the dividend. For example, 1000/1,000,000 is equal to $10^3/10^6 = 10^{-3}$.

Any large or small numbers with decimals can be written in scientific notation. The age of the earth, 4,500,000,000 years, is expressed as 4.5×10^9 years. The exponent indicates the number of places the decimal point has been moved. For a positive exponent, the decimal point has been moved to the left, by nine places in the example above. If the exponent is negative, the decimal has been moved to the right. For example, $0.00132 = 1.32 \times 10^{-3} = 13.2 \times 10^{-2}$.

Thus to perform an awkward looking calculation such as $2,200,000 \times 0.03/0.0002$, we have $(2.2 \times 10^6) \times (3. \times 10^{-2})/(2. \times 10^{-4}) = (6.6 \times 10^4)/(2. \times 10^{-4}) = 3.3 \times 10^8$.

B. CONVERSION FACTORS

Length
1 kilometer (km) = 0.621 mile (1 mi = 1.610 km).
1 meter (m) = 39.4 inches = 3.28 feet (1 ft = 0.305 m).
1 centimeter (cm) = 0.394 inch (1 in = 2.54 cm).
1 angstrom (Å) = 10^{-8} cm.

Area
1 square kilometer (km^2) = 0.386 square miles (1 mi^2 = 2.590 km^2).
1 square meter (m^2) = 10.8 square feet (1 ft^2 = 9.29×10^{-2} m^2).
1 square centimeter (cm^2) = 0.155 square inch (1 in^2 = 6.452 cm^2).

Volume
1 cubic kilometer (km^3) = 10^9 m^3 =
0.24 cubic mile (1 mi^3 = 4.17 km^3)
1 cubic meter (m^3) = 10^6 cm^3 = 35.3 cubic feet = 264 U.S. gallons.
1 cubic centimeter (cm^3) = 6.1×10^{-2} cubic inch (1 in^3 = 16.39 cm^3).

Mass
1 metric ton = 10^6 g = 2205 pounds = 1.1 U.S. ton.
1 kilogram (kg) = 2.205 pounds (1 lb = 0.454 kg).
1 gram (g) = 3.53×10^{-2} ounce = 2.2×10^{-3} pound (1 lb = 454 g).

Temperature
To convert degrees centigrade (°C) to degrees Fahrenheit (°F), °F =
1.8 °C + 32.
To convert °F to °C, °C = (°F − 32)/1.8.

Force
A force of one dyne will produce an acceleration of 1 cm/sec^2 when applied to a mass of 1 g.

Pressure
1 bar = 10^6 dynes/cm^2 = 14.22 lb/in^2 = 0.968 atmosphere (atm).

Density
1 gram per cubic centimeter (g/cm^3) = 62.43 lb/ft^3 = 0.036 lb/in^3.

Velocity
1 kilometer per second (km/sec) = 2237 mi/hr = 3281 ft/sec, c =
speed of light in vacuum = 2.998×10^{10} cm/sec.

Gravity
G = universal gravitational constant = 6.67×10^{-8} in cm-g-sec units.

Appendix 2
The Elements

Element	Symbol	Atomic number	Abundance in weight % Continental crust	Whole earth
Actinium	Ac	89		
Aluminum	Al	13	8.26	0.44
Antimony	Sb	51	2×10^{-5}	
Argon	Ar	18	4×10^{-5}	
Arsenic	As	33	1.8×10^{-4}	
Barium	Ba	56	0.0425	
Beryllium	Be	4	2.8×10^{-4}	
Bismuth	Bi	83	1.7×10^{-5}	
Boron	B	5	1.0×10^{-3}	
Bromine	Br	35	2.5×10^{-4}	
Cadmium	Cd	48	2×10^{-5}	
Calcium	Ca	20	4.07	0.61
Carbon	C	6	0.020	
Cerium	Ce	58	6.7×10^{-3}	
Cesium	Cs	55	3×10^{-4}	
Chlorine	Cl	17	0.013	
Chromium	Cr	24	0.010	0.01
Cobalt	Co	27	2.5×10^{-3}	0.20
Copper	Cu	29	5.5×10^{-3}	
Dysprosium	Dy	66	5.2×10^{-4}	
Erbium	Er	68	2.8×10^{-4}	
Europium	Eu	63	1.2×10^{-4}	
Fluorine	F	9	0.0625	
Gadolinium	Gd	64	7.3×10^{-3}	
Gallium	Ga	31	1.5×10^{-3}	
Germanium	Ge	32	1.5×10^{-4}	
Gold	Au	79	4×10^{-7}	
Hafnium	Hf	72	3×10^{-4}	
Helium	He	2	3×10^{-7}	
Holmium	Ho	67	1.5×10^{-4}	
Hydrogen	H	1	0.14	
Indium	In	49	1×10^{-5}	
Iodine	I	53	5×10^{-5}	
Iridium	Ir	77	1×10^{-7}	
Iron	Fe	26	4.84	35.39
Krypton	Kr	36		
Lanthanum	La	57	2.5×10^{-3}	
Lead	Pb	82	1.25×10^{-3}	
Lithium	Li	3	2.0×10^{-3}	
Lutetium	Lu	71	8×10^{-5}	
Magnesium	Mg	12	1.87	17.00
Manganese	Mn	25	0.08	0.09

*Estimated elemental abundances taken from tables by B. Mason, *Principles of Geochemistry*, John Wiley & Sons, Inc. New York, 1966, and K. Krauskopf, *Introduction to Geochemistry*, McGraw-Hill, Inc., New York, 1967.

Element	Symbol	Atomic number	Abundance in weight %	
			Continental crust	Whole earth
Mercury	Hg	80	8×10^{-6}	
Molybdenum	Mo	42	1.5×10^{-4}	
Neodymium	Nd	60	2.8×10^{-3}	
Neon	Ne	10		
Nickel	Ni	28	7.5×10^{-3}	2.70
Niobium	Nb	41	2.0×10^{-3}	
Nitrogen	N	12	2.0×10^{-3}	
Osmium	Os	76	5×10^{-7}	
Oxygen	O	8	46.63	27.79
Palladium	Pd	46	1×10^{-6}	
Phosphorus	P	15	0.13	0.03
Platinum	Pt	78	1×10^{-6}	
Polonium	Po	84		
Potassium	K	19	2.11	0.07
Prasodymium	Pr	59	6.5×10^{-4}	
Protactinium	Pa	91		
Radium	Ra	88		
Radon	Rn	86		
Rhenium	Re	75	1×10^{-7}	
Rhodium	Rh	45	5×10^{-7}	
Rubidium	Rb	37	9×10^{-3}	
Ruthenium	Ru	44	1×10^{-6}	
Samarium	Sm	62	7.3×10^{-4}	
Scandium	Sc	21	2.2×10^{-3}	
Selenium	Se	34	5×10^{-6}	
Silicon	Si	14	28.93	12.64
Silver	Ag	47	7×10^{-6}	
Sodium	Na	11	2.30	0.14
Strontium	Sr	38	0.0375	
Sulfur	S	16	0.026	2.74
Tantalum	Ta	73	2×10^{-4}	
Tellurium	Te	52	1×10^{-6}	
Terbium	Tb	65	1.1×10^{-4}	
Thallium	Tl	81	4.5×10^{-5}	
Thorium	Th	90	9.6×10^{-4}	
Thulium	Tm	69	2.5×10^{-5}	
Tin	Sn	50	2×10^{-4}	
Titanium	Ti	22	0.48	0.04
Tungsten	W	74	1.5×10^{-4}	
Uranium	U	92	2.7×10^{-4}	
Vanadium	V	23	0.0135	
Xenon	Xe	54		
Ytterbium	Yb	70	3.0×10^{-4}	
Yttrium	Y	39	3.3×10^{-3}	
Zinc	Zn	30	7×10^{-3}	
Zirconium	Zr	40	0.0165	

Appendix 3
Physical Data on the Earth

Average radius	6371 km
Equatorial radius	6378 km
Polar radius	6357 km
Flattening	1/298.2
Equatorial circumference	40,077 km
Area	5.10×10^8 km^2
Volume	1.083×10^{12} km$^3 = 1.083 \times 10^{27}$ cm^3
Mass	5.977×10^{27} g
Average density	5.517 g/cm^3

B. PARTS OF THE EARTH

Average Radius and Thickness

	Radius to Top (km)	Depth to Top (km)	Thickness (km)	% of Earth's Radius
Core			3473	54.5
Inner	1300	5071	1300	20.4
Outer	3473	2898	2173	34.1
Mantle			2881	45.2
Lower	5671	700	2198	34.5
Upper	6354	17	683	10.7
Crust	6371		17	0.3
Continental		−0.8	35	0.5
Oceanic		4	8	0.1
Lithosphere				
Continental			110 − 150	~2.1
Oceanic			80	1.3

Area

	Area (km²)	% of Earth's Surface
Land	1.49×10^8	29.2
Oceans and seas	3.61×10^8	70.8
Ice sheets	1.56×10^7	3.1
Continental shelves	2.84×10^7	5.6
Continental crust	1.77×10^8	34.7
Oceanic crust	3.33×10^8	65.3

Volume, Mass, and Average Density

	Volume (km³)	Volume (%)	Mass (g)	Mass (%)	Average Density (g/cm³)
Core	1.75×10^{11}	16.2	1.88×10^{27}	31.5	10.72
Inner	9.20×10^9	0.9			
Outer	1.66×10^{11}	15.3			
Mantle	8.99×10^{11}	83.0	4.07×10^{27}	68.1	4.53
Lower	5.88×10^{11}	54.3			
Upper	3.11×10^{11}	28.7			
Crust	8.87×10^9	0.82	2.51×10^{25}	0.42	2.83
Continental	6.21×10^9	0.57	1.74×10^{25}	0.29	2.8
Oceanic	2.66×10^9	0.25	7.71×10^{24}	0.13	2.9
Oceans and Seas	1.37×10^9	0.13	1.41×10^{24}	0.023	1.03
Ice sheets	2.5×10^7	0.0023	2.3×10^{22}	4×10^{-4}	0.90
Atmosphere			5×10^{21}	8×10^{-5}	

Appendix 4
Summary of Plate Tectonic Interactions

Type of Plate Boundary	Divergent	Conservative	Convergent
Relative motion	Plates move apart	Plates slide past each other in contact	Plates approach each other, one goes down into mantle
Name of process	**Sea Floor Spreading**	**Transform Faulting**	**Subduction**
Topographic expression	Midoceanic ridges	Fault ridges and troughs, oceanic fracture zones	Oceanic trenches, island arcs, volcanic mountain chains; continental collision mountain belts (non-volcanic)
Type of magmatism	Basaltic volcanism	None	Volcanism and plutonism of intermediate average composition (andesite and granodiorite). Very little in continental collision mountain belts.
Source of magma	Partial melt of ultramafic mantle (first stage of double distillation)	None	Partial melt of basalt layer covering downgoing slab (second stage of double distillation)
Effects on size of plates	Increases area of plates involved	Neither adds to nor subtracts from area of plates	Decreases area of at least one of the plates involved

Sumary of Plate Tectonic Interactions (*continued*)

Generates	Oceanic Crust and Lithosphere	Faults	Continental Crust and Lithosphere
Earthquakes	Small and numerous	Large and destructive	Largest and most destructive
Other effects	Linear magnetic anomalies are created that permit measurement of rates and history of motion, moves continents apart.	Major strike-slip (lateral) faults	Creates island arcs, volcanic mountain chains; continental collision mountain belts (non-volcanic)
Examples	Mid-Atlantic Ridge, East Pacific Rise	San Andreas Fault (on land) Eltanin Fracture Zone (submarine)	Tonga, Japan Trenches and arcs (island arcs, oceanic-oceanic lithosphere colliding) Andes Mts. (volcanic mountain chain, oceanic-continental lithosphere collision) Himalayas, Alps (fold mountain belts, continental-continental lithosphere collision)

Ablation   Those processes, particularly melting and sublimation, by which a glacier loses material (wastes away).

Abyssal Plain   Flat area of the ocean floor where excessive sedimentation has resulted in the burial of the original submarine topography.

Aeolian Sandstone   A sandstone deposited by action of the wind (for example, sand dunes).

Agglomerate   A coarse grained pyroclastic rock consisting of unsorted volcanic ejecta up to more than a meter in diameter.

Alkaline Igneous Rock   A rock containing a relatively high proportion of alkaline metals such as sodium and potassium.

Alluvial Fan   A roughly fan-shaped body of alluvium deposited on land when a stream reaches a sudden decrease in slope and loses much of its transport power.

Alluvium   A general term for the clastic sedimentary deposits of streams.

Alpha Particle   A particle, equivalent to the nucleus of a helium atom, emitted from an atomic nucleus during radioactive decay. The alpha particle is composed of two protons and two neutrons and carries a positive charge of two.

Anion   A negatively charged atom or group of atoms.

Anthracite   Coal of the highest metamorphic rank. It is a hard black substance containing greater than 90% carbon.

Apatite   Mineral consisting of calcium phosphate together with fluorine, hydroxyl, or carbonate ions in variable proportions.

Aquifer   A permeable rock unit through which groundwater can flow and from which groundwater can be readily extracted.

Arc-trench Gap   Geographic element in arc-trench systems that separates the trench from the arc. In most cases the gap is 100 km or more wide.

Artesian Well   A well that has been drilled into a confined aquifer, in which the water is under sufficient pressure for it to rise above the local groundwater table and possibly even flow out on the surface.

Asteroid   A small planetary mass in orbit around the sun. Most asteroid orbits lie between the orbits of Mars and Jupiter.

Asthenosphere   A portion of the upper mantle of the earth, below the lithosphere, where rock strength is low enough to allow a slow, steady flow of rock mass in response to isostatic or plate tectonic motion.

Atoll   A low island consisting of a shallow lagoon enclosed by a semi-circular coral reef.

Atomic Number   The number of protons in an atomic nucleus; each chemical element is uniquely characterized by its atomic number.

Atomic Weight   More appropriately called **atomic mass,** it is essentially equal to the number of neutrons plus the number of protons in an atomic nucleus.

Banded Iron Formation   A rock consisting of alternating thin bands of chert and iron minerals, generally oxides.

Batholith   A large body of coarse grained intrusive igneous rock, having an area of more than 100 km.^{2}

Bauxite The hydrated aluminum oxide minerals that are the most important ore of aluminum metal.

Bed A tabular depositional layer in a sedimentary rock body.

Bedload The amount of material carried along the bed of a moving body of water, usually a stream or river. The material moves by sliding, rolling, or bouncing.

Benioff Zone An inclined zone of earthquake activity that marks the location of the down-going lithospheric slab during subduction.

Beta Particle A charged particle, equivalent to an electron emitted from an atomic nucleus during radioactive decay.

Bituminous Coal Coal of low metamorphic rank that contains a high percentage of relatively volatile hydrocarbons (bitumens) that can be distilled off by heat.

C

Capillary Action The tendency of a fluid to rise within narrow passages as a result of surface tension.

Cation A positively charged atom or group of atoms.

Chalcopyrite $CuFeS_2$ (mineral), the most important economic source of copper metal.

Cirque A steep walled, half bowl-like hollow on a mountain, formed by frost action and plucking at the head of a glacier.

Clastic Sediments Fragmental rock material that has been mechanically transported and deposited in a sedimentary environment.

Cleavage The tendency of a mineral to break along planar surfaces that are parallel to internal planes of atoms.

Colluvium A general term for the deposits formed by mass wasting processes.

Compressional Waves The primary (P) seismic waves, so called because they reach a seismograph first. The P waves are propagated by alternate compression and expansion of the medium (as are sound waves).

Competence In reference to streams, the largest particle the stream is capable of moving along its bed under a given set of flow circumstances. The term is also applicable to glaciers or wind.

Connate Water Water trapped in the pore spaces of a sedimentary rock at the time it was deposited.

Continental Rise The more gently sloping portion of the shelf margin that lies seaward of the continental slope; formed by sediment infill in the angle between the continental slope and the deep ocean floor.

Continental Shelf The submerged margin of a continental landmass forming a shallow extension of the land; of variable width, the shelf is commonly considered to be terminated at the first sharp break downward into the ocean deep.

Continental Slope The relatively steep (about 5°) slope that lies seaward of the continental shelf.

Core The inner spherical mass of the earth consisting of a fluid outer shell and a solid inner mass; thought to be composed mainly of metallic iron and nickel.

Coriolis Effect The tendency of material in motion on the earth's surface to be deflected due to the rotation of the earth. The magnitude of deflection depends on the latitude and velocity of the moving material.

Correlation The establishment of the time relationship between two or more geologic (rock) formations or geologic events that are geographically separated.

Craton The central stable region of continental landmass, composed chiefly of Precambrian igneous and metamorphic rocks.

Crust The outermost layer of the earth, above the Moho seismic discontinuity, consisting of the mafic oceanic crust and the felsic continental crust.

Curie Point The temperature above which magnetic minerals lose their magnetism. Conversely, the temperature below which spontaneous magnetic ordering may take place.

D

Delta A roughly fan-shaped body of alluvium deposited at the mouth of a stream.

Diapir A mass of material, less dense than the surrounding material, that has risen to its position through buoyancy.

Dike A tabular rock body having a discordant (cross cutting) relationship with surrounding rock bodies; most commonly referring to a thin tabular igneous intrusive.

Dip Inclination of a geologic surface, measured in degrees of arc below the horizontal.

Drumlin A streamlined, elongate hill composed of glacial till. The long axis parallels the ice flow and the steeper end faces upstream.

E

Ecology The study of the relationship between organisms and their environment.

Ejecta Material ejected by a volcano.

Electron A fundamental constituent of material; an elementary particle with a single negative charge and a mass of approximately 10^{-27} g; found in circumnuclear orbits in atoms.

Energy The capacity to do work. It may exist in and be transformed among many distinct forms such as thermal energy, kinetic energy, and potential energy; equivalent to mass according to Einstein's equation $E = mc^2$.

Epicenter The point on the earth's surface that lies directly above the focus or origin of an earthquake.

Erosion A general term describing the geologic process of degradation and movement of disaggregated rock material at the earth's surface by water, wind, or ice.

Eruption The ejection from a volcano of lava, pyroclastic material, and/or volcanic gases.

Esker An elongate, sinuous ridge composed of stratified glacial drift, thought to form beneath the surface of a continental glacier. May be over 100 km in length up to about 30 m in height.

Evaporite A nonclastic sedimentary rock or mineral precipitated from aqueous solutions as a result of evaporation.

Extrusive Igneous Rock An igneous rock that has crystallized after eruption onto the earth's surface.

F

Fabric (rock) The orientation of minerals in a rock.

Facies A term used to distinguish a portion of a rock unit with a distinctive group of characteristic (mineral assemblage or fossil assemblage for example) that differs from other portions of the same general rock unit.

Fan A fan-shaped, gently sloping sedimentary deposit formed at locations where there is a sharp decrease in gradient and hence a sharp drop in the sediment carrying capacity of the transporting medium.

Fault A large scale earth fracture that results in the displacement of rock masses on one side of the fault plane relative to those on the other side.

Felsic An adjective used to describe high silica igneous rocks in which light colored minerals, mostly feldspar and quartz, predominate.

Fission A nuclear process involving the breakup of an atomic nucleus to form lighter elements.

Floodplain The flat portion of a river valley that is subject to periodic inundation during episodes of excessive runoff.

Focus (of an earthquake) The point within the earth where the seismic energy of an earthquake is released.

Folding The large scale bending of rock strata during regional deformation in the earth's crust.

Fossil Fuel Remains of organisms that have become trapped in rocks under low-oxygen conditions and can be burned to produce energy.

Fracture Zone The topographic expression of a transform fault, consisting of parallel narrow ridges and troughs.

Frost Action A process of physical weathering where the pressure exerted by freezing water disintegrates rock material.

Fusion A nuclear process involving the formation of heavier elements from the combination of lighter elements.

G

Gamma Ray Extremely short-wavelength electromagnetic radiation emitted from the nucleus of an atom during radioactive decay.

Geosyncline A hypothetical large scale linear trough filled with sediments.

Geothermal Energy Energy obtained from the heat of the earth's interior, usually through the utilization of steam formed by contact of groundwater with hot rock.

Glacier A large ice mass, consisting of recrystallized snow, originating and flowing under the influence of gravity on a land surface.

Graded Bedding A type of sedimentary rock bedding in which each layer shows a progressive change in particle size, usually from coarse at the bottom to fine at the top.

Granitic A general term for coarse grained intermediate to felsic igneous rock containing predominantly feldspar and quartz. The term covers a much broader range of rock types than granite alone.

Greenhouse Effect Heating of the earth's atmosphere due to the partial retention by carbon dioxide and water vapor of solar heat energy re-radiated from the earth's surface.

Greenstone A field classification term given to altered mafic volcanic rocks. Belts of these green colored rocks are often found in Precambrian shield areas.

H

Hotspot Hypothetical concentrations of heat in the mantle beneath local areas of intraplate volcanism such as the Hawaiian Islands.

Hydrocarbon Naturally occurring substance of organic origin, consisting primarily of hydrogen and carbon atoms in various combinations.

Hydrogen Bond The strong chemical bond between adjacent H_2O molecules in water; an electrostatic bond in which each water molecule is able to form four hydrogen bonds, two bonds using its two hydrogens and two bonds linking its unpaired electrons with hydrogens from two adjacent H_2O molecules.

Hydrolysis The chemical reaction between water and another compound.

Hydrosphere The discontinuous envelope of water at or near the surface of the earth, consisting of the ocean and other surface water.

I

Intermediate A term applied to igneous rocks lying in composition between felsic and mafic.

Intrusive Rock An igneous rock that originates by solidification of magma that has intruded into older bedrock and not risen to the surface.

Ion An electrically charged atom or group of atoms formed by the gain or loss of one or more electrons (*see* Cation; Anion).

Isostasy The geologic concept that all large masses of the earth's crust tend toward floating equilbrium with respect to the surrounding crustal rock masses.

Isotope Any of two or more distinct forms of a chemical element having the same number of protons but a different number of neutrons.

J

Joint A fracture in a rock body along which no appreciable movement has occurred.

K

Kame A conical hill composed of stratified glacial drift.

Kettle Lake A lake filling a depression in glacial drift, formed by melting a buried or partially buried mass of ice.

Lignite   A low rank form of coal that is more compact than peat.

Lithification   The process by which unconsolidated sediments are converted to solid rock

Lithosphere   The term applied to that portion of the earth's crust and upper mantle above the asthenosphere and comprising the rigid material of the plates.

Loess   Loosely consolidated sediments formed from wind-deposited silt.

Low Velocity Zone   A zone in the earth's upper mantle in which the seismic velocity, particularly shear wave velocity, is lower than in the material both above and below.

M

Mafic   A term applied to low-silica igneous rocks containing an abundance of dark colored iron-magnesium silicates.

Magma   Molten silicate material originating within the earth from which igneous rocks crystallize.

Magnetic Field (earth's)   The magnetic force field surrounding the earth which acts upon all magnetic bodies and electric currents present on or in the earth. It is thought to originate in the outer core.

Magnetic Reversal   The reversal of polarity of the earth's magnetic field.

Mass Number   The number of protons plus the number of neutrons in the nucleus of an atom.

Mass Wasting   The movement of rock fragments to a lower elevation without involvement of a transporting medium; particles rolling downhill, for example.

Meander   A looplike bend in the course of a stream.

Melange   Chaotic mixture of blocks of rocks of different types and origins. Most commonly formed in the vicinity of trenches during active subduction.

Meteorite   A mass of stone or metal that has reached the earth from outer space.

Moho   The discontinuity in seismic velocity that defines the base of the earth's crust.

Mohs' Hardness Scale   A hardness scale pertaining to minerals that ranges from 1(talc) through 3(calcite), 6(feldspar), and 7(quartz) to 10 (diamond).

Molasse   A term applied to thick sequences of continental and shallow marine sedimentary rocks formed from the rapid erosion of young mountain chains.

Moraine   A constructional landform produced by an accumulation of glacial till. May be relatively flat (ground moraine) or ridgelike (end moraine, lateral moraine, and so on).

N

Natural Gas   A term for naturally occurring hydrocarbons that exist in the earth in a gaseous state. The most common type is methane (CH_4).

Neutron   An elementary particle having no electric charge and a mass approximately equal to a proton.

O

Oil Shale   Shale that contains sufficient solid hydrocarbons to be a potential source of petroleum.

Orogenesis   The process of deformation and mountain building accompanying plate motions and related movements in the lithosphere.

P

P **wave**   *See* Compressional wave.

Paleomagnetism   The natural remnant magnetism in magnetic rock bodies (for example, lava flows).

Peat   An accumulation of plant material that is partially decomposed and can be burned when dried.

Peneplain   An extensive flat surface produced by regional erosional processes.

Permeability   The ability or capacity of a rock or rock material to transmit fluids.

Petroleum   Naturally occurring liquid hydrocarbons formed from the decomposition of marine animals and plants in oxygen-poor subsurface environments.

Phenocryst   Mineral in a porphyritic igneous rock that is of conspicuously larger grain size than the surrounding (matrix) minerals.

Photosynthesis   The synthesis by growing plants of complex organic molecules from carbon dioxide, water, and nitrogen from the atmosphere and water and trace elements from the soil using sunlight as the source of energy.

Plastic Flow   Permanent deformation, without rupture, of the shape of a body.

Plate   A caplike portion of the lithosphere that acts as a rigid unit, with boundaries marked by narrow bands of earthquake activity. All plates are in motion relative to the immediately adjacent plates.

Plutonic   Term pertaining to geologic processes and rock formations occurring deep within the earth's crust.

Porosity   The amount of pore space (voids) in a rock or rock material; usually expressed as a percentage.

Porphyry   An igneous rock that contains a significant number of conspicuous phenocrysts in a fine grained matrix.

Porphyry Copper   A large, generally low grade, copper deposit closely associated with the emplacement of shallow-seated felsic igneous rocks.

Proton   An elementary particle that is a fundamental constituent of all atomic nuclei; each proton carries a positive charge equal to the negative charge on an electron.

Pyroclastic   A term applied to clastic rock material ejected from a volcano.

Radioactive Decay The spontaneous emission of a particle from an atomic nucleus, transforming the atom from one chemical element to another.

Radiometric Age Absolute age of a rock body expressed in years, calculated from the determination of the quantities of radioactive elements and their decay products.

Rank (of coal) Scale indicating the degree to which the more volatile components of coal have been driven off by heat and pressure within the earth.

Recharge Rate Rate at which the ground water in aquifers is replenished.

Regolith Loose rock material, of any origin, that forms most of the land surface and overlies the consolidated "bed" rock. That part of the regolith capable of supporting the growth of rooted plants is termed soil.

Rift A zone in the earth's crust along which divergence takes place, the lithosphere on either side moving apart.

S Wave *See* Shear wave

Saltation A process by which particles move in intermittent jumps when driven along a surface by wind or by water.

Schistosity The foliation apparent in schists and similar rocks due to the parallel arrangement of platy mineral grains.

Seamount An isolated underwater volcanic structure.

Seismic Sea Wave Wave form generated within ocean water by large scale, short term disturbances of the sea floor associated with earthquake activity.

Shear Waves The secondary (S) seismic waves, so called because they reach a seismograph after the P waves. S waves are propagated by the oscillation of particles at right angles to the direction of propogation and can be transmitted only by solids.

Sheet Wash Erosion by sheets of water flowing downslope, as distinct from water flowing in channels.

Sill A tabular intrusive igneous rock body whose surfaces are concordant (parallel) with the enclosing rock structure.

Stack A small pillarlike rock mass off a steep sided sea shore. Stacks are remnant features formed by wave erosion of headlands.

Stock An intrusive igneous rock body of roughly equidimensional outcrop form that has an exposed outcrop area of less than 100 km².

Strike The orientation in the horizontal plane of a linear feature; its direction or trend.

Subduction The process of consumption of lithospheric material at convergent plate boundaries.

Sublimation The transition of a substance from the solid to the vapor state without passing through an intermediate liquid state.

Submarine Canyon A V-shaped underwater canyon with a tributary system that resembles a river cut canyon and begins on the continental shelf.

Surface Wave Seismic waves that are propagated along the earth's surface.

These waves are slower than S and P waves but have larger amplitude and can cause serious damage.

T

Talus   A pile or sheet of loose rock fragments that accumulates at the base of a steep slope.

Tar Sand   Sand impregnated with tarry solid hydrocarbon that can be used as a source of petroleum.

Terrace   A relatively flat, benchlike surface roughly parallel to a stream. Usually produced when the stream cuts downward through its own deposits.

Thrust Fault   A fault, inclined at a small angle to the horizontal, in which one side has moved upwards relative to the other side.

Till   Unstratified and relatively unsorted glacial drift deposited directly from glacial ice.

Tillite   Lithified till.

Transform Fault   A plate boundary fault parallel to the direction of relative plate motion, so that the slip on the fault is horizontal. Links other kinds of plate boundaries.

Trap   Geologic structure in which petroleum liquids and/or natural gas have accumulated.

Trench (oceanic)   A steep sided elongate depression on the deep sea floor that is the location of active subduction.

Triple Junction   Point at which three lithospheric plates meet and from which radiate three plate boundaries.

Tsunami   Japanese word for seismic sea wave.

Turbidity Current   A rapid flow of sediment laden water downslope in subaqueous environments.

U

Ultramafic   A term applied to igneous rocks composed mainly of iron-magnesium silicates.

Unconformity   A surface between rock units marking a lack of continuity or gap in the geologic record.

Uniformitarianism   The concept that the physical and chemical laws governing geologic processes have operated uniformly throughout earth history.

Uraninite   (UO_2) Naturally occurring uranium oxide mineral that represents the main ore constituent in many uranium deposits.

V

Varve   A pair of thin sedimentary layers that contrast due to seasonal changes affecting sedimentation. Each varve represents one year; varves are most common in glacial lakes.

Viscosity   The resistance of a fluid to flow.

Volcano-plutonic Arc Linear belt within which volcanic and plutonic igneous rocks are formed. They are closely related in time and place to the subduction of oceanic lithosphere.

W

Water Table The surface that separates the zone saturated with ground water from the unsaturated zone above. The depth to the water table is equal to the level of standing water in a well.